T0260962

MULTIMEDIA
SEMANTICS

MULTIMEDIA SEMANTICS

METADATA, ANALYSIS AND INTERACTION

Raphaël Troncy
Centrum Wiskunde & Informatica, Amsterdam, The Netherlands

Benoit Huet
EURECOM, France

Simon Schenk
WeST Institute, University of Koblenz-Landau, Germany

WILEY

A John Wiley & Sons, Ltd., Publication

This edition first published 2011
© 2011 John Wiley & Sons Ltd.

Registered office
John Wiley & Sons Ltd, The Atrium, Southern Gate, Chichester, West Sussex, PO19 8SQ, United Kingdom

For details of our global editorial offices, for customer services and for information about how to apply for permission to reuse the copyright material in this book please see our website at www.wiley.com.

The right of the author to be identified as the author of this work has been asserted in accordance with the Copyright, Designs and Patents Act 1988.

All rights reserved. No part of this publication may be reproduced, stored in a retrieval system, or transmitted, in any form or by any means, electronic, mechanical, photocopying, recording or otherwise, except as permitted by the UK Copyright, Designs and Patents Act 1988, without the prior permission of the publisher.

Wiley also publishes its books in a variety of electronic formats. Some content that appears in print may not be available in electronic books.

Designations used by companies to distinguish their products are often claimed as trademarks. All brand names and product names used in this book are trade names, service marks, trademarks or registered trademarks of their respective owners. The publisher is not associated with any product or vendor mentioned in this book. This publication is designed to provide accurate and authoritative information in regard to the subject matter covered. It is sold on the understanding that the publisher is not engaged in rendering professional services. If professional advice or other expert assistance is required, the services of a competent professional should be sought.

Library of Congress Cataloging-in-Publication Data

Troncy, Raphaël.
  Multimedia semantics : metadata, analysis and interaction / Raphaël Troncy, Benoit Huet, Simon Schenk.
    p. cm.
  Includes bibliographical references and index.
  ISBN 978-0-470-74700-1 (cloth)
  1. Multimedia systems. 2. Semantic computing. 3. Information retrieval. 4. Database searching. 5. Metadata.
I. Huet, Benoit. II. Schenk, Simon. III. Title.
  QA76.575.T76 2011
  006.7 – dc22

                                                                                      2011001669

A catalogue record for this book is available from the British Library.

ISBN: 9780470747001 (H/B)
ISBN: 9781119970224 (ePDF)
ISBN: 9781119970231 (oBook)
ISBN: 9781119970620 (ePub)
ISBN: 9781119970637 (mobi)

Set in 10/12pt Times by Laserwords Private Limited, Chennai, India

Contents

Foreword

I am delighted to see a book on multimedia semantics covering metadata, analysis, and interaction edited by three very active researchers in the field: Troncy, Huet, and Schenk. This is one of those projects that are very difficult to complete because the field is advancing rapidly in many different dimensions. At any time, you feel that many important emerging areas may not be covered well unless you see the next important conference in the field. A state of the art book remains a moving, often elusive, target. But this is only a part of the dilemma. There are two more difficult problems. First multimedia itself is like the famous fable of an elephant and blind men. Each person can only experience an aspect of the elephant and hence has only understanding of a partial problem. Interestingly, in the context of the whole problem, it is not a partial perspective, but often is a wrong perspective. The second issue is the notorious issue of the semantic gap. The concepts and abstractions in computing are based on bits, bytes, lists, arrays, images, metadata and such; but the abstractions and concepts used by human users are based on objects and events. The gap between the concepts used by computer and those used by humans is termed the semantic gap. It has been exceedingly difficult to bridge this gap. This ambitious book aims to cover this important, but difficult and rapidly advancing topic. And I am impressed that it is successful in capturing a good picture of the state of the art as it exists in early 2011. On one hand I am impressed, and on the other hand I am sure that many researchers in this field will be thankful to editors and authors for providing all this material in compact, yet comprehensible form, in one book.

The book covers aspects of multimedia from feature extraction to ontological representations to semantic search. This encyclopedic coverage of semantic multimedia is appearing at the right time. Just when we thought that it is almost impossible to find all related topics for understanding emerging multimedia systems, as discussed in use cases, this book appears. Of course, such a book can only provide breadth in a reasonable size. And I find that in covering the breadth, authors have taken care not to become so superficial that the coverage of the topic may become meaningless. This book is an excellent reference sources for anybody working in this area. As is natural, to keep such a book *current* in a few years, a new edition of the book has to be prepared. Hopefully, all the electronic tools may make this feasible. I would definitely love to see a new edition in a few years.

I want to particularly emphasize the closing sentence of the book: *There is no single standard or format that satisfactorily covers all aspects of audiovisual content descriptions; the ideal choice depends on type of application, process and required complexity.* I hope that serious efforts will start to develop such a single standard considering all rich metadata in smart phones that can be used to generate meaningful extractable, rather than human generated, tags. We, in academia, often ignore obvious and usable in favor of obscure and complex. We seem to enjoy creation of new problems more than solving challenging existing problems. Semantic multimedia is definitely a field where there is need for simple tools to use available data and information to solve rapidly growing multimedia data volumes. I hope that by pulling together all relevant material, this book will facilitate solution of such real problems.

Ramesh Jain
Donald Bren Professor in Information & Computer Sciences,
Department of Computer Science Bren School of Information and Computer Sciences,
University of California, Irvine.

List of Figures

List of Tables

List of Contributors

Thanos Athanasiadis
Image, Video and Multimedia Systems Laboratory, National Technical University of Athens, 15780 Zographou, Greece

Yannis Avrithis
Image, Video and Multimedia Systems Laboratory, National Technical University of Athens, 15780 Zographou, Greece

Werner Bailer
JOANNEUM RESEARCH, Forschungsgesellschaft mbH, DIGITAL – Institute for Information and Communication Technologies, Steyrergasse 17, 8010 Graz, Austria

Rachid Benmokhtar
EURECOM, 2229 Route des Crêtes, BP 193 – Sophia Antipolis, France

Petr Berka
Faculty of Informatics and Statistics, University of Economics, Prague, Czech Republic

Susanne Boll
Media Informatics and Multimedia Systems Group, University of Oldenburg, Escherweg 2, 26121 Oldenburg, Germany

Paul Buitelaar
DFKI GmbH, Germany

Marine Campedel
Telecom ParisTech, 37–39 rue Dareau, 75014 Paris, France

Oscar Celma
BMAT, Barcelona, Spain

Krishna Chandramouli
Queen Mary University of London, Mile End Road, London, UK

Slim Essid
Telecom ParisTech, 37–39 rue Dareau, Paris, 75014 France

Thomas Franz
ISWeb – Information Systems and Semantic Web, University of Koblenz-Landau, Universitätsstraße 1, Koblenz, Germany

Lynda Hardman
Centrum Wiskunde & Informatica, Amsterdam, The Netherlands

Michael Hausenblas
Digital Enterprise Research Institute, National University of Ireland, IDA Business Park, Lower Dangan, Galway, Ireland

Michiel Hildebrand
Centrum Wiskunde & Informatica, Amsterdam, The Netherlands

Frank Hopfgartner
International Computer Science Institute, 1947 Center Street, Suite 600, Berkeley, CA, 94704, USA

Benoit Huet
EURECOM, 2229 Route des Crêtes, BP 193 – Sophia Antipolis, France

Antoine Isaac
Vrije Universiteit Amsterdam, de Boelelaan 1081a, Amsterdam, The Netherlands

Joemon M. Jose
University of Glasgow, 18 Lilybank Gardens, Glasgow G12 8RZ, UK

Florian Kaiser
Technische Universität Berlin, Institut für Telekommunkationssysteme, Fachgebiet Nachrichtenübertragung, Einsteinufer 17, 10587 Berlin, Germany

Ioannis Kompatsiaris
Informatics and Telematics Institute, Centre for Research and Technology Hellas, 57001 Thermi-Thessaloniki, Greece

Bart Lehane
CLARITY: Centre for Sensor Web Technologies, Dublin City University, Ireland

Vasileios Mezaris
Informatics and Telematics Institute, Centre for Research and Technology Hellas, 57001 Thermi-Thessaloniki, Greece

Phivos Mylonas
Image, Video and Multimedia Systems Laboratory, National Technical University of Athens, 15780 Zographou, Greece

Frank Nack
University of Amsterdam, Science Park 107, 1098 XG Amsterdam, The Netherlands

Jan Nemrava
Faculty of Informatics and Statistics, University of Economics, Prague, Czech Republic

Noel E. O'Connor
CLARITY: Centre for Sensor Web Technologies, Dublin City University, Ireland

Željko Obrenović
Centrum Wiskunde & Informatica, Amsterdam, The Netherlands

Eyal Oren
Vrije Universiteit Amsterdam, de Boelelaan 1081a, Amsterdam, The Netherlands

Georgios Th. Papadopoulos
Informatics and Telematics Institute, Centre for Research and Technology Hellas, 57001 Thermi-Thessaloniki, Greece

Tomas Piatrik
Queen Mary University of London, Mile End Road, London E1 4NS, UK

Yves Raimond
BBC Audio & Music interactive, London, UK

Reede Ren
University of Surrey, Guildford, Surrey, GU2 7XH, UK

Gaël Richard
Telecom ParisTech, 37–39 rue Dareau, 75014 Paris, France

Carsten Saathoff
ISWeb – Information Systems and Semantic Web, University of Koblenz-Landau, Universitätsstraße 1, Koblenz, Germany

David A. Sadlier
CLARITY: Centre for Sensor Web Technologies, Dublin City University, Ireland

Andrew Salway
Burton Bradstock Research Labs, UK

Peter Schallauer
JOANNEUM RESEARCH, Forschungsgesellschaft mbH, DIGITAL – Institute for Information and Communication Technologies, Steyrergasse 17, 8010 Graz, Austria

Simon Schenk
University of Koblenz-Landau, Universitätsstraße 1, Koblenz, Germany

Ansgar Scherp
University of Koblenz-Landau, Universitätsstraße 1, Koblenz, Germany

Nikolaos Simou
Image, Video and Multimedia Systems Laboratory, National Technical University of Athens, 15780 Zographou, Greece

Giorgos Stoilos
Image, Video and Multimedia Systems Laboratory, National Technical University of Athens, 15780 Zographou, Greece

Michael G. Strintzis
Informatics and Telematics Institute, Centre for Research and Technology Hellas, 57001 Thermi-Thessaloniki, Greece

Vojtěch Svátek
Faculty of Informatics and Statistics, University of Economics, Prague, Czech Republic

Raphaël Troncy
Centrum Wiskunde & Informatica, Amsterdam, The Netherlands

Vassilis Tzouvaras
Image, Video and Multimedia Systems Laboratory, National Technical University of Athens, 15780 Zographou, Greece

Thierry Urruty
University of Lille 1, 59655 Villeneuve d'Ascq, France

Miroslav Vacura
University of Economics, Prague, Czech Republic

Jacco van Ossenbruggen
Centrum Wiskunde & Informatica, Amsterdam, The Netherlands

1

Introduction

Raphaël Troncy,[1] Benoit Huet[2] and Simon Schenk[3]

[1]*Centrum Wiskunde & Informatica, Amsterdam, The Netherlands*
[2]*EURECOM, Sophia Antipolis, France*
[3]*University of Koblenz-Landau, Koblenz, Germany*

Digital multimedia items can be found on most electronic equipment ranging from mobile phones and portable audiovisual devices to desktop computers. Users are able to acquire, create, store, send, edit, browse, and render through such content at an increasingly fast rate. While it becomes easier to generate and store data, it also becomes more difficult to access and locate specific or relevant information. This book addresses directly and in considerable depth the issues related to representing and managing such multimedia items.

The major objective of this book is to gather together and report on recent work that aims to extract and represent the semantics of multimedia items. There has been significant work by the research community aimed at narrowing the large disparity between the low-level descriptors that can be computed automatically from multimedia content and the richness and subjectivity of semantics in user queries and human interpretations of audiovisual media – the so-called semantic gap.

Research in this area is important because the amount of information available as multimedia for the purposes of entertainment, security, teaching or technical documentation is overwhelming but the understanding of the semantics of such data sources is very limited. This means that the ways in which it can be accessed by users are also severely limited and so the full social or economic potential of this content cannot be realized.

Addressing the grand challenge posed by the semantic gap requires a multi-disciplinary approach and this is reflected in recent research in this area. In particular, this book is closely tied to a recent Network of Excellence funded by the Sixth Framework Programme of the European Commission named 'K-Space' (Knowledge Space of Semantic Inference for Automatic Annotation and Retrieval of Multimedia Content).

By its very nature, this book is targeted at an interdisciplinary community which incorporates many research communities, ranging from signal processing to knowledge

Multimedia Semantics: Metadata, Analysis and Interaction, First Edition.
Edited by Raphaël Troncy, Benoit Huet and Simon Schenk.
© 2011 John Wiley & Sons, Ltd. Published 2011 by John Wiley & Sons, Ltd.

representation and reasoning. For example, multimedia researchers who deal with signal processing, computer vision, pattern recognition, multimedia analysis, indexing, retrieval and management of 'raw' multimedia data are increasingly leveraging methods and tools from the Semantic Web field by considering how to enrich their methods with explicit semantics. Conversely, Semantic Web researchers consider multimedia as an extremely fruitful area of application for their methods and technologies and are actively investigating how to enhance their techniques with results from the multimedia analysis community. A growing community of researchers is now pursuing both approaches in various high-profile projects across the globe. However, it remains difficult for both sides of the divide to communicate with and learn from each other. It is our hope that this book will go some way toward easing this difficulty by presenting recent state-of-the-art results from both communities.

Whenever possible, the approaches presented in this book will be motivated and illustrated by three selected use cases. The use cases have been selected to cover a broad range of multimedia types and real-world scenarios that are relevant to many users on the Web: photos on the Web, music on the Web, and professional audiovisual media production process. The use cases introduce the challenges of media semantics in three different areas: personal photo collection, music consumption, and audiovisual media production as representatives of image, audio, and video content. The use cases, detailed in Chapter 2, motivate the challenges in the field and illustrate the kind of media semantics needed for future use of such content on the Web, and where we have just begun to solve the problem.

Nowadays it is common to associate semantic annotations with media assets. However, there is no agreed way of sharing such information among systems. In Chapter 3 a small number of fundamental processes for media production are presented. The so-called canonical processes are described in the context of two existing systems, related to the personal photo use case: CeWe Color Photo Book and SenseCam.

Feature extraction is the initial step toward multimedia content semantic processing. There has been a lot of work in the signal processing research community over the last two decades toward identifying the most appropriate feature for understanding multimedia content. Chapter 4 provides an overview of some of the most frequently used low-level features, including some from the MPEG-7 standard, to describe audiovisual content. A succinct description of the methodologies employed is also provided. For each of the features relevant to the video use case, a discussion will take place and provide the reader with the essential information about its strengths and weaknesses. The plethora of low-level features available today led the research community to study multi-feature and multi-modal fusion. A brief but concise overview is provided in Chapter 4. Some feature fusion approaches are presented and discussed, highlighting the need for the different features to be studied in a joint fashion.

Machine learning is a field of active research that has applications in a broad range of domains. While humans are able to categorize objects, images or sounds and to place them in specific classes according to some common characteristic or semantics, computers are having difficulties in achieving similar classification. Machine learning can be useful, for example, in learning models for very well-known objects or settings.

Chapter 5 presents some of the main machine learning approaches for setting up an automatic multimedia analysis system. Continuing the information processing flow described in the previous chapter, feature dimensionality reduction methods, supervised and unsupervised classification techniques, and late fusion approaches are described.

The Internet and the Web have become an important communication channel. The Semantic Web improves the Web infrastructure with formal semantics and interlinked data, enabling flexible, reusable, and open knowledge management systems. In Chapter 6, the Semantic Web basics are introduced: the RDF(S) model for knowledge representation, and the existing web infrastructure composed of URIs identifying resources and representations served over the HTTP protocol. The chapter details the importance of open and interlinked Semantic Web datasets, outlines the principles for publishing such linked data on the Web, and discuss some prominent openly available linked data collections. In addition, it shows how RDF(S) can be used to capture and describe domain knowledge in shared ontologies, and how logical inferencing can be used to deduce implicit information based on such domain ontologies.

Having defined the Semantic Web infrastructure, Chapter 7 addresses two questions concerning rich semantics: How can the conceptual knowledge useful for a range of applications be successfully ported to and exploited on the Semantic Web? And how can one access efficiently the information that is represented on these large RDF graphs that constitute the Semantic Web information sphere? Those two issues are addressed through the presentation of SPARQL, the recently standardized Semantic Web Query language, with an emphasis on aspects relevant to querying multimedia metadata represented using RDF in the running examples of COMM annotations.

Chapter 8 presents and discusses a number of commonly used multimedia metadata standards. These standards are compared with respect to a list of assessment criteria using the use cases listed in Chapter 2 as a basis. Through these examples the limitations of the currents standards are exposed. Some initial solutions provided by COMM for automatically converting and mapping between metadata standards are presented and discussed.

A multimedia ontology framework, COMM, that provides a formal semantics for multimedia annotations to enable interoperability of multimedia metadata among media tools is presented in Chapter 9. COMM maps the core functionalities of the MPEG-7 standard to a formal ontology, following an ontology design approach that utilizes the foundational ontology DOLCE to safeguard conceptual clarity and soundness as well as extensibility towards new annotation requirements.

Previous chapters having described multimedia processing and knowledge representation techniques, Chapter 10 examines how their coupling can improve analysis. The algorithms presented in this chapter address the photo use case scenario from two perspectives. The first is a segmentation perspective, using similarity measures and merging criteria defined at a semantic level for refining an initial data-driven segmentation. The second is a classification perspective, where two knowledge-driven approaches are presented. One deals with visual context and treats it as interaction between global classification and local region labels. The other deals with spatial context and formulates the exploitation of it as a global optimization problem.

Chapter 11 demonstrates how different reasoning algorithms upon previously extracted knowledge can be applied to multimedia analysis in order to extract semantics from images and videos. The rich theoretical background, the formality and the soundness of reasoning algorithms can provide a very powerful framework for multimedia analysis. The fuzzy extension of the expressive DL language $\mathcal{SHIN}$, f-$\mathcal{SHIN}$, together with the fuzzy reasoning engine, FiRE, that supports it, are presented here. Then, a model using explicitly represented knowledge about the typical spatial arrangements of objects is presented. Fuzzy constraint reasoning is used to represent the problem and to find a solution that provides an optimal labeling with respect to both low-level and spatial features. Finally, the NEST expert system, used for estimating image regions dissimilarity is described.

Multimedia content structuring is to multimedia documents what tables of contents and indexes are to written documents, an efficient way to access relevant information. Chapter 12 shows how combined audio and visual (and sometimes textual) analysis can assist high-level metadata extraction from video content in terms of content structuring and in detection of key events depicted by the content. This is validated through two case studies targeting different kinds of content. A quasi-generic event-level content structuring approach using combined audiovisual analysis and a suitable machine learning paradigm is described. It is also shown that higher-level metadata can be obtained using complementary temporally aligned textual sources.

Chapter 13 reviews several multimedia annotation tools and presents two of them in detail. The Semantic Video Annotation Tool (SVAT) targets professional users in audiovisual media production and archiving and provides an MPEG-7 based framework for annotating audiovisual media. It integrates different methods for automatic structuring of content and provides the means to semantically annotate the content. The K-Space Annotation Tool is a framework for semi-automatic semantic annotation of multimedia content based on COMM. The annotation tools are compared and issues are identified.

Searching large multimedia collection is the topic covered in Chapter 14. Due to the inherently multi-modal nature of multimedia documents there are two major challenges in the development of an efficient multimedia index structure: the extremely high-dimensional feature space representing the content, on the one hand, and the variable types of feature dimensions, on the other hand. The first index function presented here divides a feature space into disjoint subspaces by using a pyramid tree. An index function is proposed for efficient document access. The second one exploits the discrimination ability of a media collection to partition the document set. A new feature space, the feature term, is proposed to facilitate the identification of effective features as well as the development of retrieval models.

In recent years several Semantic Web applications have been developed that support some form of search. Chapter 15 analyzes the state of the art in that domain. The various roles played by semantics in query construction, the core search algorithm and presentation of search results are investigated. The focus is on queries based on simple textual entry forms and queries constructed by navigation (e.g. faceted browsing). A systematic understanding of the different design dimensions that play a role in supporting search on Semantic Web data is provided. The study is conducted in the context of image search and depicts two use cases: one highlights the use of semantic functionalities to support the search, while the other exposes the use of faceted navigation to explore the image collection.

In conclusion, we trust that this book goes some way toward illuminating some recent exciting results in the field of semantic multimedia. From the wide spectrum of topics covered, it is clear that significant effort is being invested by both the Semantic Web and multimedia analysis research communities. We believe that a key objective of both communities should be to continue and broaden interdisciplinary efforts in this field with a view to extending the significant progress made to date.

2

Use Case Scenarios

Werner Bailer,[1] Susanne Boll,[2] Oscar Celma,[3] Michael Hausenblas[4] and Yves Raimond[5]

[1] *JOANNEUM RESEARCH – DIGITAL, Graz, Austria*
[2] *University of Oldenburg, Oldenburg, Germany*
[3] *BMAT, Barcelona, Spain*
[4] *Digital Enterprise Research Institute, National University of Ireland, IDA Business Park, Lower Dangan, Galway, Ireland*
[5] *BBC Audio & Music Interactive, London, UK*

In this book, the research approaches to extracting, deriving, processing, modeling, using and sharing the semantics of multimedia are presented. We motivate these approaches with three selected use cases that are referred to throughout the book to illustrate the respective content of each chapter. These use cases are partially based on previous work done in the W3C Multimedia Semantics Incubator Group, MMSEM–XG[1] and the W3C Media Annotations Working Group[2] and have been selected to cover a broad range of multimedia types and real-world scenarios that are relevant to many users on the Web.

- The 'photo use case' (Section 2.1) motivates issues around finding and sharing photos on the Web. In order to achieve this, a semantic content understanding is necessary. The issues range from administrative metadata (such as EXIF) to describing the content and context of an image.
- The 'music use case' (Section 2.2) addresses the audio modality. We discuss a broad range of issues ranging from semantically describing the music assets (e.g. artists, tracks) over music events to browsing and consuming music on the Web.
- The 'video use case' (Section 2.3) covers annotation of audiovisual content in the professional audiovisual media production process.

---

[1] http://www.w3.org/2005/Incubator/mmsem/
[2] http://www.w3.org/2008/WebVideo/Annotations/

Multimedia Semantics: Metadata, Analysis and Interaction, First Edition.
Edited by Raphaël Troncy, Benoit Huet and Simon Schenk.
© 2011 John Wiley & Sons, Ltd. Published 2011 by John Wiley & Sons, Ltd.

The use cases introduce the challenges of media semantics in three different areas: personal photo collection, music consumption, and audiovisual media production as representatives of image, audio, and video content. The use cases motivate the challenges in the field and illustrate the kind of media semantics needed for future use of such content on the Web, and where we have just begun to solve the problem.

2.1   Photo Use Case

We are facing a market in which more than 20 billion digital photos are taken per year. The problem is one of efficient management of and access to the photos, and that manual labeling and annotation by the user is tedious and often not sufficient. Parallel to this, the number of tools for automatic annotation, both for the desktop but also on the Web, is increasing. For example, a large number of personal photo management tools extract information from the so-called EXIF header and add this to the photo description. These tools typically also allow the user to tag and describe single photos. There are also many Web tools that allow the user to upload photos to share them, organize them and annotate them. Photo sharing online community sites such as Flickr[3] allow tagging and organization of photos into categories, as well as rating and commenting on them.

Nevertheless, it remains difficult to find, share, and reuse photos across social media platforms. Not only are there different ways of automatically and manually annotating photos, but also there are many different standards for describing and representing this metadata. Most of the digital photos we take today are never again viewed or used but reside in a digital shoebox. In the following examples, we show where the challenges for semantics for digital photos lie. From the perspective of an end user we describe what is missing and needed for next generation semantic photo services.

2.1.1   Motivating Examples

Ellen Scott and her family had a nice two-week vacation in Tuscany. They enjoyed the sun on the Mediterranean beaches, appreciated the unrivaled culture in Florence, Siena and Pisa, and explored the little villages of the Maremma. During their marvelous trip, they took pictures of the sights, the landscapes and of course each other. One digital camera they use is already equipped with a GPS receiver, so every photo is stamped with not only the time when but also the geolocation where it was taken. We show what the Scotts would like to do.

Photo Annotation and Selection

Back home the family uploads about 1000 pictures from the family's cameras to the computer and decides to create an album for grandpa. On this computer, the family uses a nice photo management tool which both extracts some basic features such as the EXIF header and automatically adds external sources such the GPS track of the trip. With their memories of the trip still fresh, mother and son label most of the photos, supported by automatic suggestions for tags and descriptions. Once semantically described, Ellen starts to create a summary of the trip and the highlights. Her photo album software takes in all the pictures and makes intelligent suggestions for the selection and arrangement of the

---

[3] http://www.flickr.com

pictures in a photo album. For example, the album software shows her a map of Tuscany, pinpointing where each photo was taken and grouping them together, making suggestions as to which photos would best represent each part of the vacation. For places for which the software detects highlights, the system offers to add information to the album about the place, stating that on this piazza in front of the Palazzo Vecchio there is a copy of Michelangelo's *David*. Depending on the selected style, the software creates a layout and distribution of all images over the pages of the album, taking into account color, spatial and temporal clusters and template preference. In about an hour Ellen has finished a great album and orders a paper version as well as an online version. They show the album to grandpa, and he can enjoy their vacation at his leisure. For all this the semantics of what, when, where, who and why need to be provided to the users and tools to make browsing, selecting and (re)using easy and intuitive, something which we have not yet achieved.

Exchanging and Sharing Photos

Selecting the most impressive photos, the son of the family uploads them to Flickr in order to give his friends an impression of the great vacation. Of course, all the descriptions and annotations that describe the places, events, and participants of the trip from the personal photo management system should easily go into the Web upload. Then the friends can comment, or add another photo to the set. In all the interaction and sharing the metadata and annotations created should just 'flow' into the system and be recognized on the Web and reflected in the personal collection. When aunt Mary visits the Web album and starts looking at the photos she tries to download a few onto her laptop to integrate them into her own photo management software. Now Mary should be able to incorporate some of the pictures of her nieces and nephews into her photo management system with all the semantics around them. However, the modeling, representations and ways of sharing we use today create barriers and filters which do not allow an easy and semantics-preserving flow and exchange of our photos.

2.1.2   Semantic Description of Photos Today

What is needed is a better and more effective automatic annotation of digital photos that better reflects one's personal memory of the events captured by the photos and allows different applications to create value-added services. The semantic descriptions we need comprise all aspects of the personal event captured in the photos: then when, where, who, and what are the semantics that count for the later search, browsing and usage of the photos. Most of the earliest solutions to the semantics problem came from the field of content-based image retrieval. Content-based analysis, partly in combination with user relevance feedback, is used to annotate and organize personal photo albums. With the availability of time and location from digital cameras, more recent work has a stronger focus on combinations with context-based methods and helps us solve the when and where question. With photo management systems on the Web such as Picasa[4] and management tools such as iPhoto[5], the end user can already see that face recognition and event detection by time stamps now work reasonably well.

---

[4] http://picasa.google.com
[5] http://www.apple.com/ilife/iphoto/

One recent trend is to use the 'wisdom of the crowd' in social photo networks to better understand and to annotate the semantics of personal photo collections: user information, context information, textual description, viewing patterns of other users in the social photo network. Web 2.0 showed that semantics are nothing static but rather emergent in the sense that they change and evolve over time. The concrete usage of photos reveals much information about their relevance to the user. Overall, different methods are still under development to achieve an (automatic) higher-level semantic description of digital photos, as will be shown in this book.

Besides the still existing need for better semantic descriptions for the management of and access to large personal media collections, the representation of the semantics in the form of metadata is an open issue. There is no clear means to allow the import, export, sharing, editing and so on of digital photos in such a way that all metadata are preserved and added to the photo. Rather they are spread over different tools and sites, and solutions are needed for the modeling and representation.

2.1.3   Services We Need for Photo Collections

Services and tools for photo collections are many and varied. There are obvious practices connected to digital photos such as downloading, editing, organizing, and browsing. But also we observe the practice of combining photos into collages, creating physical products such as T-shirts, mugs or composing photo albums. This means that personal photos might have quite an interesting social life and undergo different processes and services that might be running on different platforms and on different sites. Here are some examples of these services and processes for photos (see also Chapter 3).

- Capturing: one or more persons capture an event, with one or different cameras with different capabilities and characteristics.
- Storing: one or more persons store the photos with different tools on different systems.
- Processing: post-editing with different tools that change the quality and perhaps the metadata.
- Uploading: some persons make their photos available on Web (2.0) sites (Flickr). Different sites offer different kinds of value-added services for the photos (PolarRose).
- Sharing: photos are given away or access provided to them via email, websites, print, etc.
- Receiving: photos from others are received via MMS, email, download, etc.
- Combining: Photos from different sources are selected and reused for services like T-shirts, mugs, mousemats, photo albums, collages, etc.

So, media semantics are not just intended for one service. Rather, we need support for semantics for all the different phases, from capture to sharing. This challenges the extraction and reasoning as well as the modeling and exchange of content and semantics over sites and services.

2.2   Music Use Case

In recent years the typical music consumption behavior has changed dramatically. Personal music collections have grown, favored by technological improvements in networks, storage, portability of devices and Internet services. The quantity and availability of songs have reduced their value: it is usually the case that users own many digital music files that they have only listened to once or even never. It seems reasonable to think that by providing listeners with efficient ways to create some form of personalized order in their collections, and by providing ways to explore hidden 'treasures' within them, the value of their collection will dramatically increase.

Also, notwithstanding the many advantages of the digital revolution, there have been some negative effects. Users own huge music collections that need proper storage and labeling. Searching within digital collections requires new methods of accessing and retrieving data. Sometimes there is no metadata – or only the file name – that informs about the content of the audio, and that is not enough for an effective utilization and navigation of the music collection. Thus, users can get lost searching in the digital pile of their music collection. Yet, nowadays, the Web is increasingly becoming the primary source of music titles in digital form. With millions of tracks available from thousands of websites, deciding which song is the right one and getting information about new music releases is becoming a problematic task.

In this sense, online music databases, such as Musicbrainz and All Music Guide, aim to organize information (editorial, reviews, concerts, etc.), in order to give the consumer the ability to make a more informed decision. Music recommendation services, such as Pandora and Last.fm, allow their users to bypass this decision step by partially filtering and personalizing the music content. However, all these online music data sources, from editorial databases to recommender systems through online encyclopedias, still exist in isolation from each other. A personal music collection, then, will also be isolated from all these data sources. Using actual Semantic Web techniques, we would benefit from linking an artist in a personal music collection to the corresponding artist in an editorial database, thus allowing us to keep track of her new releases. Interlinking these heterogeneous data sources can be achieved in the 'web of data' Semantic Web technologies allow us to create. We present several use cases in the music domain benefiting from a music-related web of data.

2.2.1 Semantic Description of Music Assets

Interlinking music-related datasets is possible when they do not share a common ontology, but far easier when they do. A shared music ontology should address the different categories highlighted by Pachet (2005):

- **Editorial metadata** includes simple creation and production information. For example, the song 'C'mon Billy', written by P.J. Harvey in 1995, was produced by John Parish and Flood, and the song appears as track 4 on the album *To Bring You My Love*). Editorial metadata also includes artist biographies, album reviews, and relationships among artists.
- **Cultural metadata** is defined as the information that is implicitly present in huge amounts of data. This data is gathered from weblogs, forums, music radio programs, or even from web search engine results. This information has a clear subjective component as it is based on personal opinions. Cultural metadata includes, for example, musical genre. It is indeed usual that different experts cannot agree in assigning a concrete genre to a song or to an artist. Even more difficult is a common consensus of a taxonomy of

musical genres. Folksonomy-based approaches to the characterization of musical genre can be used to bring all these different views together.

- **Acoustic metadata** is defined by as the information obtained by analyzing the actual audio signal. It corresponds to a characterization of the acoustic properties of the signal, including rhythm, harmony, melody, timbre and structure. Alternatively, music content can be successfully characterized according to several such music facets by incorporating a higher-level semantic descriptor to a given audio feature set. These semantic descriptors are predicates that can be computed directly from the audio signal, by means of the combination of signal processing, machine learning techniques, and musical knowledge.

Most of the current music content processing systems operating on complex audio signals are mainly based on computing low-level signal features (i.e. basic acoustic metadata). These features are good at characterizing the acoustic properties of the signal, returning a description that can be associated with texture, or at best, with the rhythmical attributes of the signal. Alternatively, a more general approach proposes that music content can be successfully characterized according to several musical facets (i.e. rhythm, harmony, melody, timbre, structure) by incorporating higher-level semantic descriptors.

Semantic Web languages allow us to describe and integrate all these different data categories. For example, the Music Ontology framework (Raimond et al. 2007) provides a representation framework for these different levels of descriptors. Once all these different facets of music-related information are integrated, they can be used to drive innovative semantic applications.

2.2.2 Music Recommendation and Discovery

Once the music assets are semantically described and integrated, music-related information can be used for personalizing such assets as well as for filtering and recommendations. Such applications depend on the availability of a user profile, stating the different tastes and interests of a particular user, along with other personal data such as the geographical location.

Artist Recommendation

If a user is interested in specific artists and makes this information available in his profile (either explicitly or implicitly), we can explore related information to discover new artists. For example, we could use content-based similarity statements to recommend artists. Artists who make music that sounds similar are associated, and such associations are used as a basis for an artist recommendation. This provides a way to explore the 'long tail' of music production.

An interesting feature of Semantic Web technologies is that the data being integrated is not 'walled' – music-related information does not have to live only with other music-related information. For example, one could use the fact that a member of a particular band was part of the same movement as a member of another band to provide an artist recommendation. One could also use the fact that the same person drew the album cover

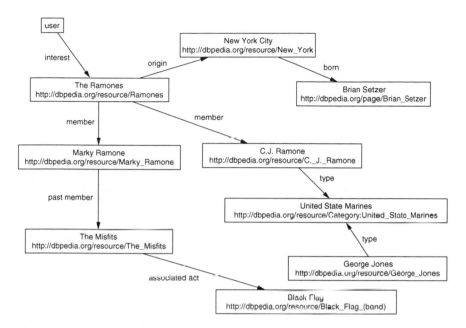

Figure 2.1 Artist recommendations based on information related to a specific user's interest.

art for two artists as a basis for recommendation. An example of such artist recommendations is given in Figure 2.1.

Personalized Playlists

A similar mechanism can be used to construct personalized playlists. Starting from tracks mentioned in the user profile (e.g. the tracks most played by the user), we can explore related information as a basis for creating a track playlist. Related tracks can be drawn from either associated content-based data (e.g. 'the tracks sound similar'), editorial data (e.g. 'the tracks were featured on the same compilation' or 'the same pianist was involved in the corresponding performances') or cultural data (e.g. 'these tracks have been tagged similarly by a wide range of users').

Geolocation-Based Recommendations

A user profile can also include further data, such as a geographical location. This can be used to recommend items by using attached geographical data. For example, one could use the places attached with musical events to recommend events to a particular user, as depicted in Figure 2.2.

2.2.3 Management of Personal Music Collections

In this case we are dealing with (or extending, enhancing) personal collections, whereas in the previous one (recommendation) all the content was external, except the user profile.

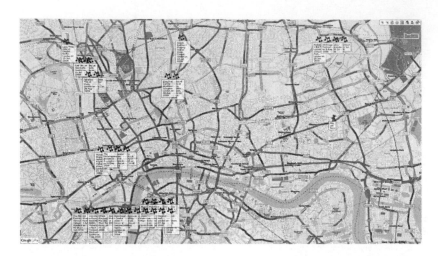

Figure 2.2   Recommended events based on artists mentioned in a user profile and geolocation.

Managing a personal music collection is often only done through embedded metadata, such as ID3 tags included in the MP3 audio file. The user can then filter out elements of her music collection using track, album, artist, or musical genre information. However, getting access to more information about her music collection would allow her to browse her collection in new interesting ways. We consider a music collection as a music-related dataset, which can be related with other music-related datasets. For example, a set of audio files in a collection can be linked to an editorial dataset using the mo:available_as link defined within the music ontology.

```
@prefix ex: <http://zitgist.com/music/track/>.

ex:80ee2332-9687-4627-88ad-40f789ab7f8a mo:available_as
    <file:///Music/go-down.mp3>.
ex:b390e4d2-788c-4e82-bcf9-6b99813f59e6 mo:available_as
    <file:///Music/dog-eat-dog.mp3>.
```

Now, a music player having access to such links can aggregate information about this personal music collection, thus creating a tailored database describing it. Queries such as 'Create me a playlist of performances of works by French composers, written between 1800 and 1850' or 'Sort European hip-hop artists in my collection by murder rates in their city' can be answered using this aggregation. Such aggregated data can also be used by a facet browsing interface. For example, the */facet* browser (Hildebrand et al. 2006) can be used, as depicted in Figure 2.3 – here we plotted the Creative Commons licensed part of our music collection on a map, and we selected a particular location.

2.3   Annotation in Professional Media Production and Archiving

This use case covers annotation of content in the professional audiovisual media production process. In the traditional production process annotation of content (often called

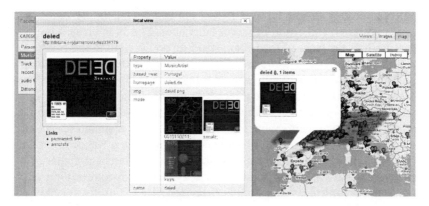

Figure 2.3 Management of a personal music collection using aggregated Semantic Web data by GNAT and GNARQL.

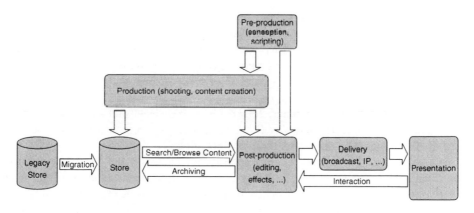

Figure 2.4 Metadata flows in the professional audiovisual media production process.

documentation) is something that happens at the very end of the content life cycle – in the archive. The reason for annotation is to ensure that the archived content can be found later. With increasing reuse of content across different distribution channels, annotation of content is done earlier in the production process or even becomes an integral part of it, for example when annotating content to create interaction for interactive TV applications. Figure 2.4 shows an overview of the metadata flows in the professional audiovisual media production process.

2.3.1 Motivating Examples

Annotation for Accessing Archive Content

Jane is a TV journalist working on a documentary about some event in contemporary history. She is interested in various kinds of media such as television and radio broadcasts, photos and newspaper articles, as well as fiction and movies related to this event. Relevant media content are likely to be held by different collections, including broadcast archives, libraries and museums, and are in different languages. She is interested in details of the media items, such as the name of the person standing on the right in this photo, or the exact place where that short movie clip was shot as she would like to show what the place looks like today. She can use online public access catalogs (OPACs) of many different institutions, but she prefers portals providing access to many collections such as Europeana.[6]

A typical content request such as this one requires a lot of annotation on the different content items, as well as interoperable representations of the annotations in order to allow exchange between and search across different content providers. Annotation of audiovisual content can be done at various levels of detail and on different granularity of content. The simple case is the *bibliograhic* (also called *synthetic*) documentation of content. In this case only the complete content item is described. This is useful for media items that are always handled in their entirety and of which excerpts are rarely if ever used, such as movies. The more interesting case is the *analytic* documentation of content, decomposing the temporal structure and creating time-based annotations (often called *strata*). This is especially interesting for content that is frequently reused and of which small excerpts are relevant as they can be integrated in other programs (i.e. mainly news and sports content).

In audiovisual archives metadata is crucial as it is the key to using the archive. The total holdings of European audiovisual archives have been estimated at 10^7 hours of film and 2×10^7 hours of video and audio (Wright and Williams 2001). More than half of the audiovisual archives spend more than 40% of their budget on cataloging and documentation of their holdings, with broadcast archives spending more and film archives spending less (Delaney and Hoomans 2004). Large broadcast archives document up to 50% of their content in a detailed way (i.e. also describing the content over time), while smaller archives generally document their content globally.

As will be discussed in this book (most notably in Chapters 4 and 12), some annotation can be done automatically, but high-quality documentation needs at least a final manual step. Annotation tools as discussed in Chapter 13 are needed at every manual metadata creation step and as a validation tool after automatic metadata extraction steps. These tools are used by highly skilled documentalists and need to be designed to efficiently support their work. Exchanging metadata between organizations needs semantically well-defined metadata models as well as mapping mechanisms between them (cf. Chapters 8 and 9).

Production of Interactive TV Content

Andy is part of the production team of a broadcaster that covers the soccer World Cup. They prepare an interactive live TV transmission that allows the viewers with iTV-enabled set-top boxes and mobile devices to switch between different cameras and to request additional information, such as information on the player shown in close-up, a clip showing the famous goal that the player scored in the national league, or to buy merchandise of the team that is just about to win. He has used the catalog of the broadcaster's archive to retrieve all the additional content that might be useful during the live transmission and has been able to import all the metadata into the production system. During the live transmission he will use an efficient annotation tool to identify players and events. These

---

[6] http://www.europeana.eu

annotations will allow the production system to offer the viewer the appropriate additional content at the right time.

In such a scenario, retrieval (cf. Chapter 14) is a highly relevant and efficient annotation tool specifically designed for live scenarios. The interoperability between the archive catalog and the production system can only be ensured by using interoperable metadata models (cf. Chapter 8).

2.3.2   Requirements for Content Annotation

In the following we discuss in more detail the main requirements coming from two important scenarios that can be derived from the examples above: broadcast archive documentation (Bailer et al. 2007) and content annotation in interactive TV production (Neuschmied et al. 2007).

Segmentation and structuring. Segmentation of programs into smaller units (e.g. stories, shots, sub-shots) is a crucial step in documentation as further annotation will be in the context of these segments. The structure should be visualized as a tree and on the timeline and it needs to be editable in both visualizations.

Annotation on every segment. Any kind of annotation should be possible on arbitrary segments. This seems obvious for descriptive metadata (content synopsis, named entities) but also applies to production metadata as this might also be different for news stories. In some cases this information may even change for every shot, for example if a news story contains a shot taken from another broadcaster. This requirement has important consequences for the underlying metadata model as it has to be flexible enough to allow segmentation into arbitrary segments and associated annotations.

Support of controlled vocabulary. For annotation of named entities and production metadata, the use of controlled vocabularies such as classifications schemes, thesauri and ontologies is crucial. This not only includes vocabularies that are part of the multimedia metadata standard that is used for representation but also established in-house vocabularies of large organizations that often have been grown over many years. The annotation tool has to provide support for responsibilities of different user groups; in many cases the annotator is only allowed to use the vocabulary while modifications of the vocabulary are in the hands of specific staff. Ideally, the controlled vocabularies should be multilingual.

Metadata interoperability. Broadcast organizations have their own content management systems, in the cases of larger broadcasters often developed in house and using proprietary data models. Thus annotation tools must support metadata formats that enable interoperability with different systems. As production is often outsourced, annotation during production might require one to interface with systems of different organizations involved. In the case of annotation for iTV production, parts of the annotation have to be represented in a format that can be interpreted by the consumer's set-top box.

Related material. Annotation of links to related material (e.g. other audiovisual material, websites) should be possible for each segment. In the archive use case this is relevant in order to put the document content into its historical context. When automatic analysis tools are used this external resource serves as a valuable source of information, for example to extract named entities mentioned in these resources, or to get section headings of newspaper articles (Messina et al. 2006). In the iTV use case this related material is usually called 'additional content', and establishing links between the audio-visual material and this content is the core purpose of annotation. In contrast to the documentation use case, the additional content also includes graphical and interactive elements, such as buttons and menus. The annotation defines which additional content is accessible in which context of the audiovisual content and defines the result of the end user's interactions.

Integration of automatic content analysis. Automatic content analysis and semantic information extraction tools can be used to support the human user in annotation. However, the results of these tools will never be perfect and thus need validation (and possibly correction) by the user. As discussed in Messina et al. (2006), a way to integrate these novel technologies into an archive documentation workflow is in the manual documentation step. Thus this validation functionality needs to be integrated in an annotation tool. A specific type of automatic content analysis result that is relevant in archive applications is the result of an automatic quality analysis (Schallauer et al. 2007). In the case of annotation for iTV production, fast content analysis tools are invoked during annotation in live production scenarios.

Rights. Rights metadata is usually handled (at least in larger organizations) by specifically trained staff and not by archive documentalists. While the metadata model needs to support rights metadata, it is excluded from the professional annotation tool for this reason.

We can see the diverse and complex requirements for professional annotation. Content annotation and archive documentation at broadcasters and cultural heritage institutions have always been done by specially trained and highly qualified staff. It thus becomes clear that, in contrast to tools targeted at end users, completeness in terms of functionality and the effectiveness of the tool are more important than simplicity (cf. Delany and Bauer 2007).

2.4   Discussion

It is evident from the use cases introduced above that on the Web scale the well-known algorithms and methods usually do not work out as desired (Hausenblas 2008). Whilst certainly the research in this direction has only started, we can build on experiences and insights from hypermedia research (Hardman 1998). There are several issues at hand, such as (automatic) metadata extraction vs. (manual) annotation[7] of multimedia resources. Furthermore, we currently have an ecosystem that is almost entirely built upon proprietary application programming interfaces (APIs) – though there exist efforts to overcome this such as DataPortability.[8] However, linked data seem to us to be a more promising and effective approach. As we have discussed in (Hausenblas and Halb 2008; Hausenblas et al. 2009), explicit handling and addressing of multimedia fragments and their inter-

---

[7] See, for example, http://www.w3.org/2005/Incubator/mmsem/XGR-image-annotation/ for the photo case.
[8] http://www.dataportability.org/

linking seem to be one of the next big steps in creating a Web-scale, open and smart multimedia ecosystem on the Web.

Acknowledgements

The authors would like to thank Erik Mannens for providing input and material that helped shape this chapter.

3

Canonical Processes of Semantically Annotated Media Production

Lynda Hardman,[1] Željko Obrenović[1] and Frank Nack[2]
[1]*Centrum Wiskunde & Informatica, Amsterdam, The Netherlands*
[2]*University of Amsterdam, Amsterdam, The Netherlands*

While many multimedia systems allow the association of semantic annotations with media assets, there is no agreed way of sharing these among systems. We identify a small number of fundamental processes of media production, which we term canonical processes, which can be supported in semantically aware media production tools. The processes are identified in conjunction with a number of different research groups within the community and were published in 2008 in a special issue of the *Multimedia Systems* journal (Vol. 14, No. 6). Our effort to identify canonical processes is part of a broader effort to create a rigorous formal description of a high-quality multimedia annotation ontology compatible with existing (Semantic) Web technologies. In particular, we are involved with work on specifying the structure of complex semantic annotations of non-textual data. This has resulted in the Core Ontology for Multimedia (COMM),[1] (Arndt et al. 2008) based on the MPEG-7 standard (MPEG-7 2001) and expressed in terms of OWL (Bechhofer et al. 2004).

We begin with a brief description of all canonical processes. Then, in order to illustrate canonical processes and connect them to the themes of this book, we describe two existing systems related to the personal photo management use case: CeWe Color Photo Book and SenseCam. CeWe Color Photo Book allows users to design a photo book on a home computer and have it printed by commercial photo finishers. SenseCam introduces

---

[1] http://multimedia.semanticweb.org/COMM/

Multimedia Semantics: Metadata, Analysis and Interaction, First Edition.
Edited by Raphaël Troncy, Benoit Huet and Simon Schenk.
© 2011 John Wiley & Sons, Ltd. Published 2011 by John Wiley & Sons, Ltd.

different approaches to structuring, searching, and browsing personal image collections, and tries to automatically locate important or significant events in a person's life.

3.1 Canonical Processes

The definition of the canonical processes of media production is the result of discussions with many researchers and practitioners in the community. This was initiated in a workgroup on 'Multimedia for Human Communication' at a Dagstuhl seminar 05091,[2] with a follow-up workshop at ACM Multimedia 2005 on 'Multimedia for Human Communication – From Capture to Convey'.[3] Our goal with these discussions and the open call for papers for the special issue of the *Multimedia Systems* journal was to establish community agreement on the model before presenting it here.

The special issue of the *Multimedia Systems* journal systematizes and extends this discussion, providing more elaborate identification of canonical processes and their mapping to real systems. In the special issue, which we summarize in this chapter, we identified and defined a number of canonical processes of media production and gave their initial formal description. The special issue is also accompanied with other papers that describe existing systems, the functionality they support and a mapping to the identified canonical processes. These companion system papers are used to validate the model by demonstrating that a large proportion of the functionality provided by the systems can be described in terms of the canonical processes, and that all the proposed canonical processes are supported in more than a single existing system. In this way we also wanted to emphasize that building a useful model must always include the agreement of a significant part of the community. The agreement on particular elements of the model and discussion about usability of the model for a particular domain are as important as – sometimes even more important than – detailed and rigorous formalization. In this special issue we present not 'yet another model', but an approach to building such a model, and the benefits of having a model that a significant part of the community agrees on. Our model is the result of long discussions in the multimedia community, and we present not only the result of this discussion, but also a number of system papers that discuss the model in a particular domain.

Based on an examination of existing multimedia systems, we have identified nine canonical processes of media production. Every process introduced into our model has at least several instances in existing systems. Our model, therefore, does not contain processes that are specific to particular systems. In the following sections we give a detailed description of these nine processes:

- *premeditate,* where initial ideas about media production are established;
- *create media asset,* where media assets are captured, generated or transformed;
- *annotate,* where annotations are associated with media assets;
- *package,* where process artifacts are logically and physically packed;
- *query,* where a user retrieves a set of process artifacts;
- *construct message,* where an author specifies the message they wish to convey;

---

[2] http://www.dagstuhl.de/en/programm/kalender/semhp/?semnr = 05091
[3] http://www.cwi.nl/~media/conferences/mhc05/mhc05.html

- *organize,* where process artifacts are organized according to the message;
- *publish,* where final content and user interface are created;
- *distribute,* where final interaction between end users and produced media occurs.

For each process, we give a brief explanation and state its inputs, outputs and the actors involved. While we give a name to each of the processes, these are meant to be used in a very broad sense. The textual description of each one specifies the breadth of the process we wish to express. Often in real life composite processes are also implemented that combine several canonical processes.

3.1.1   Premeditate

Any media creation occurs because someone has made a decision to embark on the process of creating – whether it be image capture with a personal photo camera, drawing with a drawing tool, professional news video, an expensive Hollywood film or a security video in a public transport system. In all cases there has been premeditation and a decision as to when, how and for how long creation should take place.

In all these cases what is recorded is not value-free. A decision has been made to take a picture of this subject, conduct an interview with this person, film this take of the chase scene or position the security camera in this corner. There is already a great deal of semantics implicitly present. Who is the 'owner' of the media to be created? Why is the media being created? Why has this location/background been chosen? Whatever this information is, it should be possible to collect it, preserve it and attach it to the media that is to be created. For this we need to preserve the appropriate information that can, at some later stage, be associated with one or more corresponding media assets.

3.1.2   Create Media Asset

After a process of premeditation, however long or short, at some point there is a moment of media asset creation. A device can be used to collect images or sound for a period of time, be it photo or video camera, scanner, sound recorder, heart-rate monitor, or MRI.

Note that in this process, we do not restrict the creation of a media asset to newly recorded information. Media assets can also be created in other ways. For example, images can be created with image editing programs or generated by transforming one or more existing images. The essence is that a media asset comes into existence and we are not interested in the method of creation *per se*. If the method is considered significant, however, then this information should be recorded as part of the annotation.

3.1.3   Annotate

The annotate process allows extra information to be associated with any existing process artifact. The term 'annotation' is often used to denote a single human user adding metadata to facilitate search. Here we view annotation as the broader process of adding more easily machine-processable descriptions of the artifact.

The annotations need not be explicitly assigned by a user, but may be assigned by an underlying system, for example by supplying a media asset as input to a feature analysis algorithm and using the extracted result to annotate the media asset. We make no distinction between whether annotations are selected from an existing vocabulary or machine-generated. If deemed relevant, the identity of the human assigner or the algorithm can be recorded in the annotation (Arndt et al. 2008).

We do not prescribe the form of annotations, but require that they can be created and associated with one or more artifacts. We also do not impose limitations on the structure of annotations, due to the high diversity of annotation formats in practice. In most semantically rich systems, however, the structure of an annotation may include a reference to a vocabulary being used, one of the terms from the vocabulary plus a value describing the media asset.

The annotation can refer to any artifact as a whole, but the annotation could also be more specific. In this case, an anchor mechanism is needed to refer to the part of the media asset to which the annotation applies (Halasz and Schwartz 1994). An anchor consists of a media-independent means of referring to a part of the media asset and a media-dependent anchor value that specifies a part of the media asset. For example, for an image this could be an area, for an object in a film a time-dependent description of an area of the image. For further discussion on anchor specifications, see Hardman (1998, p. 53).

3.1.4   Package

The process of packaging provides a message-independent grouping of artifacts. The output of this process is aimed at authors and developers, to help them to maintain process artifacts, and is unrelated to the final presentation organization. This process, for example, can assign a group of related media items and annotations an identity so that it can be retrieved as a unit. One of the simplest forms of packaging is putting related files in one directory, where the directory path provides an identity for the package.

We make a distinction between physical and logical packaging, where physical packaging reflects the organization of units in a database or file system, and logical packaging defines logical relations among items. For example, a Synchronized Multimedia Integration Language (SMIL) presentation is logically one presentation unit, but links media components physically packaged in many files in a distributed environment. On the other hand, a multimedia database can physically be packaged in one file, but contain many logical units.

3.1.5   Query

The query process selects a number of artifacts from a repository of artifacts. Up until now the processes we describe have concentrated on creating, storing and describing primarily media assets. These are needed for populating a media repository. Note that our definition of media repository does not necessarily imply the existence of a complex storage infrastructure, but we assume that systems have a repository where they keep media assets and other artifacts, in the simplest case a hierarchically organized file directory structure. Once there is a repository of artifacts it can be queried for components whose associated media assets correspond to specified properties.

We do not wish to use a narrow definition of the term 'query', but intend to include any interface that allows the artifacts to be searched, using query languages of choice or (generated) browsing interfaces that allow exploration of the content of the archive. It is worth noting that many systems that provide advanced query interfaces also provide support for other processes. For example, browser interfaces can, in addition to a simple query interface, organize intermediate results to present them to a user for feedback, and create temporary presentations that are then published and distributed to the user.

A query of the system may be in terms of media assets, or in terms of the annotations stored with the media assets. A query needs to specify (indirectly) the annotation(s) being used, and includes techniques such as query by example. The mechanisms themselves are not important for the identification of the process.

3.1.6 Construct Message

A presentation of media assets, such as a film or an anatomy book, is created because a human author wishes to communicate something to a viewer or reader. Constructing the message which lies behind the presentation is most often carried out by one or more human authors. When a viewer watches a film or a reader reads a book then some part of the intended original message of the author will hopefully be communicated. In order to give different processes access to the underlying intent, we include an explicit process which brings a processable form of the message into the system. Just as capturing a media asset is input into the system, so is the specification of the message an author wishes to convey.

In some sense, there is no input into the construct message process. However, the real input is the collection of knowledge and experience in the author her/himself. The output of the process is a description of the intended message. For example, a multimedia sketch system, such as described in Bailey et al. (2001), allows an author to gradually build up a description of the message. For the message to be machine processable the underlying semantics need to be expressed explicitly.

A query implicitly specifies a message, albeit a simple one, that an author may want to convey, since otherwise the author would not have been interested in finding those media assets. The query is, however, not itself the message that the author wishes to convey.

In general, we give no recommendation here for the expression of the semantics of the message. We expect that it contains information regarding the domain and how this is to be communicated to the user, but we do not assign anything more than a means of identifying a particular message.

3.1.7 Organize

While querying allows the selection of a subset of media assets, it imposes no explicit structure on the results of one or more queries. The process of organization creates a document structure that groups and orders the selected media assets for presentation to a user. How this process occurs is, again, not relevant, but may include, for example, the linear relevance orderings provided by most information retrieval systems. It also includes the complex human process of producing a linear collection of slides for a talk; creating multimedia documents for the web; ordering shots in a film; or even producing a static two-dimensional poster.

The process of organization is guided by the message (the output of the construct message process). The organization depends on the message and how the annotations of the process artifacts relate to the message. For example, annotations concerning dates could be used to order assets temporally. The resulting document structure may reflect the underlying domain semantics, for example a medical or cultural heritage application, but is not required to. The structure may be color-based or rhythm-based, if the main purpose of the message is, for example, aesthetic rather than informative.

In the arena of text documents, the document structure resulting from organization is predominantly a hierarchical structure of headings and subheadings. The document structure of a film is a hierarchical collection of shots. For more interactive applications, the document structure includes links from one 'scene' to another. In a SMIL document, for example, par and seq elements form the hierarchical backbone of the document structure we refer to here (SMIL 2001).

3.1.8 Publish

The output of the organize process is a prototypical presentation that can be communicated to an end user. This serves as input to the publication process which selects appropriate parts of the document structure to present to the end user. The publication process takes a generic document structure and makes refinements before sending the actual bits to the user. These may include selecting preferred modalities for the user and displayable by the user's device. The resulting presentation can be linear (non-interactive, for example, a movie) or non-linear (interactive, for example, web presentation).

Publication can be seen as taking the document structure from the internal set of processes and converting it (with potential loss of information) for external use. Annotations may be added to describe the published document, for example the device or bandwidth for which the publication is destined. Annotations and alternative media assets may be removed to protect internal information or just reduce the size of the data destined for the user.

3.1.9 Distribute

Created content has to be, synchronously or asynchronously, transmitted to the end user. This final process involves some form of user interaction and requires interaction devices, while transmission of multimedia data to the user device goes through some of the transmission channels including the internet (streamed or file-based), non-networked media (such as a CD-ROM or DVD) or even analog recording media (e.g. film).

It is important to note that the term 'distribution' in our model has a much broader meaning than in classical linear production of media. It can also be used to describe interactive non-linear productions, such as games or other interactive presentations. The resulting system would implement a complex process including query, organize and publish processes in addition to distribution (end-user interaction). For example, some systems can have a final presentation where the storyline depends on user feedback. In this case, the presentation system would include the canonical processes query (to select next part of the story), organize (to organize selected items coherently), publish (to create internal document ready for presentation) and distribute (to present and expose the control interface to user).

3.2 Example Systems

In order to illustrate canonical processes on real-world examples, and to connect them to the themes of this book, we describe two existing systems related to the personal photo management use case: CeWe Color Photo Book and SenseCam. CeWe Color Photo Book allows users to design a photo book on a home computer and have it printed by commercial photo finishers. SenseCam introduces different approaches to structuring, searching and browsing personal image collections, to try to automatically locate important or significant events in a person's life. More detailed description of these systems can be found in the special issue of the *Multimedia Systems* journal.

3.2.1 CeWe Color Photo Book

The *CeWe Color Photo Book* system, described by Sandhaus et al. (2008), allows users to design a photo book on a home computer, and then have it printed by commercial photo finishers. CeWe system inputs are images generated by users, mostly images from users' digital cameras. The system uses low-level feature analysis to select, group and lay out pictures, but it also allows manual authoring enabling users to override suggested selection and layout. The system introduces several steps related to feature extraction, clustering, selection, layout, manual authoring, sorting, ordering and printing.

Feature Extraction

The CeWe Color Photo Book system uses feature extraction methods to annotate images with low-level data. Inputs to this process are digital photos with the EXIF header as creation metadata. The photos undergo a sequence of steps, such as edge detection and color histogram extraction. In the terms of the canonical processes this is a *complex annotation process* involving the CeWe software as a computing actor, outputting one or more annotation artifacts (metadata) associated with the input photos.

Clustering and Selection

Following the annotation processes, a time clustering is performed on the photo set. Inputs to these processes are the photos with associated annotations from the feature extraction step. The clustering process is based on the creation time annotation. In terms of canonical processes, this time clustering is a *complex process* which consists of several instances of a *logical package process*.

The photo clusters are the basis for the selection process which chooses the photos to be included in the photo book. Inputs to this process are the clusters of annotated photos with additional user constraints. With constraints the user can state if similar, blurred or poor print quality images should be selected, and/or if brighter images should be preferred. The user can also limit the number of pages and the number of photos per page. In terms of canonical processes, the selection process is a *complex process* with several instances of the canonical *query* and *organize* processes. Outputs of this query process are selected photo clusters, each cluster with one or more photos.

Layout

The layout process automatically arranges the photos over the pages and chooses the appropriate background for each photo book page. Inputs to this phase are the photo clusters, user preferences, as well as predefined templates for page layouts and backgrounds. User preferences include the style of the page layout and preferred backgrounds. The time clustering information is used to ensure that temporally related photos are kept together, preferably on the same photo book page. Color histogram annotations and user preferences are used to select the background for each page. In terms of canonical processes, the layout process is an instance of the canonical *organize* process. The result of this process is a preliminary photo book layout.

Manual Authoring

The automatically generated photo book layout can undergo various modifications and additions by the user. Alterations on individual photos can be made by 'filters' (such as sepia, black and white, sharpen) which we can describe as instances of the canonical process *transform* (which is an instance of the *create media asset* process). Other modifications include cropping, resizing, moving or rotating of photos, which do not change photos but modify the layout. The user can also delete unwanted photos or add additional photos and text annotations. In terms of canonical processes, this manual authoring is a complex process that, in addition to image transformations, also includes another instance of the *organize* process, as the user has the ability to reorganize images and pages. The result of the manual authoring process is the final photo book layout, which in terms of canonical processes is an instance of a *generic document structure*.

While the user is designing the photo book, he is always able to switch to a preview mode where he can see how the book will look when it is finally produced. In terms of canonical processes, this preview process can be seen as a *complex process* consisting of iterative *publish* and *distribute* processes.

Store and Order

When the user is satisfied with the photo book layout, he can store the layout (with links to the photos) in a physical file. This activity we may describe as an instance of the canonical *physical packaging* process.

To order a physical copy of the photo book, the user can send all required information over the internet or create a CD and send it by mail. This is primarily a *publish* process, with the photo album layout and the associated photos as input. Within this process the required resolution of the photos for the printing process is determined and the photos are scaled down to save bandwidth or space on a CD – that is, this process also involves a canonical *transform* process. The result of this process is a set of image files and a file describing the layout with additional annotations such as shipping and billing information.

Print

The printing process is another instance of a canonical *publish* process, where the abstract photo book layout and selected images are transformed into a physical product, the printed photo book. Inputs to this process are the transformed photos from the the CeWe software, with the order file. The printed photo book is then sent to the customer in a final *distribution* process. The canonical processes and their relation to photo book production are summarized in Table 3.1.

3.2.2   SenseCam

SenseCam (Lee et al. 2008) is a system for structuring, searching and browsing personal image collections. The main goal of the system is to locate important or significant events in a person's life. The system takes as input images created by a small wearable personal device which automatically captures up to 2500 images per day. The system processes these using feature analysis, allowing the user to browse the collection of images organized according to their novelty using different granularities (e.g. today, this week, this month). SenseCam processes images through phases related to capturing, structuring and display.

Capture/Upload

The user can wear the SenseCam device as a visual archive tool for recording images during a particular event or throughout daily life. Before actual capturing of images starts, the user is involved in a form of *premeditate* process, where her wish/motivation to record images for a period of time results in her action of wearing the camera and use of the software system to create the initial user profile. In a more personalized service, the user's motive for wearing the device could be explicitly recorded when she decides to embark on using the system, potentially leading to an adaptive interface optimized for a particular motive.

Wearing the SenseCam throughout the day automatically *captures* a number of photos, and *annotates* them with time stamps and other sensed data (such as ambient temperature and light levels), creating the initial set of media assets and metadata.

Processing and Indexing

A more intensive *annotate* process occurs during the processing time, after the user uploads the file to her computer. Photos with their initial metadata go through a series of content-based analyses to further add machine-generated *annotations*; similarities among the photos are analyzed to determine the boundaries of individual events; a landmark photo is determined from each event; and the novelty value is calculated for each event by comparing the similarities between all events on that day and all other events that occurred during the previous week. From this data, the associations among events are established, both within a day and within a week period.

Using this additional metadata, the system *packages* photos into discrete events so that they can be retrieved as the main units of searching and browsing. Most of these processes are instances of *logical packaging* as the generated metadata are separated from the actual

Table 3.1 Canonical processes and their relation to photo book production

Canonical process	CeWe system
Premeditate	**Capture:** Planning an event (e.g. a holiday trip) which is to be documented in the photo book. Premeditate potentially influenced by external decisions *Input:* Thoughts of author(s) / external decisions *Output:* Schedule/plan
Create Media Asset	**Capture:** Taking a photo *Input:* Spontaneous decision of the photographer, schedule/plan *Output:* Photo equipped with EXIF header (Creation Metadata) **Author:** Altering a photo (cropping, resizing, filtering, rotating) *Input:* Photo in the photo book *Output:* Altered photo **Creation of text annotations:** *Input:* Editor, photo book software, schedule/plan of event *Output:* Text annotation
Annotate	**Author:** Feature extraction on photos (color histograms, edge detection, ...) *Input:* Photos from the input photo set *Output:* Generated metadata
Package	**Capture:** Organizing photos in a separate folder *Input:* Photos, schedule/plan from premeditate *Output:* Folder with photos, identifier is the folder name **Author:** Automatic time clustering of photos *Input:* Photos, time metadata *Output:* Photo clusters **Storing the photo book layout on a hard disc** *Input:* Photo book description *Output:* Physical file with layout information
Query	**Author:** Selecting a subset of images from the clustered input photo set *Input:* Photo clusters, user parameters for photo selection *Output:* Altered photo clusters (subset)
Construct Message	**Capture:** Spontaneous decision to take a photo *Input:* Photographer and his ideas and thoughts *Output:* Decision to take a photo **Author:** Layout decisions for the photo book *Input:* Photographer and his ideas, thoughts, creativity *Output:* Human layout decisions
Organize	**Author:** Actual author process: organizing photos and text over the pages. Split into two steps: the first is automatically done by the CeWe software, the second refinements manually by the CeWe software user *Input:* Photos, text annotations, human layout decisions *Output:* Structured description of the photo book layout

(continued overleaf)

Table 3.1   (*continued*)

Canonical process	CeWe system
	Preparing a photo book order which includes additional annotations like shipping, payment information *Input:* Photo book layout *Output:* Photo book order, such as CD image or internet order
Publish	**Author:** Internal preview of the photo in the photo book software *Input:* Photo book layout *Output:* Rendered images of photo book pages **Print:** Turning the structural photo book description into a physical product *Input:* Photo book order *Output:* Manufactured photo book
Distribute	**Author:** Presenting the user the rendered preview images of the photo book *Input:* Rendered images of photo book pages *Output:* Screen presentation **Print:** Shipping the photo book to the customer *Input:* Manufactured photo book

stored photos; that is, they are only marked-up information that points to different parts of the photo set.

Described automatic processes add partial descriptions of the photos in order to kick-start the initial organization and presentation. Subsequently, metadata may be further enriched with user descriptions as the user provides further feedback during interactive searching and browsing at a later stage.

Accessing the Photos

To review past events, the user can visit the online SenseCam image management system, and *query* the system by explicitly selecting dates, or by typing the terms to search annotation text. The system also supports query by example. The result of a query is a structured set of pointers to the subset of the archived photos ranked by the content-based similarity measures.

Results of the user query are further *organized* for the final presentation. This involves creating an interface template that specifies the sizes of each image according to the novelty assigned to each of the selected events and associated attributes. The output of this process is the visual summary of landmark events to be displayed to the user.

In the *publish* process, the organized information is displayed on the web interface in a comic-book style layout.

It is important to note that generating the final presentation also involves a *construct message* process, where the message emphasizes the novelty of events detected in the photos. The inputs to this process are the system parameters, such as relative size of images, as well as the customized user parameters. The system parameters are determined

Table 3.2 Description of dependencies between visual diary stages and the canonical process for media production

Canonical process	SenseCam image management system
Premeditate (1)	The user deciding to use a wearable camera to capture images of a day's events or of significant events such as going to a wedding, birthday party, zoo which (s)he would like to be recalled and reviewed at some later stage *Input:* User intention/motivation to record images *Output:* Decision to wear the device and access the system; initial user profile that the user provides on the initial online registration form
Create Media Asset (2)	Images passively captured by the SenseCam wearable camera *Input:* Sensor data that triggers automatic photo capture *Output:* Raw images along with sensor file
Annotate (3)	All images automatically time-stamped on download of images from SenseCam to computer *Input:* Output of (2) *Output:* Time-stamped SenseCam images Automatically describe each image in terms of its low-level visual features such as color, texture, or edge *Input:* Time-stamped images from computer *Output:* Annotated images During user interaction, user adds text annotation to events to add value to his/her archive
Package (4)	Automatically segment the annotated images into events *Input:* Annotated images from (3) *Output:* Images organized into events
Query (5)	User selects a particular date, week, month, or a range of dates; or types in text query; or requests all similar events by clicking on Find Similar button *Input:* Images from (4) and query statement (specific date range or text query terms or an example image representing an event) *Output:* List of events and their relative uniqueness rating
Construct Message (6)	Setting display options such as the number of events to be displayed on the page; setting the speed of slideshow on each event, etc. *Input:* User's intention to modify the presentation parameters to suit her interaction/viewing style and preferences *Output:* Modified set of presentation parameters, to be used for this user once the process has occurred
Organize (7)	Creating an interface template that will emphasize the most important events in a visual manner to the user *Input:* List of events with importance values from (5) *Output:* Summary of landmark events to be prepared for display

(continued overleaf)

Table 3.2   (*continued*)

Canonical process	CeWe system
Publish (8)	Selecting appropriate events and their metadata to be presented on the screen (web interface); alternatively, generating appropriate formats in PDF file (for print-out) or on DVD (for offline interactive browsing on TV screen); generating downsized versions suitable for a mobile phone or PDA consumption *Input:* Results of process (7) and all associated presentation-specific information *Output:* Generated presentation formats
Distribute (9)	Displaying the interactive visual summary on the web interface; printing out the daily summary in PDF format; inserting the generated DVD on interactive TV and browsing with a remote controller; transmitting to a mobile phone or a PDA, etc. *Input:* Results of process (8) *Output:* Viewing on a web browser, on a printed paper, or interacting with a DVD player

by the designer at the design time, but the user can also modify some of them. For example, by default, the number of events presented on a page (whether it is for a single day or multiple days) is set at 20, but during browsing the user can adjust this value, enabling her to partially construct a message. Dependencies between visual diary stages and canonical processes for media production are summarized in Table 3.2.

3.3   Conclusion and Future Work

In this chapter we have described a small number of fundamental processes of media production, termed 'canonical processes', which can be supported in semantically aware media production tools. The processes are identified in conjunction with a number of different research groups within the community and are being published in a special issue of the *Multimedia Systems* journal. In order to introduce canonical processes and connect them to the themes of this book, we have described two existing systems related to the personal photo management use case: CeWe Color Photo Book and SenseCam.

The focus of our work is not on rigorous formalization of canonical processes, but on their identification and mapping to real systems. We do, however, see the need for future work to link our current descriptions into a higher-level ontology, and to specify the structure of annotations more precisely. Describing the processes using a foundational ontology, such as DOLCE (Gangemi et al. 2002), provides a solid modeling basis and enables interoperability with other models. DOLCE provides description templates (patterns) for the specification of particular situations and information objects. A next step in the specification of the ontologies would be to express these as specializations of the DOLCE model – in particular, the DOLCE situations.

4

Feature Extraction
for Multimedia Analysis

Rachid Benmokhtar,[1] Benoit Huet,[1] Gaël Richard[2] and Slim Essid[2]

[1]*EURECOM, Sophia Antipolis, France*
[2]*Telecom ParisTech, Paris, France*

Feature extraction is an essential step for processing of multimedia content semantics. Indeed, it is necessary to extract embedded characteristics from within the audiovisual material in order to further analyze its content and then provide a relevant semantic description. These characteristics, extracted directly from the digital media, are often referred to as low-level features. While there has been a lot of work in the signal processing research community over the last two decades toward identifying the most appropriate feature for understanding multimedia content, there is still scope for improvement. One particular effort was led by the Moving Picture Experts Group (MPEG) consortium to provide a set of tools to standardize multimedia content description. This chapter provides an overview of some of the most frequently used low-level features, including some from the MPEG-7 standard, to describe audiovisual content. This chapter begins with an overview of the prominent low-level features for image, video, audio and text analysis. A succinct description of the methodologies employed for their extraction from digital media is also provided. Traditional MPEG-7 descriptors, as well as some more recent developments, will be presented and discussed in the context of a specific application scenario described in Chapter 2 of this book. Indeed, for each of the features relevant to parental control and content filtering of video scenes related to violence, a discussion provide the reader with essential information about its strengths and weaknesses.

With the plethora of low-level features available today, the research community has refocused its efforts toward a new aim: multi-feature and multi-modal fusion. As single-feature systems provide limited performance, it is natural to study the combined use of multiple characteristics originating from the multiple modalities of documents. Here, the various aspects one should consider when attempting to use multiple features from

Multimedia Semantics: Metadata, Analysis and Interaction, First Edition.
Edited by Raphaël Troncy, Benoit Huet and Simon Schenk.
© 2011 John Wiley & Sons, Ltd. Published 2011 by John Wiley & Sons, Ltd.

various modalities together are presented. In particular, approaches for feature selection and feature fusion will be reported as well as synchronization and co-occurrence issues in the context of our chosen scenario: parental control and content filtering of video.

4.1 Low-Level Feature Extraction

4.1.1 What Are Relevant Low-Level Features?

Good descriptors should describe content with high variance and discriminance to be able to distinguish any type of media, taking into account extraction complexity, the sizes of the coded descriptions, the scalability and interoperability of the descriptors. MPEG-7 defines a set of descriptors, some of which are just structures of descriptor aggregation or localization (Eidenberger 2003).

The scenario of this chapter aims to propose an automatic identification and semantic classification of violent content involving two persons (e.g. fighting, kicking, hitting with objects) using visual, audio and textual modalities of multimedia data. As an example, the *punch concept* can be automatically extracted based on audiovisual events, for example two persons in visual data, one moving toward the other, while a punch sound and scream of pain are detected.

The visual modality can involve video shot segmentation, human body and object recognition (weapons, knives), and motion analysis. In the audio modality, signal segmentation and classification in sound categories (speech, silence, music, screaming, etc.) can determine whether or not an event is violent. In the following the MPEG-7 descriptors which we used in this work are presented.

4.1.2 Visual Descriptors

Color Descriptors

Color is the most commonly used descriptor in image and video retrieval, and has been extensively studied during the last two decades. The current version of MPEG-7 (2001) includes a number of histogram descriptors that are able to capture the color distribution with reasonable accuracy for image search and retrieval applications. However, there are some dimensions to take into account, including the choice of color space, the choice of quantization in color space, and quantization of the histogram values. MPEG-7 supports 5 color spaces, namely RGB, HSV, YCrCb, HMMD, and monochrome. The HMMD space is composed of the hue (H), the maximum (max) and the minimum (min) of RGB values, and the difference between the max and the min values. The descriptors are preferentially defined in non-linear spaces (HSV, HMMD) closely related to human perception of color. We now present technical details of each color descriptor.

1. The **dominant color descriptor (DCD)** specifies a set of dominant colors in an image, where a small number of colors are enough to characterize the color information (Cieplinski 2000). In Idrissi et al. (2002), the authors demonstrate that on average we need between four and six colors to build an image histogram model. The DCD is computed by quantization of the pixel color values into a set of N dominant colors c_i,

presented as the following vector components:

$$DCD = \{(c_i, p_i, v_i), s\} \quad i = \{1, 2, \ldots, N\} \tag{4.1}$$

where p_i is the percentage of pixels of the image, v_i the variance associated with c_i, and s is the spatial coherence (i.e. the average number of connecting pixels of a dominant color using a 3×3 masking window).

2. The **color layout descriptor (CLD)** is a compact representation of the spatial distribution of colors (Kasutani and Yamada 2001). It is computed using a combination of grid structure and dominant descriptors, as follows (Figure 4.1):

   • *Partitioning:* the image is divided into an (8×8) block to guarantee the resolution or scale invariance (Manjunath et al. 2001).
   • *Dominant color selection:* for each block, a single dominant color is selected. The blocks are transformed into a series of coefficient values using dominant color descriptors or average color, to obtain $CLD = \{Y, Cr, Cb\}$ components, where Y is the coefficient value of luminance, and Cr and Cb for chrominance.
   • *Discrete cosine transform (DCT):* the three components (Y, Cb, Cr) are transformed into three sets of DCT coefficients.
   • *Non-linear quantization:* a few low-frequency coefficients are extracted using zigzag scanning and quantized to form the CLD for a still image.

3. The **color structure descriptor (CSD)** encodes local color structure in an image using a structuring element of (8×8) dimension as shown in Figure 4.2. The CSD is computed by visiting all locations in the image, and then summarizing the frequency of color occurrences in each structuring element on four HMMD color space quantization possibilities: 256, 128, 64 and 32 bin histograms (Messing et al. 2001).

4. The **scalable color descriptor (SCD)** is defined as the hue–saturation–value (HSV) color space with fixed color space quantization. The Haar transform encoding is used to reduce the number of bins of the original histogram with 256 bins to 16, 32, 64, or 128 bins (MPEG-7 2001).

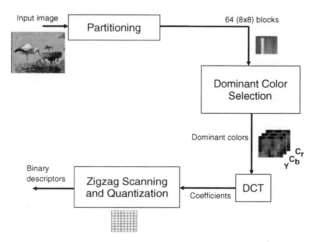

Figure 4.1 Color layout descriptor extraction.

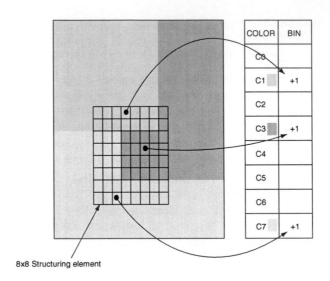

8x8 Structuring element

Figure 4.2   Color structure descriptor structuring element.

5. The **group of frame/group of picture descriptor (GoF/GoP)** is obtained by combining the SCD of the individual frames/images comprised in the group (Manjunath et al. 2002). Three GoF/GoP descriptors are proposed: average, median and intersection scalable color histograms. The average histogram is obtained by averaging the corresponding bins of all the individual histograms. For the median histogram, each GoF/GoP histogram bin value is computed by taking the median of the corresponding bin values of all the individual histograms. The intersection histogram is computed by taking the minimum value of the corresponding bins of all the individual histograms.

Texture Descriptors

Texture, like color, is a powerful low-level descriptor for content-based image retrieval. It refers to the homogeneity properties or not of the visual patterns. In MPEG-7, three texture descriptors are proposed: homogeneous texture, texture browsing, and edge histogram.

1. The **homogeneous texture descriptor (HTD)** characterizes the regional texture using local spatial frequency statistics. The HTD is extracted by Gabor filter banks. According to some results in the literature (Manjunath and Ma 1996; Ro et al. 2001), the optimal numbers for angular and directional parameters are 6 frequency times and 5 orientation channels, resulting in 30 channels in total, as shown in Figure 4.3. Then we compute the energy e and energy deviation d for each channel (Manjunath et al. 2002; Xu and Zhang 2006).

   After filtering, the first and second moments (avg, std) in 30 frequency channels are computed to give the HTD as a 62-dimensional vector,

   $$HTD = [avg, \text{std}, e_1, ..., e_{30}, d_1, ..., d_{30}]. \tag{4.2}$$

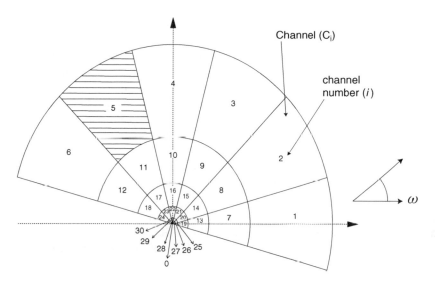

Figure 4.3   HTD frequency space partition (6 frequency times, 5 orientation channels).

The Gabor filters have the capacity to realize filtering close to that realized by our visual perception system. These are sensitive in both orientation and frequency (Manjunath et al. 2001).

2. The **texture browsing descriptor (TBD)** relates to a perceptual characterization of texture, similar to human visual characterization in terms of regularity, coarseness, and directionality (Manjunath et al. 2002). A texture may have more than one dominant direction and associated scale. For this reason, the specification allows a maximum of two different directions and scale values. The TBD is a give-dimensional vector expressed as

$$TBD = [r, d_1, d_2, s_1, s_2]. \tag{4.3}$$

Here regularity r represents the degree of periodic structure of the texture. It is graded on a scale from 0 to 3: irregular, slightly regular, regular, and highly regular. The directionalities (d_1, d_2) represent two dominant orientations of the texture: d_1 denotes the primary dominant orientation and d_2 the secondary. The directionality is quantized to six values, ranging from $0°$ to $150°$ in steps of $30°$. The scales (s_1, s_2) represent two dominant scales of the texture: s_1 denotes the primary dominant scale and s_2 the secondary. The scale is quantized on four levels in the wavelets decomposition, with 0 indicating a fine-grained texture and a 3 indicating a coarse texture.

Like the HTD, the TBD is also computed based on Gabor filter banks with 6 orientations and 4 scales, resulting in 24 channels in total. The analysis is based on the following observations (Wu et al. 1999):

- Structured textures usually consist of dominant periodic patterns. The more regular the periodicity is, the stronger the structuredness is.
- A periodic or repetitive pattern, if it exists, could be captured by filtered images. This behavior is usually captured in more than one filtered output.

- The dominant scale and orientation information can also be captured by analyzing projections of the filtered images.

3. The **edge histogram descriptor (EHD)** expresses only the local edge distribution in the image. An edge histogram in the image space represents the frequency and the directionality of the brightness changes in the image. The EHD basically represents the distribution of five types of edges in each local area called a sub-image. Specifically, dividing the image into (4×4) non-overlapping sub-images. Then, for each sub-image, we generate an edge histogram. Four directional edges ($0°$, $45°$, $90°$, $135°$) are detected in addition to non-directional one. Finally, it generates an 80-dimensional vector (16 sub-images, 5 types of edges). The descriptor can be improved for retrieval tasks by adding global and semi-global levels of localization of an image (Park et al. 2000).

   Independently of the type of media, the EHD does not measure on the ranges [0.32, 0.42], [0.72, 0.81] and [0.9, 1], because the EHD simply counts edges of certain orientations in predefined spatial regions. So, 30% of allowed data are not used.

Shape Descriptors

Like color and texture descriptors, shape has an important role to play in image object retrieval. However, shape descriptors are less developed than color and texture due to shape representation complexity. In MPEG-7, three descriptors are proposed: the region-based shape descriptor, contour-based shape descriptor, and 3D shape descriptor.

1. The **region-based shape descriptor (R-SD)** expresses the distribution of all pixels constituting the shape (interior and boundary pixels). The descriptor is obtained by the decomposition of the shape into a set of basic functions with various angular and radial frequencies as shown in Figure 4.4 (12 angular and 3 radial functions), using angular radial transformation (ART) (Manjunath et al. 2002). The ART is a two-dimensional complex transform defined on a unit disk in polar coordinates. ART coefficients of order (n, m) are given by:

$$F_{nm}(V_{nm}(\rho, \theta), f(\rho, \theta)) = \int_0^{2\pi} \int_0^1 V_{nm}^*(\rho, \theta) f(\rho, \theta) \rho \partial \rho \partial \theta, \qquad (4.4)$$

where $f(\rho, \theta)$ is an image function in polar coordinates, and $V_{nm}(\rho, \theta) = A_m(\theta) R_n(\rho)$ is the ART basis function that is separable along angular A_m and radial R_n directions:

$$A_m(\theta) = \frac{1}{2\pi} \exp(jm\theta), \qquad (4.5)$$

$$R_n(\rho) = \begin{cases} 1 & n = 0 \\ 2\cos(\pi n \rho) & n \neq 0. \end{cases} \qquad (4.6)$$

Normalization by the magnitude of the largest coefficient and quantization to 4 bits are used to describe the shape.

2. The **contour-based shape descriptor (C-SD)** presents a closed two-dimensional object or region contour in an image or video sequence (Zhang and Lu 2003). Six techniques

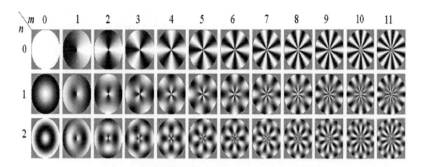

Figure 4.4 Real parts of the ART basis functions (12 angular and 3 radial functions) (MPEG-7 2001).

are proposed in MPEG experiments: Fourier-based technique, shape representation based on Zernike moments and ART, turning angles, wavelet-based technique, and curvature scale space (CSS) representation. However, for content-based image retrieval, a shape representation should satisfy some properties such as affine invariance, compactness, robustness, and low computation complexity. It was concluded that the CSS-based descriptor offered the best overall performance (Manjunath et al. 2002).

To create a CSS description of a contour shape, N equidistant points are selected on the contour, starting from an arbitrary point on the contour and following the contour clockwise. The contour is then gradually smoothed by repetitive low-pass filtering of the x and y coordinates of the selected contour points (Figure 4.5), until the contour becomes convex (no curvature-zero crossing points are found). The concave part of the contour is gradually flattened out as a result of smoothing. Points separating concave and convex parts of the contour and peaks (maxima of the CSS contour map) in between are then identified and the normalized values are saved in the descriptor. Additional global features such as *eccentricity* (equation (4.7)), *circularity* (equation (4.8)) and the *number of CSS peaks* of original and filtered contours can be used to

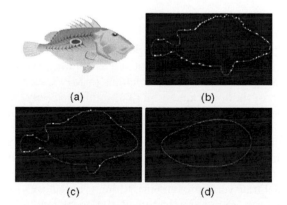

(a)   (b)

(c)   (d)

Figure 4.5 CSS representation for the fish contour: (a) original image, (b) initialized points on the contour, (c) contour after t iterations, (d) final convex contour.

form more practical descriptors (Manjunath et al. 2002):

$$\text{eccentricity} = \sqrt{\frac{i_{20} + i_{02} + \sqrt{i_{20}^2 + i_{02}^2 - 2i_{20}i_2 + 4i_{11}^2}}{i_{20} + i_{02} - \sqrt{i_{20}^2 + i_{02}^2 - 2i_{20}i_{02} + 4i_{11}^2}}}, \tag{4.7}$$

where $i_{02} = \sum_{k=1}^{M}(y_k - y_c)^2$, $i_{20} = \sum_{k=1}^{M}(x_k - x_c)^2$, $i_{11} = \sum_{k=1}^{M}(x_k - x_c)(y_k - y_c)$, M is the number of points inside the contour shape and (x_c, y_c) is the center of mass of the shape;

$$\text{circularity} = \frac{\text{perimeter}^2}{\text{area}}. \tag{4.8}$$

3. The **3D shape descriptor (3D-SD)** provides an intrinsic shape description of three-dimensional mesh models, independently of their spatial, location, size, and meshed representation. The 3D-SD is based on the shape spectrum (SS) concept, which expresses local geometrical attributes of 3D surfaces, and it is defined as shape index values – function of two principle curvatures (Koenderink 1990) – calculated over the entire mesh.
Let p be a point on the 3D surface, and k_p^1, k_p^2 be the principal curvatures associated with point p. The shape index at point p, denoted by $I_p \in [0, 1]$ is defined by

$$I_p = \frac{1}{2} - \frac{1}{\pi}\arctan\frac{k_p^1 + k_p^2}{k_p^1 - k_p^2}, \quad k_p^1 \geq k_p^2, \tag{4.9}$$

except for planar regions, where the shape index is undefined. The shape index captures information about the local convexity of a 3D surface.
The 3D-SD uses a histogram with 100 bins. Two additional variables are used to form the descriptor. The first one expresses the relative area of planar surface regions of the mesh, with respect to the entire area of the 3D mesh. The second one is the relative area of all polygonal components where reliable estimation of the shape index is not possible.

Motion Descriptors

Motion descriptions in video sequences can provide even more powerful clues as to content, introducing its temporal dimension, but are generally very expensive in terms of processes requiring a huge volume of information. In MPEG-7, four descriptors characterize various aspects of motion: camera motion, motion activity, motion trajectory, and parametric motion descriptors.

1. The **camera motion descriptor (CMD)** details what kind of global motion parameters are present at an instance in time in a scene provided directly by the camera, supporting seven camera operations (Figure 4.6), in addition to fixed: panning (horizontal rotation), tracking (horizontal transverse movement), tilting (vertical rotation), booming (vertical transverse movement), zooming (change of the focal length), dollying (translation along the optical axis), and rolling (rotation around the optical axis); see Manjunath et al. (2002).

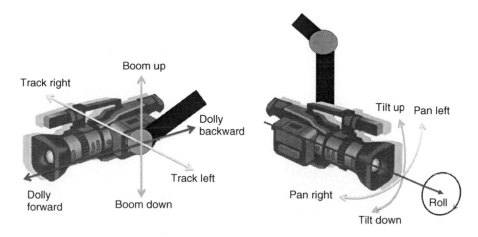

Figure 4.6 Camera operations.

2. The **motion activity descriptor (MAD)** shows whether a scene is likely to be perceived by a viewer as being slow, fast-paced, or action-paced (Sun et al. 2002). An example of *high activity* is goal scoring in soccer game scenes, while *low activity* includes TV news and interview scenes. Additional attributes can also be extracted such as the following:

- *Intensity of motion:* based on standard deviations of motion vector magnitudes. The standard deviations are quantized into five activity values. A high (low) value indicates high (low) activity.
- *Direction of motion:* expresses the dominant direction while a video shot has several objects with differing activity.
- *Spatial distribution of activity:* indicates the number and size of active regions in frame. For example, an interviewer's face would have one large active region, while a sequence of a busy street would have many small active regions.
- *Temporal distribution of activity:* expresses the variation of activity over the duration of the video shot.

3. The **motion trajectory descriptor (MTD)** details the displacement of objects over time, defined as a spatio-temporal localization and presented by the positions of one representative point, such as its center of mass. The trajectory is represented by a list of $N - 1$ vectors, where N is the number of frames in the video sequence (Manjunath et al. 2002). When the video is long, describing every frame would require a large amount of description data and long computing time for retrieval. Partitioning the video frames can reduce the description data size complexity (You et al. 2001).

The proposed description scheme represents each segmented motion trajectory of an object using the polynomial functions

$$\begin{cases} x(t) = x_i + v_{x_i}(t - t_i) + \frac{1}{2}a_{x_i}(t - t_i)^2, \\ y(t) = y_i + v_{y_i}(t - t_i) + \frac{1}{2}a_{y_i}(t - t_i)^2. \end{cases} \qquad (4.10)$$

The parameters i, x_i, y_i, v_i, and a denote the segment index, the initial x and y position of centroid points, the velocity of an object, and the acceleration in each segment, respectively. The trajectory model is a first- or second-order approximation of the spatial positions of the representative point through time, as shown in Figure 4.7.

4. The **parametric motion descriptor (PMD)** addresses the global motion of objects in video sequences, by one of the following classical parametric models (Manjunath et al. 2002):

Models	Equation
Translational (2 parameters)	$\begin{cases} v_x(x, y) = a_1 \\ v_y(x, y) = a_2 \end{cases}$
Rotation/scaling (4 parameters)	$\begin{cases} v_x(x, y) = a_1 + a_3 x + a_4 y \\ v_y(x, y) = a_2 - a_4 x + a_3 y \end{cases}$
Affine (6 parameters)	$\begin{cases} v_x(x, y) = a_1 + a_3 x + a_4 y \\ v_y(x, y) = a_2 + a_5 x + a_6 y \end{cases}$
Perspective (8 parameters)	$\begin{cases} v_x(x, y) = (a_1 + a_3 x + a_4 y)/(1 + a_7 x + a_8 y) \\ v_y(x, y) = (a_2 + a_5 x + a_6 y)/(1 + a_7 x + a_8 y) \end{cases}$
Quadratic (12 parameters)	$\begin{cases} v_x(x, y) = a_1 + a_3 x + a_4 y + a_7 xy + a_9 x^2 + a_{10} y^2 \\ v_y(x, y) = a_2 + a_5 x + a_6 y + a_8 xy + a_{11} x^2 + a_{12} y^2 \end{cases}$

Here $v_x(x, y)$ and $v_y(x, y)$ represent the x and y displacement components of the pixel at coordinates (x, y). The descriptors specify the motion model, the time interval, the coordinate and the value of parameters a_i.

To estimate global motion in a frame, a regular optimization method can be used, where the variables are the parameters of the chosen motion model, and the motion compensation error is the objective function to be minimized,

$$E = \sum_i (I'(x_i', y_i') - I(x_i, y_i))^2, \tag{4.11}$$

between the motion transformed $I'(x_i', y_i')$ and the reference image $I(x_i, y_i)$. Note that the sum is made up of common pixels.

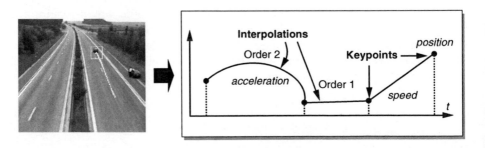

Figure 4.7   Motion trajectory representation (one dimension).

Visual Descriptor Properties

The color descriptors contain a small number of bins, suitable for fast indexing and fast queries. However, CLD and SCD descriptors have a more complex computation process than DCD and SCD descriptors, thus increasing their computational time.

While MPEG-7 standards accommodate different color spaces, most of the color descriptors are constrained to one or a limited number of color spaces to ensure interoperability.

HTD (62 bins) and EDH (80 bins) descriptors are simple and quick to extract, but imply time-consuming browsing. Conversely, TBD (5 bins) is more complex and time-consuming to extract, but provides faster browsing.

C-SD is usually simple to acquire and generally sufficiently descriptive. It is invariant to orientation, scale and mirroring of the object contour (Manjunath et al. 2002). Unfortunately, two disadvantages have to be mentioned: first, C-SD does not capture global features, implying poor retrieval accuracy of the CSS method for curves which have a small number of concavities or convexities; second, it is computationally demanding at the extraction stage. The R-SD descriptor takes into account all pixels within a shape region. It has a small size but is sensitive to segmentation noise and requires even more computation for extraction.

Finally, motion descriptors provide coherent, useful and complementary descriptions. In particular, the MTD size is roughly proportional to the number of key points and depends on the coordinates of the center of mass.

In order to represent the visual characteristics associated with the *punch* concept in scenes of violence, as described in the scenario, the human body or silhouette pattern can be described using its contour and region shape. Some extremely violent actions that are inappropriate for children (e.g. beatings, gunfire and explosions) result in bleeding, which can be identified using color descriptors (Nam et al. 1998).

Most action scenes involve rapid and significant movements of persons and objects. So, motion activity and trajectory to detect the person in movement are used. We can also compute the orientation of arms and legs to draw certain inferences based on their motion pattern over time (Datta et al. 2002). However, the potential existence of weapons is not easy to detect as the object might not be visible due to occlusion. Other descriptors can be introduced such as audio and text, to detect the punch sound and scream of pain, which can provide an efficient descriptions of violent scenes (Mühling et al. 2007; Theodoros et al. 2006).

4.1.3 Audio Descriptors

Concerning audio descriptors, there exists a fairly consensual view about the ideal features for traditional speech tasks such as automatic speech recognition and speaker recognition. However, the greater variability of general audio signals has led researchers to propose numerous features for the various tasks of audio classification (Essid 2005; Peeters 2004). Clearly, there is no consensual set, even for a given task (such as musical instrument recognition or musical genre identification).

A large number of these features are included in the MPEG-7 standard (Herrera et al. 2003), but many others exist. The features considered are classically termed *instantaneous*

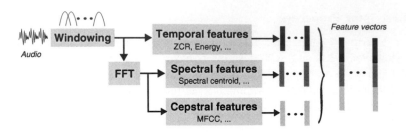

Figure 4.8 Schematic diagram of instantaneous feature vector extraction.

features when they are computed on short (or windowed) overlapping time segments (see Figure 4.8). Some of these features are computed using longer analysis windows and their value is usually replicated on all short segments in order to have all features (those from short or long windows) in a single data vector associated with a short analysis window. Conversely, the features are classically called *global features* when they are computed on the whole signal (e.g. one feature value for a given signal).

The features proposed in the literature can be roughly classified into four main categories: *temporal*, *spectral*, *cepstral*, and *perceptual*.

Temporal Features

These features are directly computed on the time domain waveform. The advantage of such features is that they are usually straightforward to compute. Some of the most common temporal features are further described below.

1. The **time-domain envelope** (also called the energy envelope) is a reliable indicator for silence detection – which is a useful component for many audio segmentation tasks – or for determining video clip boundaries. A simple implementation relies on the computation of the root mean square of the mean energy of the signal $x(n)$ within a frame i of size N:

$$e(i) = \sqrt{\frac{1}{N}\sum_{n=0}^{N-1} x(n)^2}.$$

2. The **zero crossing rate (ZCR)** is given by the number of times the signal amplitude crosses the zero value (or equivalently the number of times the signal amplitude undergoes a change of sign). It can be computed over short or long windows and is given (for a frame i of size N) by

$$z_{cr}(i) = \frac{1}{2}\sum_{n=1}^{N-1} |\mathrm{sign}[x(n)] - \mathrm{sign}[x(n-1)]|,$$

where $\mathrm{sign}[x(n)]$ returns the sign of the signal amplitude $x(n)$. It is a very popular feature since it can, in a simple manner, discriminate periodic signals (small ZCR values) from noisy signals (high ZCR values). This feature has been successfully used in speech/music discrimination problems for clean signals (Scheirer and Slaney 1997); see Figure 4.9 for an example.

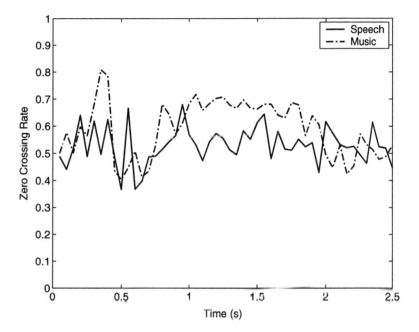

Figure 4.9 Zero crossing rate for a speech signal and a music signal. The ZCR tends to be higher for music signals.

3. **Temporal waveform moments** allow representation of different characteristics of the shape of the time domain waveform. They are defined from the first four central moments,

$$\mu_i = \frac{\sum_{n=0}^{N-1} n^i x(n)}{\sum_{n=0}^{N-1} x(n)},$$

and include:

- the *temporal centroid* which describes the center of gravity of the waveform and is is given by $S_c = \mu_1$;
- the *temporal width* which describes the spread around the mean value and is given by $S_w = \sqrt{\mu_2 - \mu_1^2}$;
- the *temporal asymmetry* S_a which represents the waveform asymmetry around its mean and is defined from the temporal skewness by $S_a = \frac{2\mu_1^3 - 3\mu_1\mu_2 + \mu_3}{S_w^3}$;
- and the *temporal flatness* S_f which represents the overall flatness of the time-domain waveform and is defined from the spectral kurtosis by $S_f = \frac{-3\mu_1^4 + 6\mu_1\mu_2 - 4\mu_1\mu_3 + \mu_4}{S_w^4} - 3$.

  Note that these moments can also be computed on the amplitude envelope or on the spectrum (see the spectral features section below). These features have been successfully used for musical instrument recognition tasks (Essid et al. 2006a).

4. **Amplitude modulation features** characterize the tremolo of a sustained sound when measured in the frequency range 4–8 Hz, and the graininess or roughness of

a sound when in the range 10–40 Hz. Eronen (2001) defines six coefficients: the AM frequency, AM strength, AM heuristic strength, product of tremolo frequency and tremolo strength, as well as the product of graininess frequency and graininess strength.

5. **Autocorrelation coefficients** can be interpreted as the signal spectral distribution in the time domain. In practice, it is common to only consider the first M coefficients which can be obtained as

$$R_i(k) = \frac{\sum_{n=0}^{N-k-1} x(n).x(n+k)}{\sqrt{\sum_{n=0}^{N-k-1} x^2(n)}.\sqrt{\sum_{n=0}^{N-k-1} x^2(n+k)}}.$$

These features were also successfully used in musical instrument recognition tasks (Brown 1998; Essid et al. 2006a).

Spectral Features

Due to the importance of the frequency content in audio perception, a number of features are directly deduced from a spectral representation of the signal. Most features are obtained from the *spectrogram* which is a 3D representation of a sound where energy in each frequency band is given as a function of time. More precisely, the spectrogram represents the modulus of the discrete Fourier transform (DFT) computed on sliding windows with a 50% overlap. Typically, the DFT $X_i(k)$ of the ith frame of audio signal $x(n)$ is given by

$$X_i(k) = \sum_{n=0}^{N-1} x(n)e^{-2j\pi kn/N}, \tag{4.12}$$

and the spectrogram is then obtained by the matrix where each column is the modulus of the DFT of an audio signal frame:

$$SPEC = [\|X_0\| \|X_1\| \ldots \|X_L\|]. \tag{4.13}$$

Similarly to the temporal features, a very high number of proposals have been made. The spectral features described herein are amongst the most popular in recent work tackling audio classification tasks:

1. **Spectral moments** describe some of the main spectral shape characteristics. They include the spectral centroid which was shown to be perceptually related to sound brightness (Iverson and Krumhansl 1994; Krumhansl 1989; McAdams 1993), the spectral width, spectral asymmetry and spectral flatness. They are computed in the same way as the temporal moments features by replacing the waveform signal $x(n)$ by the Fourier frequency components $X_i(k)$ of the signal (see Figure 4.10 for an example involving two music signals). These features have also proved very efficient in tasks related to percussive signal transcription (Gillet and Richard 2004), music genre recognition (Tzanetakis and Cook 2002), and music mood recognition (Liu et al. 2003).

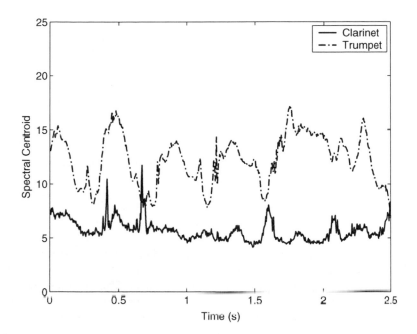

Figure 4.10 Spectral centroid variation for trumpet and clarinet excerpts. The trumpet produces brilliant sounds and therefore tends to have higher spectral centroid values.

2. **Energy in frequency bands** is often desirable to search for information in specific frequency bands. This may be achieved by computing energy or energy ratios in predefined frequency bands. The number of bands, the shape of the prototype filter and the overlap between bands greatly vary between studies (e.g. 3, 6, and 24 bands in Tanghe et al. (2005), Gillet and Richard (2004), and Herrera et al. (2003) respectively for a task of drum track transcription). The bands can also be equally spaced on the frequency axis or placed according to logarithmic or perceptual laws (see Essid et al. (2006a) for energy parameters in octave bands).
3. **Other spectral shape features:** Many other parameters have been suggested to describe the spectral shape of an audio signal and some of them are part of the ISO-MPEG-7 standard (MPEG-7 2001). They include, in particular, the following:
   - *Amplitude spectral flatness:* an alternative option for the spectral flatness parameter is to compute the ratio between the geometric and arithmetic means of the spectral amplitude (globally or, for example, in several octave bands):

$$Sf(i) = \frac{\prod_k |X_i(k)|^{1/K}}{\frac{1}{K} \sum_k |X_i(k)|},$$

   where $|X_i(k)|$ is the spectral amplitude of the kth component for the ith frame of signal. Two related features can also be computed, *spectral slope* and *spectral decrease*, which measure the average rate of spectral decrease with frequency; more details can be found in Peeters (2004).

- *Spectral roll-off* is defined as the frequency under which a predefined percentage (typically between 85% and 99%) of the total spectral energy is present. This parameter was in particular used in tasks such as music genre classification (Tzanetakis and Cook 2002).
- *Spectral flux* characterizes the dynamic variation of the spectral information. It is either computed as the derivative of the amplitude spectrum or as the normalized correlation between successive amplitude spectra.
- *Spectral irregularity features* which aim at a finer information description linked to the sound partials (e.g. individual frequency components of a sound). Several approaches have also been proposed to estimate these features (Brown et al. 2000; Peeters 2004).

Cepstral Features

Such features are widely used in speech recognition and speaker recognition. This is duly justified by the fact that such features allow the contribution of the filter (or vocal tract) in a source-filter model of speech production to be estimated. They can also be interpreted as a representation of the spectral envelope of the signal which justifies their use in general audio indexing applications such as audio summarization (Peeter et al. 2002) or music retrieval by similarity, since many audio sources also obey a source-filter model. The most common features are the *mel-frequency cepstral coefficients* (MFCC). However, other derivations of cepstral coefficients have also been proposed, for example the *linear prediction cepstral coefficients* and the *perceptual linear prediction cepstral coefficients* (Hermansky 1990; Hönig et al. 2005). Cepstral coefficients are also very commonly used jointly with their first and second derivatives. We shall now present more details for each cepstral descriptor.

MFCC are obtained as the inverse discrete cosine transform (type III) of the energy $\tilde{X}_m$ in predefined frequency bands:

$$c_n(i) = \sqrt{\frac{2}{N_f}} \sum_{m=1}^{N_f} \tilde{X}_m(i) \cos\left(\frac{n(m - \frac{1}{2})\pi}{N_f}\right).$$

In practice, a common implementation uses a triangular filter bank where each filter is spaced according to a mel-frequency scale[1] (see Figure 4.11). The energy coefficients $\tilde{X}_m(i)$ are obtained as a weighted sum of the spectral amplitude components $|X_i(k)|$ (where the weights are given according to the amplitude value of the corresponding triangular filter). The number N_f of filters typically varies between 12 and 30 for a bandwidth of 16 kHz.

---

[1] The mel-frequency scale corresponds to an approximation of the psychological sensation of heights of a pure sound (e.g. a pure sinusoid). If several analytical expressions exist, a common relation between the mel scale mel(f) and the Hertz scale f is given by Furui (2001):

$$\text{mel}(f) = 1000 \log_2\left(1 + \frac{f}{1000}\right). \tag{4.14}$$

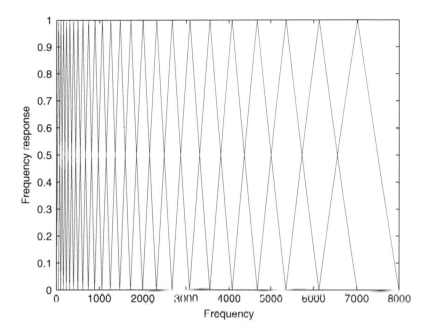

Figure 4.11 Frequency response of a mel triangular filterbank with 24 subbands.

Perceptual Features

Human perception studies have allowed us to better understand the human hearing process; however, the results of these studies are far from having been fully exploited. The MFCC features described above integrate some aspects of human perception (e.g. the integration of energy in frequency bands distributed according to the mel scale), but many other perceptual properties could be used. To illustrate this category, three perceptual features are described below:

1. **Loudness** (measured in sones) is the subjective impression of the intensity of a sound in such a way that a doubling in sones corresponds to a doubling of loudness. It is commonly obtained as the integration of the specific loudness $L(m)$ over all subbands:

$$L_T = \sum_{m=1}^{M_{erb}} L(m),$$

each band being defined on the Equivalent Rectangular Bandwidth psychoacoustic scale (Moore et al. 1997). The specific loudness can be approximated (Peeters 2004; Rodet and Jaillet 2001) by

$$L(m) = E(m)^{0.23},$$

where $E(m)$ is the energy of the signal in the mth band.

2. **Sharpness** can be interpreted as a spectral centroid based on psychoacoustic principle. It is commonly estimated as a weighted centroid of specific loudness (Peeters 2004).

3. **Perceptual spread** is a measure of the timbral width of a given sound and is measured as the relative difference between the specific loudness and the total loudness:

$$S_p = \left(\frac{L_T - \max_n(L(n))}{L_T} \right)^2.$$

Some Other More Specific Features

For some indexing problems, the specific nature of the audio signal may be considered. For example, for monophonic musical instrument recognition, it is possible to consider specific features that would be very difficult (if not impossible) to estimate in polyphonic music. Such features include the following:

- The **odd to even harmonic energy ratio** measures the proportion of energy that is carried by odd harmonics compared to the energy carried by the even harmonics. This is especially interesting in characterizing musical instruments such as the clarinet that exhibit more energetic odd partials than even partials (Peeters 2004). It is computed as

$$OEHR = \frac{\sum_{m=1,3,5,\dots,N_h-1}^{N_H} A^2(m)}{\sum_{m=2,4,6,\dots,N_h}^{N_H} A^2(m)}, \tag{4.15}$$

where $A(m)$ is the amplitude of the mth partial and N_h is the total number of partials (assuming here that N_h is even).
- **Octave band signal intensities** are an interesting alternative to the odd to even harmonic energy ratio. They are meant to capture in a rough manner the power distribution of the different harmonics of a musical sound without resorting to pitch-detection techniques. Using a filterbank of overlapping octave band filters, the log energy of each subband sb, and the log of the energy ratio of each subband to the previous subband $sb - 1$, can be measured to form two descriptors that have proven efficient for musical instrument classification for example (see Essid et al. (2006a) for more details).
- The **attack duration** is defined as the duration between the beginning of a sound event and its stable part (which is often considered as its maximum). It is also a very important parameter for sound identification although it is clearly difficult to estimate in a mixture of different sounds.

As another example, for speech emotion recognition, it is also common to extract very specific features such as the three parameters below which have proved to be efficient in a fear-type emotion recognition system (Clavel et al. 2008):

- **Formants**[2] **parameters** play a prominent role in phoneme identification and especially in vowel identification. Formants are characterized by their center frequency and their bandwidth. It is therefore possible to deduce several parameters from the range and dynamic of variation of the first few formants.

---

[2] One specificity of speech sounds is the presence of spectral resonances, the so-called formants.

- The **harmonic to noise ratio** can be used to estimate the amount of noise in the sound. It is measured as the ratio between the energy of the harmonic component and the energy of the noise component. Both components can be obtained by a periodic/aperiodic decomposition (Yegnanarayana et al. 1998).
- **Fundamental frequency (or pitch)** is defined as the inverse of the period of a periodic sound (such as voiced sounds in speech or monophonic music signals). A number of parameters can be deduced for the purpose of audio indexing tasks including the *range of variation* of the fundamental frequency, the *duration of voiced segments* in speech or the *mean fundamental frequency* of a given audio segment. Other popular features linked to the fundamental frequency include *jitter* and *shimmer*. Jitter illustrates a dynamic aspect of vocal quality change. It characterizes the amount of frequency modulation of the fundamental frequency of a produced sound or succession of voiced sounds. It is given by

$$JS = \frac{\sum_{n=2}^{N-1} |2 * B_n - B_{n-1} - B_{n+1}|}{\sum_{n=1}^{N} B_n},$$

where B_n is the estimated fundamental period at time n. Shimmer, which is related to a measure of amplitude modulation, is often used jointly with jitter since the two phenomena usually appear simultaneously.

Audio Descriptor Properties

Most of the low-level features described above can be reliably extracted from the audio signal in the sense that a small perturbation of the input signal will lead to a small change of the features' value. However, some of the features integrate higher-lever characteristics, especially when they are linked to a signal model (e.g. spectral irregularity, attack duration, or formant parameters). The choice of an appropriate feature set for a given task will therefore rely on a trade-off between robust low-level features and model-based features which are often very efficient but which may be more difficult to extract.

In terms of complexity, time domain features are often the simplest to compute but all features presented can be extracted at a very reasonable cost. It is, however, important to notice that model-based features such as the odd to even harmonic energy ratio or the harmonic to noise ratio may become significantly more complex to extract due to the underlying approach for the estimation of the harmonic content of the audio signal.

In an application of violent scene detection, the task will mainly consist of detecting rapid and abrupt changes of the audio content. To that end, an appropriate feature set will include for example the time-domain envelope and features representing the spectral envelope of the time-varying signal (such as the MFCC or spectral moments). To capture the rapid evolution of these parameters (e.g. to detect gunshots), it is very common to use the first and second derivatives of the selected features (Chen et al. 2006; Clavel et al. 2005). It may also be important to be able to detect threat or fear emotion in speech for which more specific parameters may be needed such as pitch features or formant parameters (Clavel et al. 2008).

4.2   Feature Fusion and Multi-modality

When processing multi-modal documents, typically videos, it is natural to try to combine the features extracted from the different modalities (e.g. the audio and image streams) in order to achieve a thorough analysis of the content. This process is known as *feature fusion*, also referred to as *early fusion*, which we define generically as the process whereby features extracted from distinct 'contexts' are compiled into a common feature representation. In fact, different types of contexts may be considered, especially:

- *temporal* contexts, when fusion acts on feature vectors extracted over different time windows, which is also known as *temporal integration*;
- *spatial* contexts, when fusion acts on different space regions, typically different image regions and possibly also different audio channels;
- *cross-modal* contexts, when features extracted from different data streams (images, audio, texts) are combined.

Note that any combination of the previous contexts is possible. Thus, in a video classification scenario, for example, one may consider the fusion of features extracted from different image regions with a sequence of audio features extracted over a video shot duration. The aim of fusion is to provide a richer and more thorough description of the multimedia objects, that would ease discriminating between them, especially by capturing the dependencies between the different feature vectors being combined. In fact, these dependencies are often very informative about class memberships (think of the relations between different image regions, audio and image streams in videos, or the temporal evolution of the features, etc.).

Now let us consider the fusion of a set of feature vectors $(\mathbf{x}_t)_{1 \leq t \leq T}$ extracted from T different contexts. Various feature fusion techniques can be employed depending both on the contexts and the type of information coded in each $\mathbf{x}_t$. We distinguish two common configurations:

1. The case where each feature vector $\mathbf{x}_t$ contains instances of the same scalar features (in the same order), typically measured at different time or space positions indexed by t; for example $(\mathbf{x}_t)_{1 \leq t \leq T}$ can be a sequence of MFCC vectors measured over T successive analysis windows or a sequence of colour histograms computed over T video frames; we will refer to this type of fusion as *homogeneous fusion*.
2. The case where each $\mathbf{x}_t$ is a feature vector obtained from a different stream of data, typically an audio stream and an image stream, which we will refer to as *cross-modal fusion*.

Obviously, both types of fusion can be combined. We will see that while some fusion techniques can be applied generically, some others are specific to one of the two previous configurations.

4.2.1   Feature Normalization

Prior to fusion, *feature normalization* is often necessary both to cope with the different dynamic ranges of components of the feature vectors (which may be particularly important when a large number of features is used) and the potential mismatch between the feature probability distribution of the training and testing sets.

If specific normalization techniques can be designed for a given type of signal – such as vocal tract normalization for speech signals (Eide and Gish 1996; Pitz and Ney 2005) – most common methods are rather generic.

The most straightforward technique, here called *min–max normalization*, involves mapping the range of variation of each feature in a predefined interval (e.g. $[-1, +1]$). Although simple, this has proved successful for large sets of features with heterogeneous dynamics. However, *mean–variance normalization* is often preferred since it allows centered and unit variance features to be obtained. Such a normalization is obtained as

$$\hat{x}_d = \frac{x_d - \mu(x_d)}{\sigma(x_d)}, \tag{4.16}$$

where $\mu(x_i)$ and $\sigma(x_i)$ are respectively the mean and standard deviation of the feature x_d.

The above technique normalizes the first- and second-order moments of the features. It is, however, possible to normalize the probability distribution function by means, for example, of histogram equalization (Molau et al. 2001). Such approaches aim to transform the test data so that their distribution matches the distribution of the training data. New directions are also explored to integrate the temporal dimension into the normalization process. Often referred to as filtering techniques, they have recently been of interest in the field of robust speech recognition (Xiao et al. 2007).

4.2.2 Homogeneous Fusion

We focus on the situation where the index t is a time index.[3] This is generally known as temporal integration (Meng 2006).

The most widespread temporal integration approach tends to be the use of the first statistical moments, especially the mean and standard deviation, estimated on the feature vector trajectory $(\mathbf{x}_t)_{1 \le t \le T}$ (Joder et al. 2009; Scaringella and Zoia 2005; Wold et al. 1996). An efficient summary of the sequence is thus obtained with the advantage of a significant computational load reduction, compared to the case where every single observation $\mathbf{x}_t$ is evaluated by the classifiers. However, the integration of the whole sequence of observations (over the signal duration) may result in less accurate classification. In fact it has been shown (Joder et al. 2009) that it is advantageous to combine feature fusion with late fusion (i.e. fusion of the classifiers' decision) in order to achieve good performance, hence considering T-length integration windows with T much smaller than the signal length.

More sophisticated temporal integration techniques have been proposed by researchers to address the fact that simple trajectory statistics are invariant to permutations of the feature vectors $\mathbf{x}_t$, which means that they are unable to capture the dynamic properties of the signals. Proposals include the concatenation of successive vectors (Slaney 2002), the use of a filterbank to summarize the periodogram of each feature (McKinney and Breebart 2003), the use of autoregressive models to approximate the feature temporal dynamics

---

[3] There is little loss of generality in so doing, as the same techniques still apply to other 'homogeneous' scenarios, although with different interpretations.

Athineos et al. (2004); Meng (2006), and the extraction of new spectral hyper-features from the original sequence $(\mathbf{x}_t)_{1 \leq t \leq T}$ (Joder et al. 2009). These approaches tend to provide little in the way of effective improvements to the classification accuracy and are generally far more complex.

4.2.3   Cross-modal Fusion

For the sake of conciseness, we focus on a specific type of cross-modality, namely the audiovisual scenario where we suppose two streams of features are extracted from each video: a stream of audio features $\mathcal{X}_a \in \mathbb{R}^{D_a}$ and a stream of visual features $\mathcal{X}_v \in \mathbb{R}^{D_v}$. This scenario is found in many applications, ranging from audiovisual speech processing (Potamianos et al. 2004) and biometric systems (Rua et al. 2008) to video and music indexing systems (Gillet and Richard 2005; Gillet et al. 2007; Iyengar et al. 2003).

We will suppose that the audio and visual features are extracted at the same rate – which is often obtained by downsampling the audio features or upsampling the video features as in Rua et al. (2008) – and that they share the same length.

Feature Concatenation

The most basic audiovisual feature fusion approach consists of concatenating the audio and visual feature vectors, $\mathbf{x}_a$ and $\mathbf{x}_v$ ($\mathbf{x}_a \in \mathcal{X}_a$ and $\mathbf{x}_v \in \mathcal{X}_v$), to form $\mathbf{x}_{av} = [\mathbf{x}_a, \mathbf{x}_v]$. However, the dimensionality of the resulting representation is often too high, which negatively affects the learning of the classifiers. Consequently, approaches to reduce the dimensionality of the problem have been proposed.

Feature Transformation

After feature concatenation, one might need to reduce the resulting potentially large feature space dimensionality. A common approach is to use transformation techniques such as Principal component analysis (PCA) or linear discriminant analysis (LDA). The principle of PCA, also known as the Karhunen–Loève transform, is to find a linear transformation of the initial correlated feature set to obtain a reduced uncorrelated feature set (of dimension $M < D$) that best represents the data in a least-squares sense. In other words, this can be viewed as a projection of the initial data $\mathbf{x}$ on the new coordinate axes (called principal components) for which the variances of $\mathbf{x}$ on these axes are maximized – for example, the first axis corresponds to the maximal variance, the second axis corresponds to the maximal variance in the direction orthogonal to the first axis, and so on (Hyvärinen et al. 2001). In practice, PCA can be obtained by computing an eigenvalue decomposition (EVD) of the covariance matrix $\mathbf{R}_{\mathbf{xx}}$ of the data $\mathbf{x}$ such that

$$\mathbf{R}_{\mathbf{xx}} = \mathbf{U}\boldsymbol{\Gamma}\mathbf{U}^{\mathsf{T}}, \tag{4.17}$$

where the diagonal elements of $\boldsymbol{\Gamma}$ contain the eigenvalues sorted in decreasing order. The transformed data $\mathbf{y}$ is then obtained by applying $\mathbf{U}^{\mathsf{T}}$ to the data $\mathbf{x}$:

$$\mathbf{y} = \mathbf{U}^{\mathsf{T}}\mathbf{x}. \tag{4.18}$$

It can then be easily verified that the covariance matrix $\mathbf{R_{yy}}$ of the transformed data is given by $\mathbf{R_{yy}} = \mathbf{U^T R_{xx} U^T} = \mathbf{\Gamma}$. Then, it follows that the components of the transform data $\mathbf{y}$ are uncorrelated (since $\mathbf{R_{yy}}$ is diagonal) and that the first few components (the so-called principal components) concentrate the variance of the original data $\mathbf{x}$. Note that the PCA exploits the redundancy in $\mathbf{x}$ (or mutual correlation of the features) to obtain a reduced set of uncorrelated features. It is then useless to apply the PCA to independent elements. It is also worth to remark that PCA is particularly well adapted when combined with a Gaussian model with diagonal covariance.

A number of extensions of the classical PCA have already been proposed, including linear approaches such as weighted PCA, normalized PCA and supervised PCA (Koren and Carmel 2004) and non-linear approaches based, for example, on neural networks.

Feature Selection

As will be described in more depth in Chapter 5, feature selection aims to select only useful descriptors for a given task and discard the others. When applied to the feature vectors $\mathbf{x}_{av}$, feature selection can be considered as a feature fusion technique whereby the output will hopefully retain the 'best of $\mathbf{x}_u$ and $\mathbf{x}_v$', that is, a subset of most relevant audiovisual features (with respect to the selection criterion).

Audiovisual Subspaces

Many techniques have been suggested to map the audio and visual feature vectors to a low-dimensional space where the essential information from both streams is retained. This includes techniques widely used for feature transformation, such as PCA (Chibelushi et al. 1997), independent component analysis (Smaragdis and Casey 2003) and LDA (Chibelushi et al. 1997), but also various methods aiming to effectively exploit the dependencies between the audio and visual features, with the assumption that the type of association between these feature streams is useful for the classification task.

The latter methods find two mappings f_a and f_v (that reduce the dimensions of $\mathbf{x}_a$ and $\mathbf{x}_v$) such that a dependency measure $S_{av}(f_a(\mathcal{X}_a), f_v(\mathcal{X}_v))$ is maximized. Various approaches can be described, using this same formalism. Darrell et al. (2000) choose the mutual information (Cover and Thomas 2006) as a dependency measure and find single-layer perceptrons f_a and f_v projecting the audiovisual feature vectors to a two-dimensional space. Other approaches use linear mappings to project the feature streams. Canonical correlation analysis (CCA) aims to find two unit-norm vectors $\mathbf{t}_a$ and $\mathbf{t}_v$ such that

$$(\mathbf{t}_a, \mathbf{t}_v) = \arg \max_{(\mathbf{t}_a, \mathbf{t}_v) \in \mathbb{R}^{D_a} \times \mathbb{R}^{D_v}} \mathrm{corr}(\mathbf{t}_a' \mathcal{X}_a, \mathbf{t}_v' \mathcal{X}_v). \tag{4.19}$$

Yet another technique (known to be more robust than CCA) is co-inertia analysis (CoIA). This involves maximizing the covariance between the projected audio and visual features:

$$(\mathbf{t}_a, \mathbf{t}_v) = \arg \max_{(\mathbf{t}_a, \mathbf{t}_v) \in \mathbb{R}^{D_a} \times \mathbb{R}^{D_v}} \mathrm{cov}(\mathbf{t}_a' \mathcal{X}_a, \mathbf{t}_v' \mathcal{X}_v). \tag{4.20}$$

We refer the reader to Goecke and Millar (2003) for a comparative study of these techniques and to Rua et al. (2008) for an interesting example of their use in audiovisual speaker verification.

4.3   Conclusion

In this chapter, the most prominent multi-modal low-level features for multimedia content analysis have been presented. Given the parental control and content filtering of video scenarios, the individual characteristics have been placed in a realistic context, allowing their strengths and weaknesses to be demonstrated. Recently, researchers in multimedia analysis have started to give particular attention to multi-feature, multi-modal multimedia analysis. A brief but concise overview has been provided here. Both feature selection and feature fusion approaches were presented and discussed, highlighting the need for the different features to be studied in a joint fashion. The alternative to fusion at the feature level (often referred to as early fusion) is classifier fusion (late fusion) where the combination of cues from the various features is realized post classification. This topic, among others, will be discussed in the next chapter.

5

Machine Learning Techniques for Multimedia Analysis

Slim Essid,[1] Marine Campedel,[1] Gaël Richard,[1] Tomas Piatrik,[2]
Rachid Benmokhtar[3] and Benoit Huet[3]
[1]*Telecom ParisTech, Paris, France*
[2]*Queen Mary University of London, London E1 4NS, UK*
[3]*EURECOM, Sophia Antipolis, France*

Humans have natural abilities to categorize objects, images or sounds and to place them in specific classes in which they will share some common characteristics, patterns or semantics. For example, in a large set of photographs a natural classification strategy could be to organize them according to their similarity or the place they were taken (e.g. all photographs of Michelangelo's *David* would be grouped together). However, this is a somewhat tedious task and it would therefore be desirable to have automatic tools to obtain such classification. Whilst localization technologies (e.g. GPS for outdoors and RF-based approaches for indoors) can tell us where the photograph was taken, they do not necessarily tell us what the photo depicts. Thus it would appear natural that machine learning can be useful, for example, in learning models for very well-known objects or settings (such as Michelangelo's *David* or other landmark images (Van Gool et al. 2009)).

More generally, the fundamental task of a classification algorithm is therefore to put data into groups (or classes) according to similarity measures with or without a priori knowledge about the classes considered. The approaches are termed *supervised* when a priori information (such as the labels) is used. They are otherwise called *unsupervised*.

For the supervised case, the most straightforward and traditional approach to automatic classification is based on a training phase in which class models are built (see Figure 5.1). For this phase, a training database gathering a representative (and ideally very large) set

Multimedia Semantics: Metadata, Analysis and Interaction, First Edition.
Edited by Raphaël Troncy, Benoit Huet and Simon Schenk.
© 2011 John Wiley & Sons, Ltd. Published 2011 by John Wiley & Sons, Ltd.

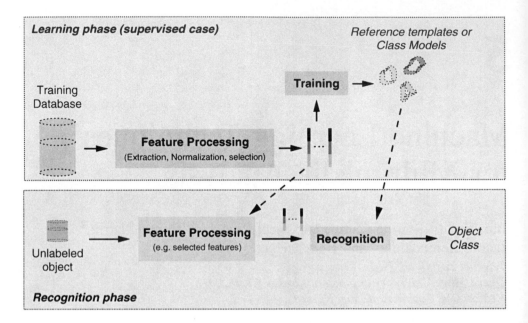

Figure 5.1   Schematic architecture for an automatic classification system (supervised case).

of examples for each class is used. A set of features is then computed for each instance
of each class of the training database. These features are then used either as reference
templates or to build a statistical model for each class. If the number of initial features is
large, it is then particularly useful to rely on feature transformation techniques to obtain
efficient features of smaller dimension. Finally, in the recognition phase, the same set of
features is computed and is compared either to the templates or to the model to determine
what the most probable classes of the current object are.

For example, in a problem of music instrument recognition in polyphonic music (Essid
et al. 2006a), a class represents an instrument or a group of instruments (e.g. brass).
In a photograph classification problem, the class can be defined by the location (e.g.
Pisa, Sienna) or a specific characteristic (e.g. pictures with blue skies). For violent scene
detection, the problem can be considered as a three-class classification task where each
class is related to the intensity of violence in the scene (scenes with extreme violence,
violent scenes, other neutral scenes).

The aim of this chapter is not to give an exhaustive overview of existing machine
learning techniques but rather to present some of the main approaches for setting up
an automatic multimedia analysis system. Following the description of a large set of
appropriate features for multimedia analysis (see Chapter 4), this chapter is organized as
follows. Section 5.1 is dedicated to feature dimensionality reduction by automatic feature
selection algorithms. Then Section 5.2 is devoted to the classification approaches including
both supervised and unsupervised techniques. The fusion (or consensus) approaches are
then described in Section 5.3. Some conclusions are presented in Section 5.4.

5.1   Feature Selection

In many classification tasks, a very large number of potentially useful features can be considered – for example, several hundreds of features are used in the musical instrument recognition system presented in Essid et al. (2006a) and in the fear-type emotion recognition system proposed in Clavel et al. 2008. Though it is sometimes practicable to use all these features for the classification, it is obviously sub-optimal to do so, since many of them may be redundant or, even worse, noisy owing to non-robust extraction procedures. Thus, feature selection or transformation (see Chapter 4) become inevitable in order to reduce the complexity of the problem – by reducing its dimensionality – and to retain only the information that is relevant in discriminating the possible classes, hence yielding a better classification performance.

Feature selection is an interesting alternative to feature transform techniques such as PCA (see Section 4.2.3) as the latter present the inconvenience of requiring that all candidate features be extracted at the test stage (before the transform found during training is applied to them). Moreover, PCA does not guarantee that noisy features will be eliminated (since noisy features may exhibit high variance) and the transformed features are difficult to interpret, which is a major drawback if one expects to gain some understanding of the classes.

In feature selection (FS), a subset of D_s features is selected from a larger set of D candidates with the aim of achieving the best classification accuracy.[1] The task is very complex: not only is it impractical to perform the exhaustive subset search because of the extremely high combinatorics involved, especially when d is also to be found, but also it is costly to evaluate the classification accuracy for each candidate feature subset. Therefore feature selection is generally solved in a sub-optimal manner by introducing two main simplifications:

• Instead of using the classification accuracy, a simpler feature selection criterion is preferred that uses the initial set of features intrinsically. This is referred to as a *filter approach*, as opposed to the *wrapper approach* where the classifier is used in the selection process and the criterion is directly related to the classification accuracy.
• Brute-force search is avoided by leveraging a near-optimal search strategy, as will be described hereafter.

5.1.1   Selection Criteria

What selection criteria can be used with filter approaches (where one does not rely on any classifier to select the features)? The most commonly used criteria tend to be chosen among the following categories:

---

[1] At least, this is our aim for most multimedia indexing tasks; accuracy may not be the only concern in other areas (Liu and Motoda 2000).

- *Discriminant power*, whereby features that provide the best class separability are preferred; frequently used separability measures include variants of inter-class scatter to intra-class scatter ratios (Theodoridis and Koutroumbas 1999), for instance Fisher's discriminant ratio (when individual scalar features are evaluated) and various probabilistic distances such as the divergence or Bhattacharryya distance (Duda et al. 2001).
- *Redundancy*, although potentially useful for consolidating class separability (Guyon and Elisseeff 2003), may produce negative side effects, especially when d is small. Redundant features should generally be avoided, both for efficiency reasons, and for selecting complementary features that span over the diverse properties of the classes to properly discriminate them.
- *Information and dependence* refer to criteria that select features which are strongly related to the classes considered, either using a dependence measure, such as the mutual information between features and classes (Zaffalon and Hutter 2002), or an information gain measure that characterizes how the uncertainty about the classes is reduced by the features (Liu and Motoda 2000).

5.1.2 Subset Search

If an exhaustive feature subset search is envisaged, the search space size is 2^D, which clearly indicates that this option is too time-consuming for realistic D values (often amounting to several hundreds). This leads one to resort to one of the various sub-optimal search methods that aim at solutions as close as possible to the optimal one, in a more or less sophisticated manner. Any search method will adopt a *search direction* and a *search strategy*.

Choosing a search direction consists in choosing a feature subset generation procedure, generally in a sequential way. Basically, this is achieved by answering the question: 'should one grow a feature subset from any empty solution or should one rather start with the original complete set of features and progressively reduce it?' The first option is referred to as *forward generation* where, based on the selection criterion, one or more relevant features are added at each step, while the latter is named *backward generation* where, at each step, one or more features are discarded. Both directions can actually be explored through *bidirectional generation* or *sequential floating search* with the aim of avoiding getting stuck in local optima (Liu and Motoda 2000; Theodoridis and Koutroumbas 1999). Another option is *random generation* which chooses random search directions as the search progresses (Liu and Motoda 2000).

Beyond the search direction issue, a near-optimal search strategy is still to be defined, which can be chosen among the following options.

- *Heuristic* search strategies rely on heuristics in performing the search, making a strong trade-off between search speed and optimality of the solution. They explore a restricted number of search paths with the risk of getting stuck in local optima.
- *Stochastic* search strategies make the whole search process non-deterministic by testing random selections of features (using random generation); instances of such strategies include simulated annealing and genetic algorithms (Theodoridis and Koutroumbas

1999). More optimal solutions are expected to be found using such strategies, often at the cost of a greater computational load; see Debuse and Rayward-Smith (1997), Lanzi (1997), Demirekler and Haydar (1999), Essid et al. (2006b), and Oh et al. (2002) for examples.

5.1.3 Feature Ranking

Another simplistic and yet popular feature selection approach that is worth mentioning, reduces the task to one of ranking each feature. Each individual feature is first scored – independently of the others – using some criterion (say, a separability criterion). Then the features are sorted with respect to their scores and the D_s top-ranked elements are retained for the classification.

This approach is clearly sub-optimal compared with the previous search strategies, which does not prevent it from yielding satisfactory performance in practice. Its main advantage is obviously its low complexity.

5.1.4 A Supervised Algorithm Example

We now present a popular feature selection algorithm that tends to provide satisfactory results for multimedia classification. It is a filter approach using a scatter matrix criterion and heuristic sequential forward search.

Let Σ_k be the covariance matrix estimated from the n_k training examples belonging to class C_k, $1 \leq k \leq K$, and defined by its elements

$$\Sigma_k(i, j) = \frac{(\mathbf{x}_i^k - \boldsymbol{\mu}^k)^t (\mathbf{x}_j^k - \boldsymbol{\mu}^k)}{n_k - 1}, \quad 1 \leq i, j \leq D_s, \tag{5.1}$$

with $\boldsymbol{\mu}^k = \frac{1}{n_k} \sum_{i=1}^{n_k} \mathbf{x}_i^k$. Let $\pi_k = \frac{n_k}{n}$ be the estimate of $P(C = C_k)$ and $\boldsymbol{\mu} = \frac{1}{n} \sum_{k=1}^{n} \mathbf{x}_k$ be the empirical mean of all the examples. Finally, let

$$\mathbf{S}_w = \sum_{k=1}^{K} \pi_k \Sigma_k$$

and

$$\mathbf{S}_b = \sum_{k=1}^{K} (\boldsymbol{\mu} - \boldsymbol{\mu}_k)(\boldsymbol{\mu} - \boldsymbol{\mu}_k)^t.$$

$\mathbf{S}_w$ is the intra-class scatter matrix and $\mathbf{S}_b$ is the inter-class scatter matrix. Good class separability is achieved by maximizing the inter-class scatter while minimizing the intra-class scatter. Therefore the separability criterion (to be used as selection criterion) can be defined as

$$J_s = \text{tr}(\mathbf{S}_w^{-1} \mathbf{S}_b). \tag{5.2}$$

The selection is then performed following Algorithm 1 in the next page.

Algorithm 1 Forward filter algorithm

Require: $\mathbf{X} \leftarrow [\mathbf{x}_1^T, \ldots \mathbf{x}_i^T, \ldots \mathbf{x}_n^T]^T$ {Training examples}
         D_s {Target number of features}
  {Initialization}
$r \leftarrow 0$ {Iteration number}
$d_r \leftarrow 0$ {Dimension at iteration r}
$\mathcal{S}_{d_r} \leftarrow \{\}$ {Subset of features selected}
while $d_r < D_s$ **do**
  $j = 1$
  while $j < D$ **do**
    $\mathcal{S}' \leftarrow \mathcal{S}_{d_r} \bigcup (x_{i,j})_{1 \leq i \leq n}$
    Evaluate separability on $\mathcal{S}'$ using expression (5.2)
    $j \leftarrow j + 1$
  end while
  $j_0 \leftarrow$ index of the feature that maximizes J_s
  $\mathcal{S}_{d_{r+1}} \leftarrow \mathcal{S}_{d_r} \bigcup (x_{i,j_0})_{1 \leq i \leq n}$
  $d_{r+1} \leftarrow d_r + 1$
  $r \leftarrow r + 1$
end while
Ensure: $\mathcal{S}_{D_s}$ {D_s selected features}

Unsupervised Feature Selection

Context of Use
When the classes of interest are not known a priori, an unsupervised feature selection problem arises. Specific approaches are then needed which can be, as in the supervised case, either of wrapper or filter type.

Unsupervised Wrappers
Concerning wrappers, the basic algorithm presented above can be adapted to the unsupervised case. More specifically the final application is generally related to a clustering task; hence the selection of features is performed using comparison between clustering results. The main problem is that there is no *consensual* method to evaluate the efficiency of a clustering or even to compare two clustering results. Several ideas have been proposed in the literature using information theory-based criteria, such as mutual information, to compare two classification results and then decide which features should be selected to find the best clustering. In Di Gesu and Maccarone (1985) both statistical and possibilistic (fuzzy) tools are combined.

This approach is clearly dependent on the quality of the clustering methods and the evaluation criteria. This can also be very time-consuming. Other approaches propose to use genetic algorithms (Kim et al. 2002) to explore the space of feature subsets instead of using greedy approaches; they are confronted with the same problem when defining a fitness function to decide which clustering result is better or weaker than the others.

Unsupervised Filters

Filters are supposed to be faster, based on simple, mainly statistical, properties. For example, they can easily be used to reduce redundancy in the original feature set. This redundancy can be analyzed using a similarity criterion between features, such as a correlation measure (possibly a Pearson coefficient) or the maximal information compression index (Mitra et al. 2002). This criterion is used to group features, and representatives of each group make up the final set of selected features. This process is very similar to what is performed in data selection. In Campedel et al. (2007) several feature clustering approaches are presented, including a consensual approach (see Section 5.3) providing automatic estimation of the number of features to be selected as well as mining tools to study feature dependencies.

5.2 Classification

In this section, we present an overview of the best-known supervised and unsupervised approaches used in the context of multimedia data management. Section 5.2.1 is dedicated to historical algorithms based on learning by examples and statistical modeling; these algorithms are still frequently used because of their computational tractability and mathematical foundation. During the 1990s, kernel methods (e.g. support vector machines) emerged and produced competitive performances compared to biologically inspired approaches (mainly neural networks). Hence, Section 5.2.2 is dedicated to the description of the principles of kernel methods and related problems (such as model selection). Section 5.2.3 can be considered as an extension to the classification of sequences. Section 5.2.4 describes state-of-the-art biologically inspired methods based on evolutionary algorithms.

5.2.1 Historical Classification Algorithms

k-Nearest Neighbors

In the 18th century, Carl von Linné (1707–1778) attempted to classify species and produced the first taxonomies (hierarchies of classes) based on observed attributes. Nowadays, it is possible to automatically classify such data and to group data objects with respect to the similarity of their mathematical signatures. Such signatures are obtained using feature extractors, as presented in Chapter 4. The closeness of data examples can be computed with respect to this mathematical similarity. Close data examples (in the feature space) are called neighbors. If the task is to classify a new signature according to a given taxonomy, the natural way to proceed is to compare this new signature to examples associated to the given taxonomy. These examples constitute the learning (or training) set. The predicted class is often taken as the majority class among the neighbors of the new signature in the training set. Of course, a majority is not always reached and one may need to weight the contribution of each neighbor by its distance to the considered sample. This classification procedure, called k-nearest neighbors (k-NN), is fast, depending on the learning set size, and efficient, but particular care should be taken with the choice of the distance and with the number k of neighbors considered.

k-Means

Unsupervised approaches can help structure any database without reference to specific knowledge such as taxonomies. They are only based on (mainly statistical) models on the feature space. The basic idea is to find a grouping able to gather data belonging to a cluster as close as possible to its prototype. For discrete data, the problem is mathematically transcribed as finding the clusters C and prototypes Z while minimizing g defined by

$$g(C, Z) = \sum_{k=1}^{K} \sum_{l \in C_k} |\mathbf{x}_l - \mathbf{z}_k|^2,$$

where C is the clustering result composed of K clusters C_k with prototype $\mathbf{z}_k$. This optimization criterion is a sum of squares (SSQ) criterion that appeared in the late 1950s (see Fisher (1958), for example). Bock (2007) presents a very interesting history of algorithms based on SSQ.

Several optimization procedures have been proposed in the literature, one which, called k-means – also known as *nuées dynamiques* (Diday 1970) or the *Lloyd algorithm* for quantization application (Lloyd 1982) – continues in widespread use. This algorithm iteratively estimates the prototype (as the centroid of the samples associated with a cluster) and associates samples with the closest prototype using 1-NN classification. The k-means procedure stops when a given number of iterations is reached or when no change in the prototypes estimation occurs. This algorithm is linear but has two main drawbacks: the number K of clusters should be tuned by the user and different initializations of the prototypes can lead to different clustering results.

Unsupervised approaches are not restricted to k-means-like algorithms. Many more are presented in the literature (Jain and Dubes 1988; Kyrgyzov 2008), especially hierarchical algorithms that are able to construct trees of classes and fuzzy algorithms able to deal with soft (probabilistic or possibilistic) memberships.

Bayes Classification

In the foregoing we mentioned fast and very well-known algorithms for data classification. In fact k-means and k-NN are suitable and very efficient when data are distributed according to a Gaussian distribution in the clusters. Of course, signatures extracted from multimedia documents can follow completely different distribution laws and it is known that when a priori information is available, statistical modeling can be very efficient. The most famous and simple classifier is certainly the *naive Bayes* classifier. Many more detailed approaches are presented in Hastie et al. (2001).

Let X_d be the dth feature variable and C be the class variable. Using Bayes' rule, we can express the probability of observing C, given $X_1, \ldots, X_D$:

$$p(C|X_1, \ldots, X_D) = \frac{p(C)p(X_1, \ldots, X_D|C)}{p(X_1, \ldots, X_D)}.$$

If we assume that each feature X_d is conditionally independent of every other feature (naive Bayes condition), this expression can be simplified:

$$p(C|X_1, \ldots, X_D) = \frac{p(C)}{p(X_1, \ldots, X_D)} \prod_{d=1}^{D} p(X_d|C).$$

The naive Bayes classifier combines this model with a decision rule. One common rule is to pick the hypothesis that is most probable; this is known as the maximum a posteriori decision rule. The corresponding classifier is the classify function defined as follows:

$$\text{classify}(\mathbf{x} = x_1, \ldots, x_D) = \arg \max_{1 \leq k \leq K} p(C|X_1, \ldots, X_D).$$

The x_d are realizations of variables X_d and k is a realization of C. The probability $p(X_d|C)$ is related to a statistical model choice, for example the Gaussian mixture models presented below.

Gaussian Mixture Models

The Gaussian mixture model (GMM) has been widely used in the speech community since its introduction by Reynolds for text-independent speaker identification (Reynolds and Rose 1995). It has also been successful in many other classification tasks (Brown et al. 2000). In such a model, the distribution of the feature vectors is described by a Gaussian mixture density. For a given feature vector $\mathbf{x}$, the mixture density for the class C_k is given by

$$p(\mathbf{x}|C_k) = \sum_{m=1}^{M} w_{m,k} b_{m,k}(\mathbf{x}), \tag{5.3}$$

where the weighting factors $w_{m,k}$ are positive scalars satisfying $\sum_{m=1}^{M} w_{m,k} = 1$. The probability density is thus a weighted linear combination of M Gaussian component densities $b_{m,k}(\mathbf{x})$ with mean vector $\mu_{m,k}$ and covariance matrix $\Sigma_{m,k}$ given by

$$b_{m,k}(\mathbf{x}) = \frac{1}{(2\pi)^{P/2}|\Sigma_{m,k}|^{\frac{1}{2}}} e^{\left(-\frac{1}{2}(\mathbf{x}-\mu_{m,k})'(\Sigma_{m,k})^{-1}(\mathbf{x}-\mu_{m,k})\right)} \tag{5.4}$$

The parameters of the model for the class C_k, denoted by

$$\lambda_k = \{w_{m,k}, \mu_{m,k}, \Sigma_{m,k}\}_{m=1,\ldots,M},$$

are estimated using the well-known Expectation–Maximization (EM) algorithm (Dempster et al. 1977). Classification is achieved applying the maximum a posteriori decision rule, that is,

$$\hat{k} = \arg \max_{1 \leq k \leq K} \sum_{t=1}^{n} \log p(\mathbf{x}_t|C_k) \tag{5.5}$$

where K is the number of possible classes, $p(\mathbf{x}_t|C_k)$ is given in (5.3), $\mathbf{x}_t$ is the test feature vector observed at time (or space region when dealing with images) t, and n is the total number of observations considered in the decision.

5.2.2 Kernel Methods

Such simple classifiers are well known and often used in the literature because of their mathematical basis and practical use. However, there are other efficient learning approaches, based on linear discrimination and kernel theory, which are able to deal with small learning sets and robust to noise.

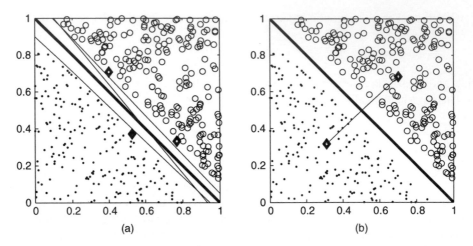

Figure 5.2 Comparison between SVM and FDA linear discrimination for a synthetic two-dimensional database. (a) Lots of hyperplanes (thin lines) can be found to discriminate the two classes of interest. SVM estimates the hyperplane (thick line) that maximizes the margin; it is able to identify the support vector (indicated by squares) lying on the frontier. (b) FDA estimates the direction in which the projection of the two classes is the most compact around the centroid (indicated by squares); this direction is perpendicular to the discriminant hyperplane (thick line).

Linear Discrimination

Linear discrimination simply tries to learn how to separate two classes of data using a hyperplane (i.e. a line in a two-dimensional feature space). Between two linearly separable classes, there are an infinite number of hyperplanes that can be used to separate them; hence several optimality criteria have been presented in the literature to find the *best* separator described by feature weights w and a bias b such as $\langle \mathbf{w}, \mathbf{x} \rangle + b = 0$.[2] The two most famous classifiers are Fisher discriminant analysis (FDA) and the support vector machine (SVM); see Cortes and Vapnik (1995). The (binary) decision function is the same and compares $f(\mathbf{x})$, defined by

$$f(\mathbf{x}) = \langle \mathbf{w}, \mathbf{x} \rangle + b,$$

to a threshold (this can simply be 0) to predict the output class.

On the one hand, FDA is regarded as seeking a direction for which the projections $f(\mathbf{x}_i), i = 1, \ldots, n$, are well separated (See Figure 5.2b). On the other hand, an SVM finds the hyperplane that maximizes the margin between the two classes (See Figure 5.2a). Both of these approaches are still in widespread use in the literature because of their strong mathematical foundations (Fisher 1958; Vapnik 1998) as well as their broad applicability (e.g. pattern analysis, classification, recognition and retrieval). It is noteworthy that SVM is also appreciated for its robustness to noisy (mislabeled) data. Moreover, plenty of SVM tools are available on the Web, for example libsvm (Chang and Lin 2001); their efficiency plays a part in the success of this algorithm.

---

[2] The operator $\langle \cdot, \cdot \rangle$ is the dot product in the feature space.

From Two-Class to Multi-class Problems

Optimization criteria used to perform two-class discrimination can be extended to multi-class problems. In practice, these direct multi-class optimization solutions are heavy and combination approaches are preferred. Two main strategies are classically adopted, called one-vs-one and one-vs-all classifications. The idea is to combine binary classifiers in order to produce multi-class decisions. In the one-vs-one strategy, every pair of classes is considered to produce a classifier. For a K-class problem, $K(K-1)/2$ unitary classifiers should be combined, using a majority decision rule. On the other hand, when dealing with a large number of classes, a one-vs-all strategy can be used: each class is discriminated from the rest of the world (represented by all the other classes); the most positive output $f(\mathbf{x})$ designates the predicted class. This implies that the classifier outputs can be compared, which is why probabilistic outputs are preferred.

Kernel Trick

The success of FDA and SVM is not restricted to linear discrimination problems. Thanks to the *kernel trick*, the frontier linearity assumption can be avoided. The trick is simple and considers that all dot products in the original feature space can be replaced by a dot product in a transformed feature space. This can be transcribed as:

$$\langle \mathbf{x}_i, \mathbf{x}_j \rangle \text{ is replaced by } \mathcal{K}(\mathbf{x}_i, \mathbf{x}_j) = \langle \phi(\mathbf{x}_i), \phi(\mathbf{x}_j) \rangle.$$

For example, when using the Euclidean distance in the original space, it is also possible to use:

$$d(\mathbf{x}_i, \mathbf{x}_j) = |\mathbf{x}_i - \mathbf{x}_j|^2 = \langle \mathbf{x}_i, \mathbf{x}_i \rangle + \langle \mathbf{x}_j, \mathbf{x}_j \rangle - 2\langle \mathbf{x}_i, \mathbf{x}_j \rangle$$

$$\text{is replaced by } d(\mathbf{x}_i, \mathbf{x}_j) = \mathcal{K}(\mathbf{x}_i, \mathbf{x}_i) + \mathcal{K}(\mathbf{x}_j, \mathbf{x}_j) - 2\mathcal{K}(\mathbf{x}_i, \mathbf{x}_j).$$

$\mathcal{K}$ is called the kernel function, and the transformation function ϕ does not need to be explicitly known. The kernel theory is not reduced to a trick; see Schoelkopf and Smola (2002) for the mathematical basis of the kernel and a discussion on the assumptions of positivity and definiteness.

Kernel Methods

The trick can be applied to FDA and SVM approaches; this means that the current problem is transformed into a linear discrimination problem in the new feature space. In fact it can also be applied to k-means, k-NN as well as PCA using the transformation of the distance expression. Moreover, it has been demonstrated in Zhou and Chellappa (2006) that the assumption that the data is Gaussian in the transformed feature space is more acceptable than in the original space. A great deal of work has been published in the literature concerning the application of kernels; see Shawe-Taylor and Cristianini (2004) for more details. Note that kernels are not only employed with vectorial representations but also with graphs. There are also methodologies to construct your kernel according to your personal data; see, for example, the studies of Vert (2008).

Model Selection

We observe that we can apply all of our well-known algorithms in the transformed feature space. However, when considering kernel approaches, the main problem is to choose the most efficient kernel. Many kernels are available in the literature; among them the Gaussian kernel is the best known, essentially because there is only one parameter to tune:

$$\mathcal{K}(\mathbf{x}_i, \mathbf{x}_j) = \exp(-\gamma|\mathbf{x}_i - \mathbf{x}_j|^2). \tag{5.6}$$

In fact there are too many choices: What are the best features? What is the best similarity measure? What is the best kernel? What are the best kernel parameters? It is quite difficult to give a unique answer and the literature proposes many, more or less practical, solutions.

One very practical way to proceed is to try and test several kernels with parameters taken from a grid of values. This can be done in a cross-validation approach (on the learning set only, of course, to avoid overfitting). The learning set is separated into N parts: all the configurations are learnt on $N - 1$ parts and tested on the remaining one. The process is iterated so that each part is seen as a test set, and the evaluation criterion is based on the mean error (and standard deviation) obtained on the test sets. The configuration corresponding to the least error is the winner. Of course this approach is only tractable with a small number of configurations. There are also other parameters estimation methods mainly based on simplified optimization criteria allowing gradient-descent approaches; see, for example, Chapelle and Vapnik (2001).

When we are dealing with supervised approaches, there exist well-established performance measurements while unsupervised methods do not present such a consensual advantage (Campedel 2007). The problem is not reduced to kernel selection, but can be extended to estimation of all other parameters. For example, when dealing with the kernel k-means algorithm (Shawe-Taylor and Cristianini 2004), the problem is not only to choose a good kernel but also to choose the best number of clusters, as well as the best way to initialize the process. Let X be the variable representing the n observations, M being a model with parameters θ. The idea is to find θ that maximizes the probability

$$p(\theta|X, M) = \frac{p(X|\theta, M)p(\theta|M)}{p(X|\theta)}.$$

The problem is how to express this idea differently in order to extract a criterion that can be used to compare several models. The answer is given as the Bayesian information criterion (BIC: Schwarz 1978), also called the minimum description length (MDL: Rissanen 1978) by the information theory community. The MDL is expressed as

$$\min_{q,\theta} - \log(p(X|\theta)) + \frac{1}{2}q\log(n),$$

where q corresponds to the parameters of the penalty term. In Kyrgyzov et al. (2007a) Kyrgyzov proposes to extend this criterion to kernel approaches in order, for example, to estimate the number of clusters for kernel k-means.

Application to Retrieval
All these classification approaches are very useful to structure any database, identify objects in images, or sounds in musical recordings. They are also the basis of user-based retrieval methodologies called *relevance feedback* approaches which are based on a binary classifier able to discriminate between what is relevant to the user and what is not. Chapter 14 of this book is dedicated to aspects of retrieval.

5.2.3 Classifying Sequences

The previous classification approaches do not exploit structural dependencies that may exist in the data. Nevertheless, these potential dependencies are very useful for classification. Intuitively, the way the different parts of an object are connected to each other is clearly important in identifying that object. Similarly, the temporal evolution of musical notes and their articulations are of great importance in recognizing the instrument that played them, just as phoneme succession is important for recognizing a spoken word, etc. The methods presented in this section take advantage of the sequential nature of the data to be classified (be it temporal or spatial). We first briefly present a very popular generative model, namely the hidden Markov model. Subsequently, we make reference to an attractive kernel tool that can be used with various discriminative approaches, particularly SVMs.

Hidden Markov Models

Hidden Markov models (HMMs) are powerful statistical models that have proved efficient for various sequence classification tasks. HMMs remain the privileged technique for speech recognition systems and have been used in other multimedia indexing applications such as musical instrument classification (Eronen 2003; Joder et al. 2009), 2D and 3D object and shape recognition (Bicego and Murino 2004; Bicego et al. 2005), and video classification (Lu et al. 2002).

In an HMM, an observed sequence of feature vectors $(\mathbf{x}_t)_{t \in 1:T}$ (using Matlab-like ranges, i.e. $1:T = \{1, 2, \ldots, T\}$) is supposed to be the realization of T random variables $X_{1:T}$ which are generated by another hidden (unobserved) sequence of discrete random variables $Q_{1:T}$ such that, for each $t \in 1:T$ (Bilmes 2002),

$$\{Q_{t:T}, X_{t:T}\} \text{ and } \{Q_{1:t-2}, X_{1:t-1}\} \text{ are independent given } Q_{t-1} \tag{5.7}$$

and

$$X_t \text{ and } \{Q_{\bar{t}}, X_{\bar{t}}\} \text{ are independent given } Q_t, \tag{5.8}$$

where $X_{\bar{t}} = X_{1:T} \backslash X_t$ (where $\backslash$ denotes the set difference operator).

Q_t are the states of the HMM, which are implicitly associated with some physical states of the objects (in our case multimedia objects) being modeled: typically some regions of visual objects, some types of audio segments, etc. This allows one to adopt distinct statistical models for the observed features depending on the state that generates them (e.g. different models for different image regions if the states are supposed to correspond to object regions). Note, however, that the states generally remain hidden and that the only observed variables are the feature vector sequence resulting from an underlying state sequence which is unobserved.

Conditions (5.7) and (5.8) have many important consequences for the probabilistic structure of an HMM (Bilmes 2002), one of which is that, given Q_t, the variable X_t is independent of every other variable in the model. This yields the factorization

$$p(\mathbf{x}_{1:T}, q_{1:T}) = p(q_1) \prod_{t=2}^{T} p(q_t | q_{t-1}) \prod_{t=1}^{T} p(\mathbf{x}_t | q_t). \tag{5.9}$$

Thus, if it is assumed that the state sequence is time-homogeneous (i.e. $p(Q_t = i | Q_{t-1} = j)$ is constant for all t), then a HMM can be completely specified by the following components:

1. an initial state probability distribution defined by a vector π whose components π_i are defined as $\pi_i = p(Q_1 = i)$;
2. a state transition matrix $\mathbf{A}$, such that $(\mathbf{A})_{ij} = p(Q_t = j | Q_{t-1} = i)$;
3. probability distributions of the observed variables X_t, $b_j(\mathbf{x}) = p(X_t = \mathbf{x} | Q_t = j)$ which are often parametric distributions described by a set of parameters $\mathcal{B}$; a widely used parametric family is the GMM family (described in Section 5.2.1).

To simplify notation, the previous HMM parameters are classically combined in a variable $\lambda = (\pi, \mathbf{A}, \mathcal{B})$. Now to exploit such models for classification, one needs to solve at least two problems:

1. *Learning.* Given a training set of N feature vector sequences $\{X_{1:T_n}\}_{1 \leq n \leq N}$ representing class C data, find the HMM model parameters λ_C that best synthesize these observations, which in a maximum likelihood sense means finding the parameters λ that maximize $p(\{X_{1:T_n}\} | \lambda)$.
2. *Classification.* Given a sequence of observations $\mathbf{x}_{1:T}$ and models λ_C, compute $p(\mathbf{x}_{1:T} | \lambda_C)$ to be able to perform maximum likelihood classification of the objects corresponding to these observations.

Efficient solutions to both problems exist, which space does not permit us to describe here. For more details, we refer the reader to the many tutorials on HMM (Bilmes 2002; Ghahramani 2001; Rabiner 1989).

Discriminative approaches Based on Alignment Kernels

Alignment kernels (Bahlmann et al. 2002; Cuturi et al. 2007; Shimodaira et al. 2002) allow for the comparison of feature vector sequences, rather than independent observations. They are a special case of sequence kernels (Campbell et al. 2006; Louradour et al. 2007; Wan and Renals 2005) that preserve the dynamic of the sequence (while general sequence kernels may be invariant to permutations of the sequence elements). Several alignment kernels have been proposed by researchers, which have proved efficient for various classification tasks, including handwriting recognition (Bahlmann et al. 2002) and musical instrument recognition (Joder et al. 2009).

With alignment kernels, feature vector sequences (possibly of different lengths) are compared after they are synchronized by means of a dynamic time warping (DTW) alignment procedure (Rabiner 1993). A p-length alignment path π is found by DTW for two feature vector sequences $\underline{\mathbf{x}} = (\mathbf{x}_1, \ldots, \mathbf{x}_n)$ and $\underline{\mathbf{y}} = (\mathbf{y}_1, \ldots, \mathbf{y}_m)$, which can be represented by the series of index pairs $(\pi_x(1), \pi_y(1)), \ldots, (\pi_x(p), \pi_y(p))$ making the correspondence between the elements of $\underline{\mathbf{x}}$ and $\underline{\mathbf{y}}$. This is done without changing the order of the feature vectors and with no repetition; see Rabiner (1993) for more details.

Given a 'static' kernel k_0, for instance the widely used Gaussian kernel given in (5.6), an efficient alignment kernel may be defined as

$$\mathcal{K}(\underline{\mathbf{x}}, \underline{\mathbf{y}}) = \sum_{\pi \in \mathcal{A}} \prod_{n=1}^{p} \frac{\frac{1}{2} k_0(\mathbf{x}_{\pi_x(n)}, \mathbf{y}_{\pi_y(n)})}{1 - \frac{1}{2} k_0(\mathbf{x}_{\pi_x(n)}, \mathbf{y}_{\pi_y(n)})},$$

where $\mathcal{A}$ is the set of all possible alignment paths between $\underline{\mathbf{x}}$ and $\underline{\mathbf{y}}$. In Joder et al. (2009) this kernel was found to perform well for SVM musical instrument classification, compared to other proposals. We believe such kernels to be very promising for other multimedia object classification tasks.

5.2.4 Biologically Inspired Machine Learning Techniques

Recent research toward improving machine learning techniques and applied optimization has been strongly influenced and inspired by natural and biological systems. No one can doubt that nature has been doing a great job maintaining life and solving complex problems for millions of years. Perceptual systems have evolved to recognize and classify complex patterns, the child inherits the parent's chromosomes, and ant colonies display swarm intelligence and find optimal paths to food sources. In each case, we can identify specific pieces of work in artificial intelligence and intelligent systems that have developed computational methods informed by these behaviors and processes. Biologically systems are attractive not only as an approach for the design of practical machine learning techniques, but also as an approach for optimization of existing methods. In this section, we focus on well-known biologically inspired systems, such as artificial neural networks, evolutionary algorithms and swarm intelligence, which have been successfully applied to machine learning domains.

Artificial Neural Networks

Artificial neural networks (ANNs), sometimes simply called neural networks, are information processing systems that generalize the mathematical models of human cognition and neural biology. Many of the key features of ANN concepts have been borrowed from biological neural networks. These features include local processing of information, distributed memory, synaptic weight dynamics and synaptic weight modification by experience. An ANN contains a large number of simple processing units, called neurons, along with their connections. Each connection generally 'points' from one neuron to another and has an associated set of weights. These weighted connections between neurons encode the knowledge of the network. ANNs became a powerful tool in machine learning, where the knowledge is represented in the form of symbolic descriptions of learned concepts. For a good overview of various artificial neural systems and their applications, see Zhang (2000). The major weakness which limits neural networks for practical applications includes their slow training phase and the divergence of training caused by the 'local extrema' in training processes. To overcome these drawbacks, efficient evolutionary optimization techniques have been applied to speed up the training process and reduce the chances of divergence.

Evolutionary Algorithms

Evolutionary algorithms (EAs) consist of several heuristics, which are able to solve optimization tasks by imitating some aspects of natural evolution. Three nearly contemporaneous sources of EA have been kept alive over three decades and have witnessed a great increase in interest during the last fifteen years: evolution strategies (Rechenberg 1965), evolutionary programming (Fogel et al. 1966) and genetic algorithms (Holland 1975). All these methods perform a random search in the space of solutions aiming to optimize some objective function in specific ways. Evolution strategies work with vectors of real numbers as representations of solutions, and emphasize behavioral changes at the level of the individual. On the other hand, evolutionary programming stresses behavioral change at the level of the species for natural evolution. The genetic algorithm (GA) is the most popular and well-studied form of evolutionary computation inspired by and named after biological processes of inheritance, mutation, natural selection, and the genetic crossover that occurs when parents mate to produce offspring. GA has been widely and successfully applied to optimization problems specifically in supervised and unsupervised classification of digital data (Bandyopadhyay and Maulik 2002).

Swarm Intelligence

Swarm intelligence is a new biologically inspired methodology in the area of evolutionary computation for solving distributed problems. It is inspired by the behavior of natural systems consisting of many agents, such as colonies of ants, schools of fish and flocks of birds. These techniques emphasize distributed processing, direct or indirect interactions among relatively simple agents, flexibility, and robustness. The two most popular methods in computational intelligence areas are ant colony optimization (ACO) and particle swarm optimization (PSO).

ACO algorithms were first proposed by Dorigo and Gambardella (1997) as a multi-agent approach to difficult combinatorial optimization problems such as the traveling salesman problem. An important and interesting behavior of ant colonies is, in particular, how ants can find the shortest paths between food sources and their nest. The characteristic of ACO algorithms is their explicit use of elements of previous solutions. One way ants communicate is by using chemical agents called pheromones. The distribution of the pheromone represents the memory of the recent history of the swarm, and in a sense it contains information which the individual ants are unable to hold or transmit. ACO algorithms became a promising research area and, by accurate definition of the problem, they have been successfully applied to optimization of classification and clustering problems (Handl et al. 2003; Piatrik and Izquierdo 2006). Furthermore, novel clustering algorithms based on the behavior of real ants can be found in recent literature, such as AntClass (Slimane et al. 1999), AntTree (Azzag et al. 2003), and AntClust (Labroche et al. 2003).

PSO was originally developed by Eberhart and Kennedy (1995), and was inspired by the social behavior of a flock of birds. In the PSO algorithm, the birds in a flock are symbolically represented as particles. These particles are considered to be flying through a problem space searching for the optimal solution. The solution is evaluated by a fitness function that provides a quantitative value of the solution's utility. The PSO algorithm has been successfully implemented in various machine learning domains, for example in classification (Chandramouli and Izquierdo 2006), in clustering (Omran et al. 2002) and in effective learning for neural networks (Song et al. 2007).

Compared to GA, both ACO and PSO are easy to implement and there are few parameters to adjust. An important feature of the aforementioned swarm-based algorithms is their memory ability and convergence to the optimal solution. It is important to understand that a stochastic system such a genetic algorithm does not guarantee success. Nevertheless, all these methods and their applications show the importance of building artifacts whose inspiration derives from biology and the natural world.

5.3 Classifier Fusion

5.3.1 Introduction

When dealing with multimedia documents, the analysis process generally provides features/attributes/descriptors derived from different media (text, audio, image). Then classifiers can be trained on these different feature sets and give some insights into the incoming documents. Classifier combination (or fusion) can then be used to (i) compare features contribution, (ii) compare classifiers' ability according to a given feature set, (iii) enhance the final classification, etc.

The classifiers may be of different nature (e.g. a combination of neural network and nearest neighbor classifier outputs) or of the same nature on different feature sets (or subsets) describing the same documents. The idea is to grant the independence of the classifier outputs in order to provide complementary results. In practice, this assumption is generally not checked. Several schemes have been proposed in the literature according to the type of information provided by each classifier as well as their training and adaptive abilities (Jain et al. 2000).

An overview of state-of-the-art research in classifier combination applied to videos is presented in Benmokhtar and Huet (2006). In this section we focus on well-known and efficient supervised approaches (the classes of interest are known by the individual classifiers) and extend the discussion to clustering ensembles.

5.3.2 Non-trainable Combiners

Combiners that are ready to operate as soon as the classifiers are trained are termed *non-trainable* – they do not require any further training. The only methods to be applied to combine these results without learning are based on the principle of voting. They are commonly used in the context of handwritten text recognition (Chou et al. 1994).

In the late 18th century Condorcet presented a theorem about how to combine votes in a jury to reach a decision by majority vote. The assumptions are: one of the two outcomes (yes/no) of the vote is correct, and each voter has an independent probability p of voting for the correct decision. The theorem asks how many voters should be included in the group. The result depends on whether p is greater or less than $1/2$. If $p > 1/2$ (each voter is more likely to vote correctly), then adding more voters increases the probability that the majority decision is correct. In the limit, the probability that the majority votes correctly approaches 1 as the number of voters increases. On the other hand, if $p < 1/2$ (each voter is more likely to vote incorrectly), then adding more voters makes things worse: the optimal jury consists of a single voter.

What is termed a 'jury' above is effectively a classifier and the outcomes can be extended to multiple classes. This can be mathematically derived for classifiers with hard or soft outputs (Achermann and Bunke 1996; Ho 1992).

5.3.3 Trainable Combiners

Contrary to the voting principle, several methods use a learning step to combine classifier outputs and consider combining classifiers as a classification problem, in particular using probabilistic methods that give an estimation or probability of the class membership, such as neural networks, GMMs, K-NN classifier, Bayesian theory, and (probabilistic) SVM.

1. **Bayesian Combination**

    Following the principles presented in Section 5.2.1, for L independent classifiers I_l, we have

    $$p(\mathbf{x} \in C_k | I_1, \ldots, I_L) = \frac{\prod_{l=1}^{L} p(I_l | \mathbf{x} \in C_k) p(\mathbf{x} \in C_k)}{p(I_1, \ldots, I_L)}.$$

    Using the maximum a posteriori decision rule, we can construct a classifier that gives the most probable class for $\mathbf{x}$, taking into account the individual classifier outputs.

2. **Decision Templates**

    The idea, proposed in Kuncheva et al. (2001) and Kuncheva (2003), is to characterize the individual classifier outputs (crispy, fuzzy or possibilistic) corresponding to a given entry $\mathbf{x}$ by a matrix, called the decision profile, $DP(\mathbf{x})$, with L rows and K columns. Each element of this matrix contains the output of each individual classifier i, for each class k considering the input $\mathbf{x}$. A weighted sum of such matrices, computed on a training set, provides (columnwise) a decision template for each class.

    When dealing with new data, their decision profile is compared to the decision template of each class using classical similarity measures; the class with the most similar decision template is chosen.

3. **Genetic Algorithms**

    In Section 5.2.4 we mentioned that biologically inspired classifiers are available. The same principles can be applied to perform classifier combination as presented in Souvannavong et al. (2005). In that paper, a binary tree-based GA is used to find the optimal classifier combination using all possible formulas with a fixed number of inputs and the appropriate operators (sum, product, min and max) and weights.

4. **Behavior Knowledge Space**

    Contrary to the preceding combination strategies, the behavior knowledge space (BKS) (Raudys and Roli 2003) is based only on the abstract output of each classifier (class level). At the training stage, a table is created (the BKS table), containing cells corresponding to each combination of L classifiers with K possible outputs. Each cell contains the list of samples characterized by this particular set of outputs and is associated with the most representative class (using statistical modeling) among its samples. When dealing with new examples, the corresponding outputs are compared to the table cells and the corresponding class is chosen. Rejection is used to deal with empty and ambiguous cells.

5.3.4 Combination of Weak Classifiers

A weak classifier is a simple classifier (albeit one performing better than random classification) that can be computed rapidly. It can be related to a feature extraction process (random subspace, AdaBoost) or to data resampling strategies (bagging, Weighted Bag-Folding). In the literature combining weak classifiers has proved to enhance the global classification accuracy at low cost; these strategies were particularly successful in pattern recognition applications (face or object detection).

1. **Random Subspace**
   During the 1970s it was demonstrated that the random subspace method (Dempster 1967) allows one to maintain a small learning error and to improve the generalization error of linear classifiers. The main idea is to combine classifiers constructed on randomly chosen features subsets. This has been derived using linear classifiers as well as decision trees.
2. **AdaBoost**
   AdaBoost appeared in the late 1990s (Freud and Schapire 1996) to provide fast face detection with a controlled false positive rate. It was received with enthusiasm because it performs feature selection and classification simultaneously at low computation cost. The cascaded classifiers are called *weak* because they are trained on primitive features and linearly combined. The combination process is iterative and tries at each step to focus the training on hard examples. When dealing with new examples, AdaBoost predicts one of the classes based on the sign of the estimated linear combination of the chosen classifier outputs. The number of iterations can be increased to improve the combination accuracy, though at the cost of increasing the risk of overfitting and makes the process more time-consuming.
3. **Bagging**
   Bagging is based on bootstrapping and aggregating concepts; it was presented by Breiman (1996). It is implemented by averaging the parameters (coefficients) of the same linear classifier built on several bootstrap replicates.[3]
4. **Weighted BagFolding**
   Proposed in Benmokhtar and Huet (2006), Weighted BagFolding (WBF) reduces both the well-known bagging instability (Skurichina and Duin 1998) – a small change in the training data produces a large change in the behavior of classifier – and the overfitting risk for AdaBoost when the number of iterations is large (Freud and Schapire 1996). WBF uses the ten folding principle to train and obtain L (classifier) models weighted by a coefficient indicating their importance in the combination. The final decision combines measures which represent the confidence degree of each model. The weighted average decision in WBF improves the precision of ten folding by giving more importance to models with small training error, as opposed to ten folding which simply averages over all model outputs.

---

[3] These replicated sets are obtained by random selection with replacement.

5.3.5   Evidence Theory

Evidence theory is based on Dempster's original work on lower and upper bounds for a set of compatible probability distributions (Dempster 1967). Shafer (1976) expanded the original work and produced what is now generally referred to as Dempster–Shafer theory. This theory has since been further extended. In particular, a non-probabilistic approach of evidence theory was presented by Smets and Kennes (Smets 1994). It gives information about uncertainty, imprecision and ignorance that allows conflicts between sources to be measured and their reliability to be interpreted. In Bloch (2003), different evidence theory-based fusion models are presented.

5.3.6   Consensual Clustering

The previously presented fusion methods are famous and efficient for combining supervised classifiers sharing the same classes. When dealing with unsupervised classifiers, the classes (clusters) obtained are different and such approaches cannot be applied. These classifiers can deal with different feature spaces that describe the same set of documents and the main idea is to construct a new indexing structure that will show up the redundancies and originalities of the individual classifications. In the following we will focus on the notion of consensus, which leads to particularly interesting tools in terms of results exploration (mining). A good review for consensual clustering (also called *cluster ensembles* in the literature) is presented in Strehl and Ghosh (2003).

In the following we make the assumption that the individual classifications are relevant for the system and that their combination will hence provide a better representation of the data. In fact these classifications can be interpreted as valuable 'points of view' about the data and their combination as a way to make consensual concepts appear.

To avoid the manipulation of the feature spaces, it is usual to exploit the co-association matrix A^p associated with each individual classification p: $A_{ij}^p = 1$ when data i and j are classified in the same group, and otherwise $A_{ij}^p = 0$. This avoids a classification-to-classification comparison in order to align the clusters.[4] Many authors, such as Jain and Dubes (1988) and Topchy et al. (2004), proposed solutions to combine classification results based on co-association matrices, but because of the complexity of their methods they cannot be applied to a large dataset. In Kyrgyzov et al. (2007b), the authors propose to exploit a least square error criterion, already used in the late 1970s (Michaud and Marcotorchino 1979), to derive a simple and fast algorithm.

The combination of classification results can be obtained by summing the individual (possibly weighted) co-association matrices. Each element of the resulting matrix A is interpreted as the probability of data i and j being classified in the same group:

$$A_{ij} = \frac{\sum_p w_p}{\sum_p w_p A_{ij}^p}.$$

---

[4] Cluster alignment can be performed using the Hungarian method (Kuhn 1955).

The consensual partition can then be obtained by finding the hard clustering (represented by the allocation matrix B) that is the closest to A in the sense of mean square error E:

$$E = \sum_i \sum_j \left(\sum_r (B_{ir} B_{rj}') - A_{ij} \right)^2 .$$

It is not possible to explore all the solutions due to the number of data we are dealing with. Kyrgyzov (2008) proposed two different approaches based on single-link clustering or mean shift density estimation, in the domain of satellite image mining. He avoids the storage of matrix A and reformulates the optimization problem using an iterative and (close to) linear algorithm.

Consensual clustering is of interest for two reasons: it makes consensual (semantical) concepts emerge from unsupervised results; and it characterizes data and cluster

Table 5.1 Classifier fusion properties

Scheme	Architecture	Trainable	Comments
Voting methods	Parallel	No	Assumes independent classifiers
Sum, mean, median	Parallel	No	Robust; assumes independent confidence estimators
Product, min, max	Parallel	No	Assumes independent features
Adaptive weighting	Parallel	Yes	Explores local expertise
Bagging	Parallel	Yes	Needs many comparable classifiers
Boosting	Parallel/ Hierarchical	Yes	improves margins; sensitive to mislabeling; overstraining risk
Random subspace	Parallel	Yes	Needs many comparative classifiers.
Weighted BagFolding	Parallel	Yes	Iterative; time-consuming training procedure
Genetic algorithm	Hierarchical	Yes	Iterative training procedure; overstraining sensitive
Behavior knowledge space	Parallel	Yes	Adversely affected by small sample size; overfitting risk; does not require any classifier dependency Scale (metric) dependent;
SVM	Parallel	Yes	iterative; slow training; non-linear; overstraining insensitive; good generalization performance
Neural network	Parallel	Yes	Sensitive to training parameters; slow training; overstraining sensitive; non-linear classification function
GMM	Parallel	Yes	Sensitive to density estimation errors
Decision template	Parallel	Yes	The templates and the metric have to be supplied by the user; the procedure may include non-linear normalizations
Evidence theory	Parallel	Yes	Fuses non-probabilistic confidences; Several combination rules
Consensual clustering	Parallel	Yes/no	Fuses clustering results; several combination criteria

relationships through connectedness scores derived from the mean co-association matrix. Consensual clustering is also of great value for dealing with incomplete data (with missing feature values).

5.3.7   Classifier Fusion Properties

In Table 5.1, we summarize the main points that we have identified to be critical when designing classifier fusion methods, taking into account two characteristics: the type of architecture (parallel/hierarchical), and the possibility (or not) of training the methods. The last column presents some comments about the methods with their advantages or disadvantages.

5.4   Conclusion

Machine learning is a field of active research that has applications in a broad range of domains. It has not been the intention of this chapter to provide an exhaustive description of machine learning techniques but rather to provide a brief overview of the most common and successful techniques applied to multimedia data indexing. The future important challenges for the machine learning community in this field are mainly linked to two major developments in multimedia data: its rapid growth in volume and diversity. Indeed, new approaches need to be developed to appropriately scale up current systems to be used with massive multimedia databases. The diversity of data both in terms of content (audio, video, text, pictures, etc.) and quality (coding artifacts, live vs studio recordings, etc.) is also continuously challenging current approaches. New types of data will also need to be better exploited in future systems. For example, there is nowadays a significant amount of rich metadata content accessible from social networks that could be used despite its inherently approximate nature. But, more generally, it seems that the major improvements in multimedia data indexing will now come from the association of machine learning techniques with high-level semantics approaches. Indeed, as exemplified in this book, this will permit us to successfully bridge the so-called *semantic gap* between the low-level description that can be obtained by machines and the high-level information that is commonly used by humans.

6

Semantic Web Basics

Eyal Oren[1] and Simon Schenk[2]
[1]*Vrije Universiteit Amsterdam, Amsterdam, The Netherlands*
[2]*University of Koblenz-Landau, Koblenz, Germany*

I have a dream for the Web [in which computers] become capable of analysing all the data on the Web – the content, links, and transactions between people and computers. A 'Semantic Web', which should make this possible, has yet to emerge, but when it does, the day-to-day mechanisms of trade, bureaucracy and our daily lives will be handled by machines talking to machines. The 'intelligent agents' people have touted for ages will finally materialize.

Sir Tim Berners-Lee, 1999 (Berners-Lee and Fischetti 1999)

The Web has grown from a tool for scientific communication into an indispensable form of communication. Although originally developed to facilitate knowledge management, the reuse of information on the Web is limited. The lack of reuse is the result of data hidden inside closed systems such as relational databases, with a rigid schema structure, a lack of universal, reusable, identifiers, and a lack of expressive and extensible schemas. The Semantic Web improves the Web infrastructure with formal semantics and interlinked data, enabling flexible, reusable, and open knowledge management systems.

The Semantic Web can be considered as an enrichment of the current Web, employing for example named-entity recognition or document classification, resulting in semantically annotated Web documents (Handschuh 2005; Kiryakov et al. 2004); on the other hand, the Semantic Web is also interpreted as an interlinked 'Web of data' (Berners-Lee 2006; Shadbolt et al. 2006) enabling ubiquitous data access and unexpected reuse and integration of online data sources.

In this chapter, we will explain the basic elements of the Web and the Semantic Web. We will approach the Semantic Web from both viewpoints: we will explain the basic RDF(S) model for knowledge representation on the Semantic Web, and explain its relation to existing Web standards such as HTML, XML, and HTTP. Next, we will explain the

Multimedia Semantics: Metadata, Analysis and Interaction, First Edition.
Edited by Raphaël Troncy, Benoit Huet and Simon Schenk.
© 2011 John Wiley & Sons, Ltd. Published 2011 by John Wiley & Sons, Ltd.

importance of open and interlinked Semantic Web datasets, outline the principles for publishing such linked data on the Web, and discuss some prominent openly available linked data collections. Finally, we will show how RDF(S) can be used to capture and describe domain knowledge into shared ontologies, and how logical inferencing can be used to deduce implicit information based on such domain ontologies.

Once these basic elements have been explained, Chapter 7 will further discuss the notion of ontologies, and introduce the expressive OWL language for building and describing these ontologies.

6.1   The Semantic Web

The Semantic Web is an ongoing evolution of the Web into a more powerful and more reusable infrastructure for information sharing and knowledge management. The current Web is a publishing platform and indeed allows us to connect with arbitrary information sources across all physical and technical boundaries. But the Web merely allows us to publish documents and links; very little consideration is given to the content or meaning of the documents or to the meaning of the links. As a consequence, the Web serves as an excellent giant document repository and, as a communication platform, enables the provision of online services, but knowledge reuse is limited because no uniform standard is available to express the meaning or intended usage of pieces of online information.

The Semantic Web (Berners-Lee et al. 2001) is a web of information that is more understandable and more usable by machines than the current Web. It can be regarded as an extension of the existing Web, whose information is mostly human-readable. Although the current Web also has some machine-usable structure such as head and body of documents, levels of heading elements, and classes of div elements, this structure has coarse granularity and little agreed meaning. As an analogy, imagine a reader from a Western country reading a Chinese text. While she may be able to identify some high-level structure in the text, she will not be able to even understand the symbols used when writing it. The Semantic Web allows for finer granularity of machine-readable information and offers mechanisms to reuse agreed meaning.

The Semantic Web can also be considered similar to a large online database, containing structured information that can be queried. But in contrast to traditional databases, the information can be heterogeneous: it does not conform to one single schema. The information can be contradictory: not all facts need be consistent. The information can be incomplete: not all facts need to be known. And resources have global identifiers allowing interlinked statements to form a global 'Semantic Web' (van Harmelen 2006).

Some of the important Semantic Web standards are shown in Figure 6.1. The fundamental data model of the Semantic Web is the Resource Description Framework (RDF); see Klyne and Carroll (2004). RDF is a language for asserting statements about arbitrary resources, such as "john knows mary". RDF uses URIs to identify all resources involved in these assertions; commonly used URI schemes include URLs (for Web resources), ISBN numbers (for books), and LSIDs for concepts in the life sciences. Sauermann and Cyganiak (2008) explain how to find and choose proper URIs when publishing RDF data. By reusing URIs for common resources such as people, places, or events, and by reusing vocabularies or ontologies, RDF statements from different sources automatically interlink to form a 'giant global graph' of statements.

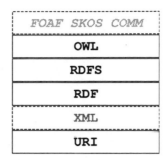

Figure 6.1 Layer cake of important Semantic Web standards.

The RDFS and OWL standards (discussed later in this and the next chapter) allow people to define ontologies or vocabularies along with their intended meaning and explicit semantics. Ontologies (formal representations of shared meaning) generally consist of class and subclass hierarchies (e.g. all men are persons), and relationship definitions (e.g. fatherOf is a transitive relation between a person and a man). The top layer in figure 6.1 shows some example ontologies and vocabularies: the FOAF "friend of a friend" vocabulary, the SKOS "simple knowledge organization system" vocabulary, and the COMM "core ontology for multimedia", defined using RDFS and OWL. Many more vocabularies and ontologies have been defined and are used in (often publicly available) Semantic Web data.

SPARQL is a language for querying RDF data. Linked Data defined a couple of best practices for publishing semantic data on the Web, which embraces Web principles. XML is included as part of the layer cake, because the other standards reuse parts of XML technology, such as XML Schema Datatypes. Moreover, RDF can be serialized in XML documents. However, RDF can very well be expressed without using XML.

In this chapter we will introduce RDF/S and Linked Data. In the following chapter we will focus on OWL, SKOS and SPARQL. COMM and other vocabularies, technologies and tools built on these foundations are discussed in the following chapters.

6.2 RDF

RDF is a formal language in the sense that a syntax, grammar, and model-theoretic semantics are defined (Hayes ed.). The semantics provide a formal meaning to a set of statements through an interpretation function into the domain of discourse. But this interpretation function is relatively straightforward and explicit: the semantics of RDF prescribe relatively few inferences to be made from given statements – there is only little implicit information in statements. RDF can thus be seen as a language for statements without specifying the meaning of these statements. More poetically, RDF can be regarded as an alphabet, allowing one to construct words and sentences, but not yet a language, since the words and sentences have not yet been given a meaning.

Such computer-usable meaning can be achieved by defining a vocabulary (a set of terms) for RDF and by specifying what should be done when such a term is encountered. Currently, two such vocabularies have been agreed and standardized. The first is RDF Schema (RDFS), which allows one to express schema-level information such as class membership, sub-class hierarchies, class attributes (properties), and sub-property hierarchies (Brickley and Guha 2004). RDFS allows simple schema information, but its expressiveness is limited. The Web Ontology Language (OWL) therefore extends RDFS (although the two are formally not completely layered) and provides terms with additional expressiveness and meaning (McGuinness and van Harmelen 2004).

This section introduces the semantics of RDF as defined in (P. Hayes ed.). We omit Datatype-Entailment (D-Entailment), XMLLiterals and container properties. We will give brief side remarks about them where appropriate. An RDF document consists of (*subject*, *predicate*, *object*) *statements* or *triples*. *Subject*, *predicate* and *object* are resources with a unique ID. Such statements can also be seen as directed node-arc-node links in a graph. An RDF document, then, is a graph (in fact a hypergraph, as predicates in turn can be used as subjects and objects of other statements), as shown in Figure 6.2. A statement (*subject*, *predicate*, *object*) means the relation denoted by *predicate* holds between *subject* and *object*. *Subject*, *predicate* and *object* are called resources or entities.

In the following, we will use RDF to model a simple scenario following the photo use case as described in Chapter 2. Ellen Scott is traveling through the Maremma, and asks her husband to take a picture of her with a marvelous view of the Mediterranean in the background. Ellen also records the geocoordinates where the photo is taken. To represent all this information in RDF, we start by naming the resources involved, that is, the real world objects, electronic items and the relations between them.

Resources in RDF can be URIs, which identify things, labels, which represent values such as numbers or strings, or existentially quantified variables called blank nodes:

Definition 6.2.1 (URI, URL) *A* Uniform Resource Identifier *(URI) is a unique identifier according to RFC 2396 (Berners-Lee et al. 1998). A* Unified Resource Locator *(URL) represents a resource by its primary access mechanism, that is, its network location. A URI can denote any resource. URIs are treated as constants in RDF. Let* $\mathbb{U}$ *be the set of all URIs.*

Example 6.2.2 *http://dbpedia.org/resource/Maremma is a* URL *denoting the Maremma.* urn:isbn:0062515861 *is a URI, which uses the URN scheme instead of http and identifies the book* Weaving the Web, *which contains the introductory quote of this chapter.*

RDF allows literals (simple data values) to appear as object of statements. These literals can by typed with a datatype, which defines a lexical space (representation), a value

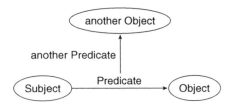

Figure 6.2   A Basic RDF Graph.

space (actual values), mappings between these two, and possibly operations on values. For example, the datatype `xsd:double` defines operations subtracting two numbers from each other, which could be implemented by a program that processes this data (if it is aware of these datatypes). RDF only defines one generic datatype (XML literal) but recommends the use of the XML Schema datatypes.

RDF allows literals to be localized to a language. Such tags can be used to adjust the (visual) display of statements: a French-speaking person could use a French label while an English-speaking person would see an English label.

Definition 6.2.3 (RDF Literal) *An RDF* literal *is one of the following:*

A plain literal *of the form* `"<string>"(@<lang>)?`*, where* `<string>` *is a string and* `<lang>` *is an optional language tag according to Alvestrand (2001). A plain literal denotes itself.*

A typed literal *of the form* `"<string>"^^<datatype>`*, where* `<datatype>` *is a URI denoting a datatype according to XML-Schema2 (Biron et al. 2004), and* `<string>` *is an element of the lexical space of this datatype. A typed literal denotes the value obtained by applying* `<datatype>`*'s lexical-to-value mapping to* `<string>`*.*

Let $\mathbb{L}$ *be the set of all literals and* $\mathbb{L}_P$ *and* $\mathbb{L}_T$ *the sets of plain and typed literals.*

Example 6.2.4 *Examples of literals are:*

```
"Ellen Scott"
"Maremma"@en
"42.416389"^^xsd:double
"11.478056"^^xsd:double
```

The first two literals denote themselves and the last two denote double precision floating point numbers. The second literal is a String with a language tag for English.

Sometimes, we want to express statements about things for which we have no name. In such cases, blank nodes can be used.

Definition 6.2.5 (Blank Node) *A* blank node *is a unique resource, which is not a URI or a literal. It can only be identified via its properties, and cannot be named directly. Even though some RDF serializations use blank node identifiers, these are just syntactic auxiliary constructs. Blank nodes are treated by RDF as existentially quantified variables. Let* $\mathbb{B}$ *be the set of all blank nodes.*

Example 6.2.6 *If we want to express the fact that the photograph in our running example has been taken at a certain point, we can use a blank node for this point – as we are only interested in the photo and the geocoordinates, we do not need to name the point.*

Now that we can name all entities involved in our example, we can build RDF statements from them.

Definition 6.2.7 (RDF Statement) *An* RDF statement *is a triple in* $\mathbb{U} \cup \mathbb{B} \times \mathbb{U} \times \mathbb{U} \cup \mathbb{L} \cup \mathbb{B}$.

6.2.1 RDF Graphs

RDF statements form RDF graphs.

Definition 6.2.8 (RDF Graph) *An* RDF graph *G is a set of RDF statements. H is a subgraph of G if $H \subseteq G$. The vocabulary V of G is the set of URIs and literals used in statements in G. For any two RDF graphs G and H, the sets of blank nodes in G and H are disjoint. Two graphs G and H are considered* equivalent, *if G can be transformed into H simply by renaming blank nodes. An RDF graph is* ground, *if it does not contain any blank nodes.*

Note that any two graphs G and H, which are not equal, can be merged into a single graph by standardizing apart the blank nodes in G and H (making sure the blank nodes have unique identifiers) and building the union of G and H.

In the following, we will use Turtle syntax (Beckett and Berners-Lee 2008) for examples. In Turtle, RDF statements are written literally as subject, predicate, object triples and are ended with a dot ".". URIs are enclosed in angle brackets. Sequences of statements with the same `subject` can be written as `subject predicate1 object1; predicate2 object2.`. Sequences of statements with the same `subject predicate` can be written as `subject predicate object1, object2.`. Blank nodes are written using square brackets. All `predicate object` pairs within square brackets have the corresponding blank node as a subject. Alternatively, Blank nodes can be assigned an id (on the syntactical level) with the prefix `_:`, for example, `_:12ht76`. A Turtle document can contain `@PREFIX` declarations, which allow URIs to be abbreviated. For example, if we define `@PREFIX foaf: <http://xmlns.com/foaf/spec/>`, `foaf:Person` expands to `http://xmlns.com/foaf/spec/Person`.[1] The `@BASE` is used to expand all abbreviated URIs which do not have a prefix. In the following we will use the imaginary namespace `ex:` in examples.

Example 6.2.9 *The following example shows how the information we have about the photo taken of Ellen could be modeled (rather straightforwardly) in RDF. Figure 6.3 shows the same example as a graph.*

```
1 # BASE and PREFIX declarations
2 @BASE <http://example.org/elen/> .
3 @PREFIX foaf: <http://xmlns.com/foaf/spec/> .
4 @PREFIX rdf: <http://www.w3.org/1999/02/22−rdf−syntax−ns#> .
5 @PREFIX geo: <http://www.w3.org/2003/01/geo/wgs84_pos#> .
6
7 ## Basic information about the photo
```

---

[1] The Friend Of A Friend vocabulary is used to describe persons and their relationships. It is modeled using RDFS and OWL.

```
 8 # Blank node written using brackets
 9 :P1000668.jpg ex:taken−near [ a geo:Point ;
10                               geo:lat  "42.416389"^^xsd:double ;
11                               geo:long "11.478056"^^xsd:double
12                            ] .
13
14 ## Basic Information about Ellen
15 :me foaf:name "Ellen Scott"
16
17 ## Information about what the photo depicts
18 :P1000668.jpg foaf:depicts :me .
19 # Blank node written using an ID
20 :P1000668.jpg foaf:depicts _:asdk65s .
21 _:asdk65s rdfs:label "Maremma"@en .
```

6.2.2 Named Graphs

Named graphs (Carroll et al. 2005) are an extension of RDF, which allows to reference RDF graphs:

Definition 6.2.10 (Named Graph) *A named graph is a tuple* (n, G) *of a URI n called the name and an RDF graph G. n denotes G.*

Note that n denotes the graph as a whole, not the extension of G. This means that statements about n do not propagate to statements in G.

We use Trig (Bizer 2007), an extension to Turtle, as syntax for named graphs. A named graph G with name ex:name can be expressed in Trig as follows:

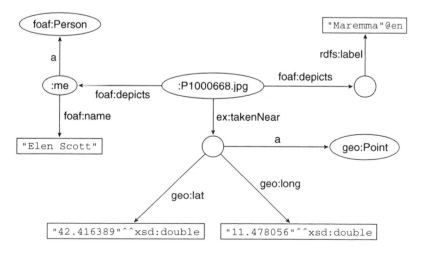

Figure 6.3 Example as a graph.

```
1 ex:name {
2 # statements in G
3 }
```

6.2.3 RDF Semantics

RDF is defined using a model-theoretic open world semantics. RDF is monotonic, that is, a statement added to an existing RDF graph cannot falsify any previously true statement. An interpretation describes the constraints on any possible world, in which all statements in the vocabulary V of an RDF graph are true. The basic RDF interpretation is very weak, as mentioned above. We will see that RDFS and OWL add stronger semantics, but are formalized similarly.

Definition 6.2.11 (Simple RDF Interpretation) *A* simple RDF interpretation $\mathcal{I}$ *of a vocabulary* V *is a tuple consisting of the following:*[2]

1. *A non-empty set* $\Delta^{\mathcal{I}}$ *called the universe of* $\mathcal{I}$;
2. *A set* $\mathbb{P}^{\mathcal{I}}$, *called the set of properties of* $\mathcal{I}$;
3. *A mapping* $\bullet_P^{\mathcal{I}} : \mathbb{P}^{\mathcal{I}} \rightarrow 2^{\Delta^{\mathcal{I}}}$ *of properties into pairs from the domain, defining the extensions of the properties;*
4. *A mapping* $\bullet_{\mathbb{U}}^{\mathcal{I}} : V \cap \mathbb{U} \rightarrow \Delta^{\mathcal{I}} \cup \mathbb{P}^{\mathcal{I}}$ *from URI references in* V *into* $\Delta^{\mathcal{I}} \cup \mathbb{P}^{\mathcal{I}}$ *defining the semantics of URIs in* V;
5. *A mapping* $\bullet_{\mathbb{L}}^{\mathcal{I}} : V \cap \mathbb{L}_T \rightarrow \Delta^{\mathcal{I}}$ *from typed literals in* V *into* $\Delta^{\mathcal{I}}$.

$\mathcal{I}$ satisfies *a ground graph* G *with vocabulary* V *(*$\mathcal{I} \models G$*), if every statement in* G *is true in* $\mathcal{I}$. *We call* $\mathcal{I}$ *a* model *of* G.

The separate treatment in 3 and 4 separates the names of predicates from their extensions and allows predicates to range over themselves without violating set-theoretic laws. While plain literals denote themselves and hence have the same meaning in every possible world, typed literals need to be interpreted. Note that an interpretation does not map blank nodes into the domain, that is, satisfaction is defined for ground graphs. We now define a semantics in the presence of blank nodes.

Definition 6.2.12 (Semantics of Blank Nodes) *Let* $\mathcal{I}$ *be an RDF interpretation of a vocabulary* V. $\mathcal{I}$ *satisfies a non-ground graph* G, *if there exists* some *instance mapping* $A : \mathbb{B} \cap V \rightarrow \Delta^{\mathcal{I}}$, *such that* $\mathcal{I}$ *satisfies* $H = G/A$, *where* G/A *denotes the graph obtained by replacing each blank node* b *in* G *with* $A(b)$.

More generally, we say that H is an *instance* of G, if $H = G/A$ for some $A : \mathbb{B} \rightarrow \mathbb{U} \cup \mathbb{L}$.

In the following, for an expression (set) X and a mapping Y, we denote by X/Y the expression (set) obtained by replacing all subexpressions (elements) x in X, which are in the domain of Y by $Y(x)$.

---

[2] As we leave out D-Entailment, we also do not treat literal values separately.

For an interpretation $\mathcal{I}$ and a URI x, we abbreviate $\bullet_{\mathbb{U}}^{\mathcal{I}}(x)$ as $x^{\mathcal{I}}$. Analogously for a typed literal x, $x^{\mathcal{I}}$ denotes $\bullet_{\mathbb{L}}^{\mathcal{I}}(x)$ and, for a blank node x, $x^{\mathcal{I}}$ denotes $(A(x))^{\mathcal{I}}$.
We now define simple RDF entailment.

Definition 6.2.13 (Simple RDF Entailment) *Let G, H be RDF graphs. G simply entails H, if for each simple RDF interpretation $\mathcal{I}: \mathcal{I} \models G \Rightarrow \mathcal{I} \models H$.*

Example 6.2.14 *For example, $\mathcal{I} = (\Delta^{\mathcal{I}}, \mathbb{P}, \bullet_{P}^{\mathcal{I}}, \bullet_{\mathbb{U}}^{\mathcal{I}}, \bullet_{\mathbb{L}}^{\mathcal{I}})$ makes each statement in the our example FOAF file true, if we additionally map both blank nodes into 1:*

- $\Delta^{\mathcal{I}} = 1, 2, 3, 4, 5, 6, 7, 8, 9, 10, 11,$ `"Ellen Scott"`, `"Maremma"@en`.

- $\bullet_{\mathbb{U}}^{\mathcal{I}} =$

URI	domain object
`:me`	*1*
`rdf:type`	*2*
`foaf:Person`	*3*
`foaf:name`	*4*
`foaf:depicts`	*5*
`rdfs:label`	*6*
`ex:takenNear`	*7*
`:P1000668`	*8*
`geo:lat`	*9*
`geo:long`	*10*
`geo:Point`	*11*

- $\mathbb{P}^{\mathcal{I}} = \{2, 4, 5, 6, 7, 9, 10\}$.

- $\bullet_{P}^{\mathcal{I}} =$

predicate	extension	
2	*1*	*3*
	1	*11*
4	*1*	`"Ellen Scott"`
5	*8*	*1*
6	*1*	`"Maremma"@en`
7	*8*	*1*
9	*1*	`"42.416389"^^xsd:double`
10	*1*	`"11.478056"^^xsd:double`

- $\mathbb{L}^{\mathcal{I}} =$

literal	value
`"42.416389"^^xsd:double`	*42.416389*
`"11.478056"^^xsd:double`	*11.478056*

The example is based on a domain consisting of integers and shows that no assumption about the exact nature of a domain is made. As we can see in the predicate extensions, no additional statements are inferred using only RDF. Also note that every RDF graph is satisfiable: every graph in RDF is true since it is not possible to express any contradictions.

An RDF vocabulary is assigned stronger semantics by imposing additional requirements on interpretations. We call a stronger interpretation with respect to a vocabulary V a *vocabulary interpretation*. The first vocabulary we define is the RDF vocabulary V_{RDF}.

Definition 6.2.15 (RDF Entailment) $\mathcal{I}$ *is called an* RDF *interpretation, if it is a simple RDF interpretation and the following additional requirements hold:*

- $\mathcal{I}$ *makes all of the following axiomatic statements true:*

```
rdf:type rdf:type rdf:Property.
rdf:subject rdf:type rdf:Property.
rdf:predicate rdf:type rdf:Property.
rdf:object rdf:type rdf:Property.
rdf:first rdf:type rdf:Property.
rdf:rest rdf:type rdf:Property.
rdf:value rdf:type rdf:Property.
rdf:_1 rdf:type rdf:Property.
rdf:_2 rdf:type rdf:Property.
...
rdf:nil rdf:type rdf:List.
```

- $x \in \mathbb{P}^{\mathcal{I}}$ *if and only if* $\langle x, rdf\!:\!Property^{\mathcal{I}} \rangle \in \bullet^{\mathcal{I}}_{P}(rdf\!:\!type^{\mathcal{I}})$.

Let G, H *be RDF graphs.* G RDF *entails* H, *if for each RDF interpretation* $\mathcal{I}: \mathcal{I} \models G \Rightarrow \mathcal{I} \models H$.

We have introduced classes, in particular the class rdf:Property of RDF properties and the class membership property rdf:type. Obviously each RDF interpretation is infinite due to the axiomatic statements (in contrast to simple RDF interpretations, which may be finite). In contrast to simple interpretations, vocabulary interpretations may be inconsistent. For example, using XMLLiterals (which basically define a literal type containing XML), it is possible to write an inconsistent RDF graph, if the literal is not valid XML. As usual, if a graph is inconsistent with respect to a vocabulary interpretation, it entails any graph in that vocabulary. These interpretations of cause are not very useful.

RDF defines containers and collections, which allow grouping resources into a bag, (closed or open-ended) sequence, or group of alternatives. Such containers can be used to state, for example, that a paper was authored by several people (in a given sequence), but they have only an intended meaning: no formal semantics is defined for containers and collections. RDF also contains a reification mechanism which enables statements about statements. RDF reification is not assigned a strong semantics either. In particular, reifying a statement does not imply that it actually exists.

6.3   RDF Schema

RDF Schema (Brickley and Guha 2004) is a vocabulary for RDF: it defines terms, their intended use, and their semantics. RDF Schema offers terms for defining classes, a sub-class hierarchy, and class-membership of resources; and for defining properties, their domain and range, and a sub-property hierarchy. As explained, RDF Schema is used to define RDF vocabularies, such as the FOAF "friend of a friend" vocabulary or the SKOS "simple knowledge organisation system" vocabulary. RDF Schema is often abbreviated as RDFS, the combination of RDF and RDF Schema as RDF(S).

RDF Schema has three basic meta-classes: rdfs:Resource, of which each resource is implicitly member; rdfs:Class, which defines sets of resources; and rdfs:Property, which defines properties (relationships) between resources. RDFS also defines some terms for informal descriptions of classes and properties, such as rdfs:comment, rdfs:label, and rdfs:seeAlso. These terms are for human-readable information only and do not carry formal semantics.

A class in RDF Schema is a named term and the collection of its members, defined using rdf:type, for example, ex:john rdf:type foaf:Person. Classes in RDFS cannot have constraints on their membership. A property in RDF Schema is a named term and its domain and range, which indicate to which class of resource this property applies (domain) and to which class of resources this property leads (range).

The RDFS semantics (Hayes ed.) defines the interpretation of class definitions, property definitions, etc. in 14 mandatory entailment rules and nine optional rules that implement stronger semantic conditions. The most important of these rules are shown in Table 6.1. The first two state that all subjects (first position of the triple) are by definition resources, and that all predicates (second position of the triple) are properties. The following two rules define the meaning of rdfs:domain and rdfs:range: if a property is used in a triple, the subject of that triple is known to belong to the property's domain and the object of that triple to the property's range. The final two rules state that types (class membership) are inherited along subclass relations (if john is a student, and all students are people, then john is a person), and that object values are inherited along subproperty relations (if john isBestFriendOf mary, and isBestFriendOf is a subproperty of knows, then also john knows mary.

Formally, the RDFS semantics is an extension of the RDF semantics. Below we give a formal definition analogous to that of RDF.

Definition 6.3.1 (RDFS Entailment) $\mathcal{I}$ *is called an* **RDFS** *interpretation, if it is an RDF interpretation and the following additional requirements hold:*[3]

Axiomatic statements $\mathcal{I}$ *makes all of the following axiomatic statements true:*

Table 6.1 Most relevant RDF(S) entailment rules

$s\,p\,o$	$\rightarrow s$ rdf:type rdfs:Resource
$s\,p\,o$	$\rightarrow p$ rdf:type rdfs:Property
$s\,p\,o$, p rdfs:domain d	$\rightarrow s$ rdf:type p
$s\,p\,o$, p rdfs:range r	$\rightarrow o$ rdf:type r
$s\,p\,o$, p rdfs:subPropertyOf p_2	$\rightarrow s\,p_2\,o$
s rdf:type c_1, c_1 rdfs:subClassOf c_2	$\rightarrow s$ rdf:type c_2

---

[3] Again, we leave out D-Entailment, XMLLiterals and container membership properties.

```
rdf:type rdfs:domain rdfs:Resource .
rdfs:domain rdfs:domain rdf:Property .
rdfs:range rdfs:domain rdf:Property .
rdfs:subPropertyOf rdfs:domain rdf:Property .
rdfs:subClassOf rdfs:domain rdfs:Class .
rdf:subject rdfs:domain rdf:Statement .
rdf:predicate rdfs:domain rdf:Statement .
rdf:object rdfs:domain rdf:Statement .
rdfs:member rdfs:domain rdfs:Resource .
rdf:first rdfs:domain rdf:List .
rdf:rest rdfs:domain rdf:List .
rdfs:seeAlso rdfs:domain rdfs:Resource .
rdfs:isDefinedBy rdfs:domain rdfs:Resource .
rdfs:comment rdfs:domain rdfs:Resource .
rdfs:label rdfs:domain rdfs:Resource .
rdf:value rdfs:domain rdfs:Resource .

rdf:type rdfs:range rdfs:Class .
rdfs:domain rdfs:range rdfs:Class .
rdfs:range rdfs:range rdfs:Class .
rdfs:subPropertyOf rdfs:range rdf:Property .
rdfs:subClassOf rdfs:range rdfs:Class .
rdf:subject rdfs:range rdfs:Resource .
rdf:predicate rdfs:range rdfs:Resource .
rdf:object rdfs:range rdfs:Resource .
rdfs:member rdfs:range rdfs:Resource .
rdf:first rdfs:range rdfs:Resource .
rdf:rest rdfs:range rdf:List .
rdfs:seeAlso rdfs:range rdfs:Resource .
rdfs:isDefinedBy rdfs:range rdfs:Resource .
rdfs:comment rdfs:range rdfs:Literal .
rdfs:label rdfs:range rdfs:Literal .
rdf:value rdfs:range rdfs:Resource .
rdfs:isDefinedBy rdfs:subPropertyOf rdfs:seeAlso .
```

Classes $x \in \bullet_C^{\mathcal{I}}(y)$ *if and only if* $\langle x, y \rangle \in \bullet_P^{\mathcal{I}}(rdf:Type^{\mathcal{I}})$;
$\quad \Delta^{\mathcal{I}} = \bullet_C^{\mathcal{I}}(rdf:Resource^{\mathcal{I}})$;
$\quad \mathbb{L} = \bullet_C^{\mathcal{I}}(rdf:Literal^{\mathcal{I}})$;
$\quad$ *Let* $\mathbb{C} = \bullet_C^{\mathcal{I}}(rdf:Class^{\mathcal{I}})$.

Domain restrictions $\langle x, y \rangle \in \bullet_P^{\mathcal{I}}(rdfs:domain^{\mathcal{I}}) \wedge \langle u, v \rangle \in \bullet_P^{\mathcal{I}}(x) \Rightarrow u \in \bullet_C^{\mathcal{I}}(y)$.

Range class restrictions $\langle x, y \rangle \in \bullet_P^{\mathcal{I}}(rdfs:range^{\mathcal{I}}) \wedge \langle u, v \rangle \in \bullet_P^{\mathcal{I}}(x) \Rightarrow v \in \bullet_C^{\mathcal{I}}(y)$.

Property hierarchies $\bullet_P^{\mathcal{I}}(rdfs:subPropertyOf^{\mathcal{I}})$ *is transitive and reflexive on* $\mathbb{P}$;
$\langle x, y \rangle \in \bullet_P^{\mathcal{I}}(rdfs:subPropertyOf^{\mathcal{I}}) \Rightarrow x, y \in \mathbb{P} \wedge \bullet_P^{\mathcal{I}}(x) \subseteq \bullet_P^{\mathcal{I}}(y)$.

Class hierarchies $x \in \mathbb{C} \Rightarrow \langle x, rdfs:Resource^{\mathcal{I}} \rangle \in \bullet_P^{\mathcal{I}}(rdfs:subClassOf^{\mathcal{I}})$;
$\bullet_P^{\mathcal{I}}(rdfs:subClassOf^{\mathcal{I}})$ *is transitive and reflexive on* $\mathbb{C}$;
$\langle x, y \rangle \in \bullet_P^{\mathcal{I}}(rdfs:subClassOf^{\mathcal{I}}) \Rightarrow x, y \in \mathbb{C} \wedge \bullet_C^{\mathcal{I}}(x) \subseteq \bullet_C^{\mathcal{I}}(y)$.

Let G, H be RDF graphs. G RDFS entails H, if for each intensional RDFS interpretation
$\mathcal{I} : \mathcal{I} \models G \Rightarrow \mathcal{I} \models H.$

The RDFS semantics are not constraints: they do not forbid statements (a constraint violation) but only infer additional information. For example, if the domain of the property ex:knows is stated to be ex:Person, then applying that property to the resource ex:john would infer that ex:john is a member of ex:Person. If another statement is given that ex:john is a member of ex:Dog, the property definition would not be violated but instead ex:john would be considered both an ex:Dog and an ex:Person. Similarly as in RDF, the semantics of RDFS is defined as such that no contradictions can be expressed, making every RDF graph satisfiable under RDFS semantics (bar literal datatypes).

Example 6.3.2 *We extend the running example with an excerpt of the FOAF ontology (Brickley and Miller 2010).*

```
foaf:Person rdfs:subClassOf foaf:Agent.
foaf:depicts rdfs:domain foaf:Image.
foaf:Image rdfs:subClassOf foaf:Document.
```

Now we can infer that

```
:me rdfs:type foaf:Agent.
:P1000668 rdfs:type foaf:Image,
:P1000668 rdfs:type foaf:Document.
```

is RDFS entailed by our example graph and V_{RDFS}.

6.4   Data Models

RDF is inspired by semi-structured data models (Abiteboul 1997; Buneman 1997) such as the OEM model (Abiteboul et al. 1997; Papakonstantinou et al. 1995). Semi-structured models are part of a long evolution history of models for data representation; we give a brief overview based on Angles and Gutierrez (2005).

The network and hierarchical models, initial representations of large-scale data, had a low abstraction level and little flexibility, as shown in Table 6.2. The relational model, described by Codd (1970) and implemented by all relational databases, introduced a higher level of abstraction by separating the logical from the physical data levels. In contrast to RDF, the relational model assumes a fixed and a priori defined data schema; furthermore, all data and schema elements use local identifiers, which hampers data reuse, integration and extensibility. Semantic models such as the entity-relationship model (Chen 1976) increase the level of abstraction and allow data modelers to include richer schema semantics such as aggregation, instantiation and inheritance.

The object-oriented data model (Atkinson et al. 1989; Copeland and Maier 1984; Kim 1990a,b) aims to overcome limitations of the relational model (type definitions limited to simple data types, all tuples must conform to schema structure, which cannot be modified during runtime, limited expressive power in, for example, inheritance or

Table 6.2   Overview of data models, from Angles and Gutierrez (2005)

model	level	data complex.	connectivity	type of data
network	physical	simple	high	homogeneous
relational	logical	simple	low	homogeneous
semantic	user	simple/medium	high	homogeneous
object-or.	logical/physical	complex	medium	heterogeneous
XML	logical	medium	medium	heterogeneous
RDF	logical	medium	high	heterogeneous

aggregation) through the principles of object-oriented design (entities modeled as objects with attribute values and relationships, encapsulation of data and behavior inside objects, message passing).

Semi-structured data is self-describing in the sense that the schema information (if available) is contained in the data itself; the data schema defines relatively loose constraints on the data or is not defined at all (Buneman 1997). Semi-structured data can be generally characterized as follows (Abiteboul 1997): it may have irregular structure, in which data elements may be heterogeneous, incomplete, or richly annotated; it may have implicit structure, in which data elements may contain structures or even plain-text that need to be post-processed; the structure may be used as a posterior data guide instead of an a priori constraining schema; the structure may be rapidly evolving; and the distinction between data and schema may not always be clear, in the presence of evolving schemas, weak constraints, and the meta-modeling capabilities of graphs and hypergraph models of semi-structured data.

RDF is based on semi-structured data models but differs in the expressiveness of its schema language, in the existence of blank nodes, and in the fact that edge labels (predicates) can be resources themselves and thus form a hypergraph. XML, with its XSD schema language, can also be considered as a semi-structured model. Important differences between RDF and XML are, on the data level, the universality of the hypergraph structure of RDF versus the tree structure of XML. On the schema level, the higher expressiveness of RDFS versus XSD with respect to class membership, class and property inheritance and conjunctive classes (Decker et al. 2000; Gil and Ratnakar 2002), allows schema authors to capture their domain model in more detail and enables expression of simple vocabulary mappings (Aleman-Meza et al. 2007). On the other hand, RDFS does not include property cardinalities (these are introduced in OWL).

6.5   Linked Data Principles

Now that we can express data in a machine-understandable way, we need to publish it 'the Web way'. The Web is extremely scalable, because it is based on a minimal set of assumptions: statelessness, accessibility of content through URLs, and hyperlinks.

Berners-Lee (2006) proposes how the same can be achieved on the Semantic Web:

1. Use URIs as names for things.
2. Use HTTP URIs so that people can look up those names.

3. When someone looks up a URI, provide useful information, using the standards (RDF, SPARQL).
4. Include links to other URIs, so that they can discover more things.

Proposal 1 is already a core of RDF. It encourages us, however, to name things. Blank nodes are not useful for lookups. Proposal 2 allows for lookups, that is, identifiers can be dereferenced in order to find more information about the resource denoted by the identifier. When someone looks up a URL, there should actually be useful information and it should refer to more resources (proposals 3 and 4). This corresponds to linking on the document-centric web. Even though the document obtained by dereferencing a URL might not completely describe a resource and all related resources (in fact such completeness is not possible in an open world), it should at least be a good starting point for further crawling.

When dereferencing a URL, the result always is a document. However, on the Semantic Web we mainly talk about non-documents. In order to have a semantically correct mechanism, we need to ensure that the URL of the document containing the description is not the same as the URL of the resource described. This can be achieved using one of the two approaches discussed in this section

6.5.1 Dereferencing Using Basic Web Look-up

HTTP URLs may contain a local part or fragment identifier, which starts with a hash "#". In the URL `http://userpages.uni-koblenz.de/~sschenk/foaf.rdf#me`, `http://userpages.uni-koblenz.de/~sschenk/foaf.rdf` identifies a document, and #me is a local identifier within this document. Hence, when an HTTP client dereferences `http://userpages.uni-koblenz.de/~sschenk/foaf.rdf#me`, it extracts the document name and in fact accesses this document.

Using fragment identifiers, the document accessed has a different URL from the resource denoted. For example, the concept representing the idea of a 'Type' in the RDF vocabulary has the URI `http://www.w3.org/1999/02/22-rdf-syntax-ns#type`. When dereferencing this URI, the document `http://www.w3.org/1999/02/22-rdf-syntax-ns` is retrieved.

An advantage of this approach is that it is technically very simple and does not need server-side infrastructure. In addition, when dereferencing multiple resources from the same ontology,[4] the ontology needs to be accessed only once.

Of cause this simplification can turn into a disadvantage in the presence of large ontologies, which are downloaded completely, even though only a small part is actually used. Moreover, it is not always feasible or desirable to include hashes in URLs.

6.5.2 Dereferencing Using HTTP 303 Redirects

The HTTP protocol allows for a multi-step communication when dereferencing a URL. With each response, the server sends an *HTTP status code* (Fielding et al. 1999).

---

[4] Before we formally introduce the term 'ontology', assume an ontology is an RDF graph using the RDFS vocabulary.

For example, when a document is requested, a server may directly return that document. The status code of the response then is "200 - OK".

Using 303 redirects when serving a description of a resource; a redirection is returned instead of an actual document. Status code '303 – See Other' is used, which means the response to the request can be found under a different URI and the response contains this different URI. The client then looks up the new URI. Again, we have achieved a separation of the URIs of a resource from the URI of the document containing the description.

Example 6.5.1 *A request for the DBpedia resource* `http://dbpedia.org/resource/` `Koblenz` *303 redirects to* `http://dbpedia.org/page/Koblenz` *when accessed with an HTML browser.*

Moreover, a user agent can send a list of accepted content types with a request to DBpedia in order to obtain either an HTML page as a human-readable representation or, for example, an RDF/XML file for machine processing. This mechanism is called content negotiation (Fielding et al. 1999; Holtman and Mutz 1998).

Example 6.5.2 *When a client requests a response in RDF/XML format, the server would redirect to* `http://dbpedia.org/data/Koblenz.rdf`.

An advantage of this approach is that it allows for arbitrary URLs and that only relevant parts of an ontology need to be retrieved. As a disadvantage, a potentially large number of HTTP requests is generated. Additionally, more control over the HTTP server and more technical know-how are required.

6.6   Development Practicalities

Many activities around the Semantic Web are led and influenced by the World Wide Web consortium (W3C). The W3C is the Web standardization organization and has published, for example, HTML, XML, WSDL, RDF(S) and OWL. The W3C leads ongoing activities around the Semantic Web. The W3C also leads education and outreach activities, through its SWEO[5] group, and has several working groups that focus on specific application areas such as the health care and life science domain (HCLS group[6]). The W3C also maintains a list of frequently asked questions[7] on the Semantic Web, a good starting point for a general overview. All W3C activities and discussion lists are public and working group participation is open to delegates from all W3C members.

In this final section of this chapter we give a short overview of available software for working with RDF. This overview is by no means complete and should only be seen as a starting point.

---

[5] `http://www.w3.org/2001/sw/sweo/`

[6] `http://www.w3.org/2001/sw/hcls/`

[7] `http://www.w3.org/2001/sw/SW-FAQ`

6.6.1 Data Stores

Concerning application development, several RDF data stores are available, both commercial, non-commercial and open source. Prominent stores include Sesame[8] (Broekstra et al. 2002), Jena[9] (Wilkinson et al. 2003), Redland[10] (Beckett 2002), YARS[11] (Harth and Decker 2005), OpenLink Virtuoso,[12] AllegroGraph,[13] Oracle 10g,[14] 4store[15] (Harris et al. 2009); see the ESW wiki[16] for an overview. Furthermore, several OWL reasoners such as Racer (Haarslev and Möller 2003), Fact++ (Tsarkov and Horrocks 2006), Kaon2 (Hustadt et al. 2004), Pellet (Sirin and Parsia 2004), HermiT (Motik et al. 2009e) and OWLIM (Kiryakov et al. 2005a) are available. As these support more expressive semantics, their architectures and usage patterns differ from "simple" RDF stores.

While in recent years scaleability of RDF storage has been an issue, state-of-the-art RDF repositories scale up to billions of statements (Virtuoso, OWLIM, 4store). The latest developments, for example 5store[17] are targeting the TeraTriples border.

The various RDF stores differ on several characteristics, such as availability and license, support for RDF(S) and OWL inferencing, support for RDF syntaxes, language bindings and APIs, scalability, etc. Historically, each store implemented its own query language, but with the standardization of the SPARQL query language and protocol most stores now support SPARQL as well. Most RDF stores use a native storage model, optimized for triples, although some are based on a relational database. RDF(S) inferencing, if supported, is usually implemented through forward-chaining during data entry: the RDF(S) rules are repeatedly evaluated and all inferred triples added to the knowledge base, until a fixpoint is reached.

Also, several logic programming frameworks include support for manipulation of RDF data, such as TRIPLE (Sintek and Decker 2002), SWI-Prolog (Wielemaker et al. 2003) and dlvhex (Eiter et al. 2006). Since these are generic rule systems, the RDF(S) semantics, and part of the OWL semantics, are implemented directly through a set of deduction rules.

6.6.2 Toolkits

In terms of tools for processing RDF, several lightweight toolkits help parse and serialize RDF from and to different syntaxes such as RDF/XML, Turtle, N3, RSS, GRDDL. These toolkits typically do not provide scalable storage and querying facilities but help process and transform RDF before storing it. We briefly list some prominent ones, more complete

---

[8] http://www.openrdf.org
[9] http://jena.sourceforge.net/
[10] http://www.librdf.org
[11] http://sw.deri.org/2004/06/yars/
[12] http://virtuoso.openlinksw.com/
[13] www.franz.com/products/allegrograph/
[14] http://www.oracle.com/technology/tech/semantic_technologies
[15] http://4store.org/
[16] http://esw.w3.org/topic/SemanticWebTools
[17] http://4store.org/trac/wiki/5store

overviews are available online:[18] in Java, Jena (Wilkinson et al. 2003), Sesame (Broekstra et al. 2002) and RDF2Go (Völkel 2005) are available, Jena and Sesame are datastores but also provide complete processing architectures; Redland (Beckett 2002) can be used in C and also has bindings in several languages such as PHP, Perl, Ruby and Python; for Python, cwm[19] and RDFLib[20] can be used; for C#, the SemWeb[21] library; for PHP, the RAP library (Oldakowski et al. 2005); and for Ruby, the ActiveRDF library (Oren et al. 2008).

---

[18] http://planetrdf.com/guide/
[19] http://www.w3.org/2000/10/swap/doc/cwm.html
[20] http://rdflib.net/
[21] http://razor.occams.info/code/semweb/

7

Semantic Web Languages

Antoine Isaac,[1] Simon Schenk[2] and Ansgar Scherp[2]

[1]*Vrije Universiteit Amsterdam, Amsterdam, The Netherlands*
[2]*University of Koblenz-Landau, Koblenz, Germany*

Having defined the Semantic Web infrastructure, which enables the creation of a web of data, two aspects remain to be addressed. The first concerns the rich *semantics* that were announced as a part of the Semantic Web vision: how can the conceptual knowledge useful for a range of applications be successfully ported to and exploited on the Semantic Web? The second aspect concerns the access to Semantic Web data: how can one query and successfully find the information that is represented on these large RDF graphs that constitute the Semantic Web information sphere?

The goal of this chapter is to make the reader familiar with relevant languages that address these two crucial matters: representing conceptual knowledge and querying RDF data. With respect to conceptual knowledge, there exist very expressive languages such as OWL, which allow formal specification of the ontologies that guide the creation of RDF data and make it amenable to automated reasoning processes. This will be the object of Section 7.2. However, we will see in Section 7.3 that OWL does not meet all conceptual representation requirements. In particular, its high level of formal precision is not appropriate for modeling a wide range of more lightweight vocabularies, where relations between concepts are not completely sharp, and which are not expected to guide the structure of RDF descriptions. Hence we will introduce the Simple Knowledge Organization System (SKOS) in Section 7.3, which is suitable for modeling such lightweight ontologies.

In Section 7.4 we will introduce SPARQL, the recently standardized Semantic Web Query language, with an emphasis on aspects relevant to querying multimedia metadata in the running examples of COMM annotations. As a practical guide, we will reference relevant implementations, list limitations of SPARQL and give an outlook on possible future extensions to address these limitations.

Multimedia Semantics: Metadata, Analysis and Interaction, First Edition.
Edited by Raphaël Troncy, Benoit Huet and Simon Schenk.
© 2011 John Wiley & Sons, Ltd. Published 2011 by John Wiley & Sons, Ltd.

7.1 The Need for Ontologies on the Semantic Web

As seen in the previous chapter, the Semantic Web requires vocabularies to be defined for the creation of useful RDF data. Consider the XML realm: there, DTDs or XML Schemas are crucial, as they allow for specifying which XML elements and attributes may appear in XML documents, and how these entities should be organized (Bray et al. 2006; Fallside and Walmsley 2004). RDF provides a more flexible way to represent knowledge. However, it is still in essence a mere description framework, which requires controlled vocabularies of some form to adequately express and share information. For example, to represent information about (audiovisual) multimedia documents, one would expect building blocks like 'program', 'sequence' and 'scene' to identify documents and their parts, and attributes such as 'genre', 'creator' or 'subject' to characterize them.

In the Semantic Web context, such vocabularies are called *ontologies*. An ontology should provide the conceptual elements required to describe entities such as encountered in a specific domain, or manipulated by a given application. An essential point, related to the intended semantic nature of the Semantic Web, is that ontologies are not restricted to the gathering of (named) building blocks for RDF graphs. They should also provide appropriate *semantics* for such building blocks. This implies, of course, that these vocabularies should emerge from careful processes, capturing and reflecting *consensus* in specific fields of human knowledge. But, beyond this, ontologies should be also exploitable by automated reasoning processes. A typical means for this, inherited from artificial intelligence research, is to rely on formalisms which allow for definitions suitable for *inference*. For example, a segment of an image again is an image and each image has at least one creator.

To achieve such goals, an appropriate ontology language[1] is necessary. In what follows we will focus on explaining the details of one specific ontology language, namely OWL, the ontology language developed by W3C.

7.2 Representing Ontological Knowledge Using OWL

RDF Schema, which was introduced in the previous chapter, is already a simple ontology language. Using it, we can introduce classes such as Document and Person, define specialization relations (e.g. Author is a sub-class of Person), and state that authorOf is a property, which has domain Document and range Author. We can then assert in RDF facts such as that Aristotle is an instance of Author who is the author of the document physics.

However, such specification may be insufficient for a whole range of semantic-intensive use cases (Heflin 2004). The W3C has therefore created the OWL ontology language, which extends the RDFS language by adding more operators, following the requirements and design principles that could be elicited from these observed use cases (Bechhofer et al. 2004).

7.2.1 OWL Constructs and OWL Syntax

OWL is defined as an extension of RDFS, with the aim of providing vocabularies to create RDF data. It is thus heavily influenced by the RDF data model, which is based on

---

[1] Such a language can actually be called a meta-language, since it enables in practice application-specific description languages to be designed.

graphs composed of nodes and directed arcs. Like RDFS, OWL distinguishes two main types of ontological resources:

- *classes*, used to define the types of a domain's entities (entities which are then called the *instances* of these classes);
- *properties*, used to relate entities together.

As a matter of fact, the designers of OWL have opted to reuse the constructs introduced by RDFS to basically characterize these entities, like `rdfs:subClassOf`, `rdfs:subPropertyOf`, `rdfs:domain` and `rdfs:range`. OWL, however, introduces new constructs to specify classes and properties with richer *axioms*.

One can specify more advanced semantic characteristics of properties, for example that `hasAuthor` and `authorOf` are the *inverse* of each other, that `hasISBN` is an *inverse-functional* property (i.e. that an object has only one possible ISBN value which identifies it) or that `hasPart` is *transitive* (every part of a part of a given object is a part of this object).

OWL also enables much richer specifications for classes. It allows classes to be defined in a complete way, or at least necessary conditions to be given for individuals to belong to a class. In the first case we can, for example, define a `Publication` as the class equivalent to the class of things having an ISBN. The second case is weaker; we can, for example, define as a necessary condition for something to be a `Journal` that it must be a publication, formally `SubClassOf(Journal Publication)`. This, however, does not imply that every publication must also be a journal.

Such definitions may involve simple class individuals or more complex *expressions* that OWL allows:

- *combinations* of classes using boolean constructs, such as *union, intersection* or *complement*;
- local *restrictions* on properties. This is an extension of the range and domain restrictions of RDFS. With the construct `owl:allValuesFrom` one can, for example, enforce that `Books` have to be written by `Persons` only. Formally, this would be written as `SubClassOf(Book, restriction(hasAuthor allValuesFrom(Person)))`.

In fact, OWL proposes a whole range of expressions to handle semantic similarity, as well as basic identity and distinction statements. As seen, one can define two classes to be equivalent. It is also possible to assert that two given classes are *disjoint*, that is, do not share individuals. For properties, the `owl:EquivalentProperty` construct can be used to declare that two properties apply to the same subjects and objects. This features are completed, at the level of individuals in general, by constructs that allow it to be stated that individuals are *different from* or *same as* other individuals.

An important OWL feature is that such identity and distinction axioms may hold perfectly between elements from different ontologies. One can therefore state that a property used in one context is semantically equivalent to a property used in a different context, or one can simply reuse it in a different ontology. This is crucial to meeting crucial interoperability requirements, such as the 'metadata interoperability' use case in Chapter 2.

Finally, it is worth mentioning that in the OWL world ontologies are themselves considered as full-fledge entities (of type `owl:Ontology`) that can themselves be described.

It is therefore possible to provide ontologies with metadata – creator, date of creation, etc. More interestingly from a Semantic Web perspective, one can state that an ontology *imports* another. This allows for sharing and reusing complete vocabularies – and their axiomatizations – in an explicit manner throughout the Semantic Web.

Concerning syntax, a first observation is that, like RDFS, OWL allows ontologies to be defined using RDF itself. It can therefore be written using the RDF XML syntax, or any of the other syntaxes mentioned in the previous chapter, like Turtle. In OWL, books can, for example, be restricted to a subclass of all objects with at most one ISBN – we then say that the link between books and ISBN number has a maximal cardinality of 1. In the XML syntax of OWL, this reads as follows:

```
<owl:Class rdf:about="#Books">
  <rdfs:subClassOf>
    <owl:Restriction>
      <owl:onProperty rdf:resource="#hasISBN" />
      <owl:maxCardinality rdf:datatype="&xsd;
        nonNegativeInteger">1</owl:maxCardinality>
    </owl:Restriction>
  </rdfs:subClassOf>
</owl:Class>
```

However, OWL also has additional syntaxes, including an XML syntax (Motik et al. 2009b), a functional abstract syntax (Motik et al. 2009d) and the popular Manchester syntax (Horridge and Patel-Schneider 2009) to make the language more accessible for human readers (Patel-Schneider et al. 2004). The same axiom would now read as `SubClassOf(Books, restriction(hasISBN minCardinality(1)))` in functional syntax. For the sake of better readability, we will use functional syntax or DL expressions in the rest of this book.

7.2.2 The Formal Semantics of OWL and its Different Layers

One of the main attractions of an ontology language like OWL is that it is grounded in formal semantics. As seen in the previous chapter, RDFS constructs come with entailment rules that can be implemented by inference engines to derive new facts from asserted ones. OWL also provides semantics for its constructs, enabling not only new information to be inferred from existing information, but also the consistency of a given RDF graph to be checked with respect to the specifications found in its underlying ontology(ies). Or, for ontology engineering purposes, enabling one to verify if the axioms found in an ontology do not raise inconsistency by themselves.

A very simple example would be to define `Shakespeare` to be both a `Person` and a `Group`. If we also add an axiom that persons and groups are disjoint – that is, that these classes cannot share any instances (formally `DisjointClasses(Person Group)`) – the ontology is inconsistent.

These mechanisms rely on a formal semantic interpretation of the ontology language constructs. In the case of OWL, one proposed interpretation is a model-theoretic one, inspired by *description logic* approaches (Baader et al. 2003). Considering a domain of interpretation (a set of resources), an OWL class can be mapped to a set of elements from this domain, its instances. Similarly, a property can be associated with a set of pairs of resources, corresponding to the subjects and objects of each statement for which the property is the predicate.

The axioms and expressions defining classes and properties are in turn interpreted as conditions that apply to the extensional interpretations of these vocabulary elements. Consider the expression DisjointClasses(Person Group) mentioned above. It can be formally interpreted as a semantic condition $EC(Person) \cap EC(Group) = \emptyset$, where $EC(c)$ denotes the interpretation of a class c. This condition can then be checked by an inference engine once the extension of the classes are (partially) known thanks to rdf:type statements, or based on other ontology axioms.

The Web Ontology Language OWL in its second version is based on the $\mathcal{SROIQ}(D)$ description logic, that is, $\mathcal{SROIQ}$ extended with datatypes. In the following we briefly introduce a fragment of OWL-2 and the fundamental reasoning problems related to DLs. For details we refer the reader to Horrocks et al. (2006) and Hitzler et al. (2009). Later in this book, we propose to extend reasoning with OWL towards fuzzy logics. Table 7.1 gives an overview of the operators used in this fragment, both in description logic syntax, which is very compact, and in OWL functional syntax. We omit operators which can be reduced to more basic ones, for example EquivalentClasses(C D) can be reduced to SubClassOf(C D), SubClassOf(D C). We also omit data properties, that is, properties ranging over literal values. The functional syntax for data properties is analogous to that for object properties, but with the prefix *Data* instead of *Object*.

Classes, roles and individuals form the vocabulary of an ontology. In OWL, they are named using URIs, or anonymous and denoted by blank nodes.

Definition 7.2.1 (Vocabulary) *A vocabulary $V = (N_C, N_P, N_I)$ is a triple where*

- N_C *is a set of URIs used to denote classes,*
- N_R *is a set of URIs used to denote roles and*
- N_I *is a set of URIs used to denote individuals.*

N_C, N_P, N_I *need not be disjoint.*

An interpretation grounds the vocabulary in objects from an object domain.

Definition 7.2.2 (Interpretation) *Given a vocabulary V, an interpretation $\mathcal{I} = (\Delta^{\mathcal{I}}, \cdot^{\mathcal{I}_C}, \cdot^{\mathcal{I}_R}, \cdot^{\mathcal{I}_I})$ is a quadruple where*

- $\Delta^{\mathcal{I}}$ *is a non-empty set called the object domain;*
- $\cdot^{\mathcal{I}_C}$ *is the class interpretation function, which assigns to each class a subset of the object domain $\cdot^{\mathcal{I}_C} : N_C \to 2^{\Delta^{\mathcal{I}}}$;*
- $\cdot^{\mathcal{I}_P}$ *is the role interpretation function, which assigns to each role a set of tuples over the object domain $\cdot^{\mathcal{I}_P} : N_P \to \Delta^{\mathcal{I}} \times \Delta^{\mathcal{I}}$;*
- $\cdot^{\mathcal{I}_i}$ *is the individual interpretation function, which assigns to each individual $a \in N_I$ an element $a^{\mathcal{I}_I}$ from $\Delta^{\mathcal{I}}$.*

Table 7.1 A core fragment of OWL2

#	Description	OWL abstract syntax	DL syntax
1	The universal concept	`owl:Thing`	$\top$
2	The empty concept	`owl:Nothing`	$\bot$
3	C is a subclass of D	`SubClassOf(C D)`	$C \sqsubseteq D$
4	Class intersection of C and D	`ObjectIntersectionOf (C D)`	$C \sqcap D$
5	Class union of C and D	`ObjectUnionOf(C D)`	$C \sqcup D$
6	C is the inverse class of D	`ObjectComplementOf(C D)`	$D \equiv \neg C$
7	The class of things which stand in relation R only with with Cs	`ObjectAllValuesFrom(R C)`	$\forall R.C$
8	The class of things which stand in relation R with at least one C	`ObjectSomeValuesFrom (R C)`	$\exists R.C$
9	The class of things which are R-related to themselves	`ObjectHasSelf(R)`	$\exists R.\text{Self}$
10	The class of things which are R-related to at least n Cs	`ObjectMinCardinality (n R C)`	$\geq nR$
11	The class of things which are R-related to at most n Cs	`ObjectMaxCardinality (n R C)`	$\leq nR)^{\mathcal{I}}$
12	The class consisting exactly of $a_1, \ldots, a_n$	`ObjectOneOf (a_1,…,a_n)`	$\{a_1, \ldots, a_n\}$
13	R is a subproperty of S	`SubObjectPropertyOf(R S)`	$R \sqsubseteq S$
14	The property obtained from chaining $R_1, \ldots, R_n$	`ObjectPropertyChain (R_1,…,R_n)`	$R_1 \circ \ldots \circ R_n \sqsubseteq S$
15	The inverse property of R	`ObjectInverseOf (R)`	R^-
16	R is asymmetric	`AsymmetricObject Property(R)`	$Asy(R)$
17	R is reflexive	`ReflexiveObject Property(R)`	$Ref(R)$
18	R is reflexive	`IrreflexiveObject Property(R)`	$Irr(R)$
19	R and S are disjoint	`DisjointObject Properties(R S)`	$Dis(R, S)$
20	a is a C	`ClassAssertion(a C)`	$a : C$
21	a and b stand in relation R	`ObjectPropertyAssertion (R a b)`	$(a, b) : R$
22	a and b do not stand in relation R	`NegativeObjectProperty Assertion (R a b)`	$(a, b) : \neg R$
23	a and b are the same	`SameIndividual(a b)`	$a = b$
24	a and b are different	`DifferentIndividuals (a b)`	$a \neq b$

Let $C, D \in N_C$, let $R, R_i, S \in N_R$ and $a, a_i, b \in N_I$. We extend the role interpretation function $\cdot^{\mathcal{I}_R}$ to role expressions:

$$(R^-)^{\mathcal{I}} = \{(\langle x, y \rangle) | (\langle y, x \rangle) \in R^{\mathcal{I}}\}.$$

We extend the class interpretation function $\cdot^{\mathcal{I}_C}$ to class descriptions:

$$\top^{\mathcal{I}} \equiv \Delta^{\mathcal{I}}$$

$$\bot^{\mathcal{I}} \equiv \emptyset$$

$$(C \sqcap D)^{\mathcal{I}} \equiv C^{\mathcal{I}} \cap D^{\mathcal{I}}$$

$$(C \sqcup D)^{\mathcal{I}} \equiv C^{\mathcal{I}} \cup D^{\mathcal{I}}$$

$$(\neg C)^{\mathcal{I}} \equiv \Delta^{\mathcal{I}} \setminus C^{\mathcal{I}}$$

$$(\forall R.C)^{\mathcal{I}} \equiv \{x \in \Delta^{\mathcal{I}} | \langle x, y \rangle \in R^{\mathcal{I}} \rightarrow y \in C^{\mathcal{I}}\}$$

$$(\exists R.C)^{\mathcal{I}} \equiv \{x \in \Delta^{\mathcal{I}} | \exists y \in \Delta^{\mathcal{I}} : \langle x, y \rangle \in R^{\mathcal{I}}\}$$

$$(\exists R.Self)^{\mathcal{I}} \equiv \{x \in \Delta^{\mathcal{I}} | \langle a, a \rangle \in R^{\mathcal{I}}\}$$

$$(\geq nR)^{\mathcal{I}} \equiv \{x \in \Delta^{\mathcal{I}} | \exists y_1, \ldots, y_m \in \Delta^{\mathcal{I}} :$$

$$\langle x, y_1 \rangle, \ldots, \langle x, y_m \rangle \subset R \wedge m \geq n\}$$

$$(\leq nR)^{\mathcal{I}} \equiv \{x \in \Delta^{\mathcal{I}} | \not\exists y_1, \ldots, y_m \in \Delta^{\mathcal{I}} :$$

$$\langle x, y_1 \rangle, \ldots, \langle x, y_m \rangle \in R \wedge m > n\}$$

$$\{a_1, \ldots, a_n\}^{\mathcal{I}} \equiv \{a_1^{\mathcal{I}}, \ldots, a_n^{\mathcal{I}}\}.$$

Class expressions are used in axioms.

Definition 7.2.3 (Axiom) *An axiom is one of the following*

- *a general concept inclusion of the form $A \sqsubseteq B$ for concepts A and B;*
- *an individual assertion of one of the forms $a . C$, $(a, b) : R$, $(a, b) : \neg R$, $a = b$ or $a \neq b$ for individuals a, b and a role R;*
- *a role assertion of one of the forms $R \sqsubseteq S$, $R_1 \circ \ldots \circ R_n \sqsubseteq S$, $Asy(R)$, $Ref(R)$, $Irr(R)$, $Dis(R, S)$ for roles R, R_i, S.*

Satisfaction of axioms in an interpretation $\mathcal{I}$ is defined as follows.

Definition 7.2.4 (Satisfaction of Axioms) *Satisfaction of axioms in an interpretation $\mathcal{I}$ is defined as follows. With $\circ$ we denote the composition of binary relations.*

$$(R \sqsubseteq S)^{\mathcal{I}} \equiv \langle x, y \rangle \in R^{\mathcal{I}} \rightarrow \langle x, y \rangle \in S^{\mathcal{I}}$$

$$(R_1 \circ \ldots \circ R_n \sqsubseteq S)^{\mathcal{I}} \equiv \forall \langle x, y_1 \rangle \in R_1^{\mathcal{I}}, \langle y_1, y_2 \rangle \in R_2^{\mathcal{I}}, \ldots,$$

$$\langle y_{n-1}, z \rangle \in R_n^{\mathcal{I}} : \langle x, z \rangle \in S^{\mathcal{I}}$$

$$(Asy(R))^{\mathcal{I}} \equiv \exists \langle x, y \rangle \in R^{\mathcal{I}} : \langle y, x \rangle \notin R^{\mathcal{I}}$$

$$(Ref(R))^{\mathcal{I}} \equiv \forall x \in \Delta^{\mathcal{I}} : \langle x, x \rangle \in R^{\mathcal{I}}$$

$$(Irr(R))^{\mathcal{I}} \equiv \forall x \in \Delta^{\mathcal{I}} : \langle x, x \rangle \notin R^{\mathcal{I}}$$

$$(Dis(R, S))^{\mathcal{I}} \equiv R^{\mathcal{I}} \cap S^{\mathcal{I}} = \emptyset$$

$$(C \sqsubseteq D)^{\mathcal{I}} \equiv x \in C^{\mathcal{I}} \rightarrow x \in D^{\mathcal{I}}$$

$$(a : C)^{\mathcal{I}} \equiv a^{\mathcal{I}} \in C^{\mathcal{I}}$$

$$((a, b) : R)^{\mathcal{I}} \equiv \langle a^{\mathcal{I}}, b^{\mathcal{I}} \rangle \in R^{\mathcal{I}}$$

$$((a, b) : \neg R)^{\mathcal{I}} \equiv \langle a^{\mathcal{I}}, b^{\mathcal{I}} \rangle \notin R^{\mathcal{I}}$$

$$a = b \equiv a^{\mathcal{I}} = b^{\mathcal{I}}$$

$$a \neq b \equiv a^{\mathcal{I}} \neq b^{\mathcal{I}}.$$

An ontology is comprised of a set of axioms. We can further differentiate between the TBox describing the terms or classes in the ontology, the RBox describing the properties and the ABox containing assertions about individuals.

Definition 7.2.5 (Ontology) *A $\mathcal{SROIQ}$ ontology O is a set of axioms as per Definition 7.2.3. A TBox (terminology) of O defines the concepts used in O. It consists only of axioms of the forms 1–12 in Table 7.1. An RBox (properties) of O defines the properties used in O. It consists only of axioms of the forms 13–19 in Table 7.1. An ABox (assertions) of O defines the individuals used in O and their relations. It consists only of axioms of the forms 20–24 in Table 7.1.*

7.2.3   Reasoning Tasks

The benefit of using a description logics based language of cause is the ability to automatically draw conclusions from the available facts. In this section we list the most common reasoning tasks for OWL2 (Motik et al. 2009c).

Definition 7.2.6 (Basic Reasoning Tasks) *Let O, O' be ontologies, and V the vocabulary of O.*

Consistency *O is consistent (satisfiable), if there exists a model of O with respect to V.*

Entailment *Let $\mathcal{I}$ be a model of O with respect to V. O entails O' ($O \models O'$) if $\mathcal{I} \models O \Rightarrow \mathcal{I} \models O'$. We say that O and O' are equivalent, if $O \models O'$ and $O' \models O$.*

Satisfiability *Let C be a class expression. C is satisfiable with respect to O if a model $\mathcal{I}$ of O with respect to V exists such that $C^{\mathcal{I}} \neq \emptyset$. We say that O is satisfiable if all class expressions in O are satisfiable. We say that O and O' are equisatisfiable, if O is satisfiable if and only if O' is satisfiable.*

Subsumption *Let C and D be class expressions. D is subsumed by C ($D \subseteq C$) with respect to O, if for each model $\mathcal{I}$ of O with respect to V the following holds: $D^{\mathcal{I}} \subseteq C^{\mathcal{I}}$. C and D are equivalent, if $D \subseteq C$ and $C \subseteq D$.*

Instance Checking *Let C be a class expression. a is an instance of C with respect to O, if for each model $\mathcal{I}$ of O with respect to V the following holds: $a^{\mathcal{I}} \in C^{\mathcal{I}}$.*

Boolean Conjunctive Query Answering *Let Q be a boolean conjunctive query of the form $\exists x_1, \ldots, x_n : A_1, \ldots, A_m$ where A_i is an atom of the form $C(s)$ or $P(s, t)$ with C a class, P a property and s and t individuals or some variable x_j. Q is an answer with respect to O if Q is true in each model of O with respect to V.*

As these basic reasoning tasks can be reduced to each other, it is sufficient for a reasoner to implement one of them (Baader et al. 2003).

A common task is to compute the full class hierarchy of an ontology, for example for visualization purposes. The full hierarchy can be computed, for example, by mutually checking subsumption for all classes in an ontology.

SPARQLDL (Sirin and Parsia 2007) is a SPARQL *entailment regime*, which uses SPARQL to express conjunctive queries in order to query OWL knowledge bases.

7.2.4   OWL Flavors

At the time of writing this book, there are a number of OWL reasoning engines available that are able to carry out such reasoning processes. Examples include Hermit (Motik et al. 2009e), RacerPro (Haarslev and Möller 2003), and Pellet (Sirin et al. 2007).

It is important to mention that several flavors of OWL have been introduced, depending on the formal expressivity and complexity required by the application at hand. Indeed, the reasoning mechanisms on which OWL inference tools rely can exhibit different levels of formal complexity, depending on the selection of OWL constructs that is considered. The more constructs are used, the less tractable inference will be.

The designers of the original OWL (Bechhofer et al. 2004) have thus proposed three sub-languages which embody different levels of compromise between expressivity and tractability: OWL-Full, OWL-DL and OWL-Lite. OWL-Full actually includes all the constructs of RDFS and the description logic $\mathcal{SHOIN}(D)$, and imposes no constraints on their use. It is, for example, permitted to consider classes themselves as instances of other classes. OWL-DL permits the use of all $\mathcal{SHOIN}(D)$ constructs, but with some constraints that prevent too complex knowledge bases to be built. Finally, the simplest flavor of OWL, OWL-Lite, imposes further constraints and limit the constructs that can be used to describe classes and properties.

In OWL2, the Full sub-language corresponds to the RDF-based semantics. The former OWL-DL corresponds to the direct semantics. Moreover, there are three dialects with reduced expressivity and desirable computational behavior (Motik et al. 2009a): OWL2-EL targets very large ontologies with a large number of classes and allows for reasoning in polynomial time. OWL2-QL allows for an implementation in LogSpace using relational database technology and hence is especially useful for implementations on top of existing databases. Finally, OWL2-RL is the fragment of OWL2 which can be implemented using forward chaining rules. It is particularly useful for relatively lightweight ontologies with a large number of individuals.

7.2.5   Beyond OWL

Readers should be aware that even the most complete flavor of OWL may yet not fit all ontology engineering requirements. Among its most noticeable shortcomings, OWL does not allow useful features such as operations on (typed) literals, such as concatenating string values or summing integers. Neither does it not enable specification of *chains of properties* – as a famous example, it is impossible to state that the *brother* of the *father* of one individual is her *uncle*.

This explains why a number of complementary knowledge representation languages coexist alongside OWL. In particular, languages that enable knowledge to be represented in the form of generic *rules* are often mentioned, such as Description Logic Programs (Grosof et al. 2003) or the Semantic Web Rule Language (Horrocks et al. 2004).

With OWL it is possible to create various kinds of ontologies that can be useful to the multimedia domain, such as the Music Ontology mentioned in Chapter 2, or the COMM ontology that will be presented in Chapter 9. However, a first application of OWL which we will discuss right now is the creation of an ontological framework to represent conceptual vocabularies that are not amenable to full-fledged Semantic Web formal ontologies.

7.3   A Language to Represent Simple Conceptual Vocabularies: SKOS

Ontology languages, such as RDFS and OWL, are a necessary part of the Semantic Web technology. With them it is possible to create structured bodies of knowledge that enable information to be described and exchanged in a flexible, yet shared and controlled way. However, ontologies are not the only source of conceptual information to be found and used on the Semantic Web.

7.3.1   Ontologies versus Knowledge Organization Systems

A number of concept schemes exist that are not meant as schemas to represent structured data about entities – defining for instance a 'subject' property that applies to 'books'. Instead, they provide mere values that will appear in such structured descriptions for these entities – for example, a 'geography' topic. Typical examples are *thesauri* (International Standards Organisation 1986) and *classification schemes* that are used to describe documents in all kinds of institutions, say, libraries and audiovisual archives. Often, these schemes result from knowledge organization efforts that largely pre-date the rise of Semantic Web technology, but are nevertheless of crucial interest for the establishment of a rich information network.[2]

These description vocabularies, often referred to as *knowledge organization systems* (KOSs), clearly belong to the knowledge level: they convey some controlled meaning for identified descriptive entities, be they 'terms', 'notations' or 'concepts'. Furthermore, they do so in a structured and controlled manner, as testified by the existence of a number of dedicated shared guidelines (International Standards Organisation 1986) and formats, such as ZThes[3] or Marc 21.[4]

Machine readability and exploitation have high priority in the design of ontologies. KOSs such as thesauri are more targeted at human users; and the range of approaches to exploit them tends to privilege domain-intuitive associations rather than complex reasoning (Isaac et al. 2009). As a result, KOSs are not easily amenable to formal semantic characterization. They focus rather on the description of lexical knowledge and simple semantic relation between concepts, such as *specialization* or mere *association*, as exemplified in the following:

```
animal
    NarrowerTerm cat
    NarrowerTerm wildcat
cat
    UsedFor domestic cat
    RelatedTerm wildcat
```

---
[2] Indeed, this is expressed by the 'support controlled vocabularies' requirement of Section 2.3.
[3] http://zthes.z3950.org/
[4] http://www.loc.gov/marc/authority/

```
BroaderTerm animal
ScopeNote used only for domestic cats
```

To port this conceptual knowledge to the Semantic Web, a first solution involves formalizing the knowledge they contain, making them full-fledge ontologies. This has already been done, and can be fruitful for projects that have a clear use for such added semantics (Hyvönen et al. 2004). Yet, it turns out to be a very difficult task, comparable to that of creating a new ontology, but at a scale larger than most ontology engineering work – KOSs routinely contain thousands of concepts.

A second, simpler solution is to represent KOS knowledge as closely as possible to what can be found in the original information sources, using an appropriate Semantic Web ontology. This of course restricts the resulting amount of formal information available, but corresponds to observed requirements and eases conversion work. The latter can even be automatized, once a *mapping* has been established between the original KOS schema and that chosen to represent KOS information on the Semantic Web (van Assem et al. 2005). One possible means for achieving such a representation is to use SKOS, the *Simple Knowledge Organization System*.

7.3.2   *Representing Concept Schemes Using SKOS*

The SKOS Context

SKOS[5] is a standard model proposed by the W3C – in the context of the Semantic Web Deployment Working Group[6] – in order to represent common KOS information in a Semantic Web compatible form, so as to:

- share information that corresponds to domain practices (i.e. close to norms such as ISO 2788);
- develop and reuse standard tools (concept-based search engines, browsers) for the exploitation of the KOSs published on the Semantic Web.

The development of SKOS was driven by requirements elicited from a numbered of practical use cases gathered from the community of KOS users (Isaac et al. 2009). As such, it benefited from the input of Semantic Web researchers, but also from information science specialists. The resulting model is presented in the SKOS Reference document (Miles and Bechhofer 2009), while a more informal overview is given in the SKOS Primer (Isaac and Summers 2009).

Describing Concepts in SKOS

SKOS proposes a concept-based model for representing controlled vocabularies. As opposed to a term-based approach, where *terms* from natural language are the first-order elements of a KOS, SKOS uses more abstract *concepts* which can have different lexicalizations. SKOS introduces a special class (skos:Concept) to be used in RDF

---

[5] http://www.w3.org/2004/02/skos/

[6] http://www.w3.org/2006/07/SWD/

data to properly type the concerned entities. To define further the meaning of these conceptual resources, SKOS features three kinds of descriptive properties.

Concept Labels

Labeling properties, for example skos:prefLabel and skos:altLabel, link a concept to the terms that represent it in language. The prefLabel value should be a *non-ambiguous term* that uniquely identifies the concept, and can be used as a descriptor in an indexing system. altLabel is used to introduce alternative terms (synonyms, abbreviations, variants) which can be used to retrieve the concept via a proper search engine, or to extend a document's index by additional keywords. It is important to note that SKOS allows the concepts to be linked to preferred and alternative labels in different languages. KOSs can thus be used seamlessly in multilingual environments, as demonstrated by the following example (using the RDF Turtle syntax introduced in the previous chapter).

```
@prefix ex: <http://example.org/animals/> .
ex:cat a skos:Concept ;
  skos:prefLabel "cat"@en ;
  skos:altLabel "domestic cat"@en ;
  skos:prefLabel "chat"@fr .
```

Semantic Relationships between Concepts

Semantic properties are used to represent the structural relationships between concepts, which are usually at the core of controlled vocabularies such as thesauri. SKOS uses skos:broader, which denotes the generalization link, skos:narrower, its reciprocal link, and skos:related, which denotes a non-hierarchical associative relationship. For instance, using skos:broader to relate two concepts (say, cat and animal) allows us to represent the fact that the first admits the second as a generalization.

```
ex:cat skos:broader ex:animal ;
  skos:related ex:wildcat .
ex:animal skos:narrower ex:cat .
```

Documentation

Even though less crucial to representing the meaning of a concept, informal documentation plays an important role in KOSs. To represent it, SKOS features documentation notes that capture the intended meaning of a concept using glosses – skos:scopeNote, skos:definition, skos:example – and management notes – skos:editorialNote, skos:changeNote and skos:historyNote.

```
 ex:cat skos:scopeNote "used only for domestic cats"@en.
```

Representing Conceptual Vocabularies

Analogously to OWL, SKOS allows vocabularies themselves to be represented and described as full-fledged individuals. It coins a skos:ConceptScheme class for this. It also introduces specific properties to represent the links between KOSs and the concepts they contain. skos:inScheme is used to assert that a concept is part of a given concept scheme, while skos:hasTopConcept states that a KOS contains a given concept as the root of its hierarchical network.

```
ex:plantVocabulary a skos:ConceptScheme ;
   skos:hasTopConcept ex:tree .
ex:tree skos:inScheme ex:plantVocabulary .
```

7.3.3 Characterizing Concepts beyond SKOS

There are more features to SKOS than the ones discussed in the above paragraph. Readers are referred to the SKOS documentation, which presents more advanced functionalities, such as grouping concepts or linking labels together (Miles et al. 2005). They should be aware, however, that the SKOS model does not provide the means to capture all the information that can be virtually attached to KOSs and their elements. Indeed, for a number of features that are not specific to the KOS domain, such as the mention of a concept's creator, SKOS recommends the use of constructs that already exist in other vocabularies – in that specific case, the Dublin Core dct:creator property. The SKOS documentation also encourages users to *extend* the SKOS constructs so as to fit their specific needs. A typical example will be to create a sub-property of skos:related to represent a specific flavor of association, such as between a cause and its consequence. Applications thus benefit from a finer representation grain while retaining compatibility with standard SKOS tools.

```
@prefix dct: <http://purl.org/dc/terms/> .
@prefix myVoc: <http://example.org/mySKOSextension/> .

myVoc:hasBodyPart rdfs:subPropertyOf skos:related .

ex:cat dct:creator <http://www.few.vu.nl/~aisaac/foaf.rdf#me> ;
   myVoc:hasBodyPart ex:whiskers .
```

Linking Concepts Together Using SKOS

SKOS allows concepts to be explicitly related within one concept scheme using semantic relations such as skos:broader. But one of the aims of SKOS is to enable the creation of richer networks. On the Semantic Web, the value of descriptions come from the links they maintain to each other, and especially to descriptions coming from a different context. Using SKOS to represent controlled vocabularies naturally allows their elements to be involved in statements that explicitly relate them to entities outside of their original context.

While a broad variety of relations are of course possible, SKOS focuses on two aspects. The first is the reuse of entities across concept schemes. In SKOS, it is perfectly legal to state that a concept belongs to several schemes at the same time, or to use a semantic relation such as skos:broader to link two concepts from different schemes.

Furthermore, SKOS proposes a range of *mapping properties* to link concepts, typically for applications where semantic interoperability has to be enabled by using semantic alignment technologies (Euzenat and Shvaiko 2007). Using the property `skos:exactMatch` or `skos:closeMatch`, it is possible to assert that two concepts have equivalent meanings. `broadMatch`, `narrowMatch` and `relatedMatch` are used to mirror the semantic relations presented in the above paragraph. These properties can be employed in situations where it is important to distinguish application-specific alignment links from carefully designed semantic relationships that specify the intrinsic meaning of considered concepts.

```
@prefix ex1: <http://example.org/animals/>
@prefix ex2: <http://example.org/moreGeneralClassification/>

ex1:felis_silvestris_catus skos:exactMatch ex2:cat.
```

The ease of creating such explicit relations across contexts with SKOS naturally stems from its relying on Semantic Web architecture: from a knowledge representation perspective, there is no fundamental difference between links within one vocabulary and links across different vocabularies. But even though this may appear obvious to the reader of this book, we recall that this is a crucial departure from the usual KOS formats, where such links can only be represented in an implicit way – typically as mere textual notes.

The Formal Semantics of SKOS

To be fully usable on the Semantic Web, SKOS comes as an OWL ontology.[7] This allows some semantic axioms to be specified which reasoners can use to derive information from asserted SKOS statements. A typical example is the property `skos:broaderTransitive` which is used to represent 'ancestor' links between concepts. To automatically produce the desired information from existing `skos:broader` statements using an OWL inference engine, this property is defined both as transitive and a super-property `skos:broader`. Consider the following example:

```
ex:cat skos:broader ex:animal.
ex:siamese_cat skos:broader ex:cat.
```

From these two statements, an OWL reasoning engine can infer the triple:

```
ex:siamese_cat skos:broaderTransitive ex:cat.
```

It is worth mentioning that it is impossible to represent all SKOS axioms using OWL alone. Some constraints would therefore require implementation by other means.

7.3.4   Using SKOS Concept Schemes on the Semantic Web

As stated, the aim of SKOS is to port existing concept schemes to the Semantic Web and to use them seamlessly in RDF descriptions. Of course this requires SKOS data

---

[7] http://www.w3.org/TR/skos-reference/skos.rdf

to be combined with other kinds of data – that is, information described using other ontologies. One typical example is when SKOS concepts are used to describe the subject of a document (its subject *indexing*). Such a thing can be done by creating an RDF description of the document concerned using a dedicated ontology, and relating this description to the one of the appropriate SKOS concept(s). Examples of ontologies that may be used, depending on the documentary context, are the Dublin Core metadata element set,[8] as in the following example.

```
<http://en.wikipedia.org/wiki/cat> dct:subject ex:cat.
```

7.4 Querying on the Semantic Web

In the previous sections and Chapter 6 we have explained how knowledge can be expressed on the Semantic Web using a layer cake of increasingly expressive languages, and how to publish it using the linked data principles. To make actual use of this knowledge, an additional language for querying knowledge on the Semantic Web is needed. This language is called SPARQL – the SPARQL Protocol And RDF Query Language. All knowledge representation languages described in the previous sections are based on RDF graphs. Hence, SPARQL is a language for querying RDF graphs, based on graph pattern matching. In addition to the actual query language, SPARQL specifies a protocol and query/result formats for interacting with SPARQL endpoint implementations on the Web (Beckett and (eds) 2008; Clark et al. 2008). SPARQL offers the same expressiveness as Datalog and SQL (Angles and Gutierrez 2008). Syntactically, SPARQL is very close to Trig, as we will see in the following.

7.4.1 Syntax

A SPARQL query consists of three main parts:

1. A query result form. Possible result forms are:
   - SELECT queries, which return variable bindings in a table;
   - CONSTRUCT queries, which return RDF by inserting variable bindings into a graph template;
   - ASK queries, which simply return true, if the graph pattern matches, and false otherwise.
2. A dataset, that is, a set of RDF graphs the graph pattern should be matched against.
3. A graph pattern.

The dataset defines the scope of query evaluation, for example a particular ontology or a set of image annotations: a list of named graphs to be used can be defined. Additionally, a default graph is defined, which is used if and only if no scoping to named graphs is done. The smallest graph pattern is a single statement pattern. By default, statement patterns are matched against the default graph. However, they can be scoped to named graphs from

---

[8] http://dublincore.org/documents/dcmi-terms/

the dataset using the keyword GRAPH. The scoping named graph can also be a variable, in which case all named graphs from the dataset are matched and the variable is bound accordingly.

Example 7.4.1 *If the dataset and graph patterns shown below are used, pattern 1 would be matched against the union of* <http://dbpedia.org> *and* gn:dbpediaMapping. *Pattern 2 would be matched against* <http://dbpedia.org> *and* <http://swoogle.com>, *binding* ?g *to the name of the named graph containing the matched statement. Pattern 3 would be matched against* <http://dbpedia.org> *only and pattern 4 would match nothing, because* gn:dbpediaMapping *is not one of the named graphs declared in the dataset using* FROM NAMED. *Note that variable names are written with a leading questionmark in SPARQL.*

```
 1  # prefixes as in Turtle
 2  PREFIX rdf:        <http://www.w3.org/1999/02/22-rdf-syntax-ns#> .
 3  PREFIX rdfs:       <http://www.w3.org/2000/01/rdf-schema#> .
 4  PREFIX owl:        <http://www.w3.org/2002/07/owl#> .
 5  PREFIX foaf:       <http://xmlns.com/foaf/0.1/> .
 6  PREFIX dbpedia:    <http://dbpedia.org/resource> .
 7  PREFIX dbpprop:    <http://dbpedia.org/property/> .
 8  PREFIX gn:         <http://sws.geonames.org/> .
 9
10  # Dataset
11  FROM <http://dbpedia.org>
12  FROM gn:dbpediaMapping
13  FROM NAMED <http://dbpedia.org>
14  FROM NAMED <http://swoogle.com>
15
16  # graph pattern 1:
17  {?s rdf:type dbpedia:Visitor_Attraction}
18  # graph pattern 2:
19  GRAPH ?g {?s rdf:type dbpedia:Visitor_Attraction}
20  # graph pattern 3:
21  GRAPH <http://dbpedia.org> {?s rdf:type dbpedia:Visitor_Attraction}
22  # graph pattern 4:
23  GRAPH gn:dbpediaMapping {?s owl:sameAs ?o}
```

Complex graph patterns are composed of simpler ones using joins, unions, left outer joins and filter expressions, which further constrain variable bindings. We will describe particularities of these connectives later in this chapter.

In the case of a SELECT query, the variable bindings produced by matching the graph pattern are projected to a tabular representation. For example, the expression SELECT ?g ?p ex:constant would return a table containing three columns with the bindings of ?g, ?p and a constant value. In the result of a SELECT query, null bindings for a subset of the variables used are allowed.

In the case of CONSTRUCT queries, the variable bindings are used to construct new RDF statements. The CONSTRUCT pattern in this case again is a graph pattern. However, it is not matched against existing data but instantiated with the variable bindings computed during graph pattern matching. All valid RDF statements created during this instantiation are returned as the query result. In this context, 'valid' means that subjects and objects of all statements are in $\mathbb{U} \cup \mathbb{B} \cup \mathbb{L}$ (as defined in the previous chapter), predicates in $\mathbb{U}$ and all variables are bound. All blank nodes in an instantiated CONSTRUCT pattern are standardized apart. The scope of a blank node is exactly the RDF graph resulting from a single instantiation of the variables in the CONSTRUCT pattern. This is in line with Definition 6.2.12, which scopes blank nodes to a single RDF graph.

In the following formal introduction we will use a syntactical subset of SPARQL, which however semantically can express all valid SPARQL SELECT and CONSTRUCT queries. For the full specification, we refer the reader to Prud'homme and Seaborne (2007).

We start by defining graph patterns. Graph patterns can contain filter expressions. For didactic reasons, we defer the definition of filter expressions.

Definition 7.4.2 (Graph Patterns) *Let $\mathbb{V}$ be the set of SPARQL variables. A* statement pattern *is an element of* $\mathbb{V} \cup \mathbb{U} \cup \mathbb{B} \cup \mathbb{L} \times \mathbb{V} \cup \mathbb{U} \times \mathbb{V} \cup \mathbb{U} \cup \mathbb{B} \cup \mathbb{L}$ *as defined in Definitions 6.2.1, 6.2.3 and 6.2.5. Statement patterns are written as statements in Turtle. A* graph pattern *is defined inductively as follows:*

- *A statement pattern is a graph pattern and a* basic graph pattern (BGP)
- *If $P_1, \ldots, P_n$ are graph patterns, $P = P_1, \ldots, P_n$ is a graph pattern. If $P_1, \ldots, P_n$ are BGPs, P is a BGP.*
- *The empty BGP {} is a graph pattern.*
- *If P is a graph pattern, $\{P\}$ is a graph pattern called a* group graph pattern.
- *If P is a graph pattern and F is a filter expression, P FILTER (F) is a graph pattern called a* filter pattern. *The scope of F is the surrounding group graph pattern.*
- *If P, Q are group graph patterns, P OPTIONAL Q is a graph pattern called an* optional pattern.
- *If P, Q are group graph patterns, P UNION Q is a graph pattern called an* alternative pattern.
- *Let $g \in \mathbb{U} \cup \mathbb{V}$. If P is a group graph pattern,* GRAPH g P *is a graph pattern called a* graph graph pattern.

Let P be a graph pattern. By $vars(P)$ we denote the set of variables used in P.

Definition 7.4.2 is a little more restrictive than Prud'homme and Seaborne (2007), which allows the same abbreviations to be used for BGPs as Turtle allows for RDF graphs. Moreover, we require the use of group graph patterns in some places, where SPARQL also allows BGPs. The additional brackets in our case make scoping more obvious. Both differences are purely syntactical. In fact, where convenient, we will use Turtle abbreviations in examples and assume that they are resolved before any further processing takes place. Moreover, we require that FILTER always occurs *after* a BGP. In contrast, Prud'homme and Seaborne (2007) allow for filter expressions to occur anywhere within a BGP. Hence, in this work we do not need to ensure FILTERs are evaluated in any special order, which makes the algebraic specification of the semantics and later the specification of extensions slightly simpler.

Example 7.4.3 *The following graph pattern selects all things which are of type* foaf:Person *and their* foaf:names. *If available, the* dbpprop:birthplace *or the* dbpprop:placeOfBirth *are selected afterwards.*

```
1 { ?person a foaf:Person. ?person foaf:name ?name }
2 OPTIONAL {
3     {?person dbpprop:birthplace ?city} UNION
4     {?person dbpprop:placeOfBirth ?city}
5 }
```

Graph patterns are matched against a dataset. The result is a set of variable bindings which can be further restricted using filter expressions.

Definition 7.4.4 (Filter Expressions) *Let* f *be a URI or one of the built-in operators listed in Section 11 of Prud'homme and Seaborne (2007).*[9]

- *Let* $v_1, \ldots, v_n \in \mathbb{V} \cup \mathbb{U} \cup \mathbb{B} \cup \mathbb{L}$. *Then* $f(v_1, \ldots, v_n)$ *is a filter expression.*
- *Let* $v_1, \ldots, v_n \in \mathbb{V} \cup \mathbb{U} \cup \mathbb{B} \cup \mathbb{L} \cup \{$filter expressions$\}$. *Then* $f(v_1, \ldots, v_n)$ *is a filter expression.*
- *If* F_1, F_2 *are filter expressions,* $(F_1$ && $F_2)$, $(F_1 \mathbin{||} F_2)$ *and* $(!F_1)$ *are filter expressions.*

As we can see, filter expressions again can be parameters of filter expressions.

Example 7.4.5 *We extend our example such that only those persons are returned who have a name starting with 'A'.*

```
1 {   { ?person a foaf:Person. ?person foaf:name ?name }
2     FILTER (REGEX(?name, "^A")) }
3 OPTIONAL {
4     {?person dbpprop:birthplace ?city} UNION
5     {?person dbpprop:place_of_birth ?city}
6 }
```

Dataset declarations, introduced intuitively at the beginning of this chapter, are defined next. We will then have everything in place to define SPARQL queries.

---

[9] STR casts any literal or URI to a string; LANG extracts only the language tag of a plain literal; LANGMATCHES checks whether a plain literal has a particular language tag; DATATYPE returns the datatype of a typed literal; BOUND checks whether a variable has been bound; sameTerm checks whether two RDF terms are the same; isIRI, isURI, isBLANK and isLITERAL check for the type of a variable; REGEX does regular expression string pattern matching; and >, <, =, <=, >=, != are the usual comparators.

Definition 7.4.6 (Dataset Declaration) *Let n be a URI, which is a name of a named graph. Then*

- *FROM n is a* dataset declaration. *We say that graph n is part of the default graph of D.*
- *FROM NAMED n is a* dataset declaration. *We say that n is a named graph in D.*
- *If D_1 and D_2 are dataset declarations, then $D_1 D_2$ is a dataset declaration.*

If D is a dataset, we denote by D_D the graphs comprising the default graph of D, by D_G the union of the graphs in D_D and by D_N the named graphs in D.

There are four types of SPARQL queries: SELECT, CONSTRUCT, ASK and DESCRIBE queries. SELECT queries return a table of variable bindings, as known from SQL. CONSTRUCT queries return RDF graphs. ASK queries return true, if the graph pattern matches the dataset. DESCRIBE queries are used to obtain a description of a single resource in RDF. We will not discuss DESCRIBE queries in this work. In the following we will omit BASE and PREFIX declarations, as they are purely syntactical.

Definition 7.4.7 (SELECT Query) *Let D be a dataset declaration and P be a group graph pattern. Let $v_1, \ldots, v_n \in \mathbb{V} \cup \mathbb{L} \cup \mathbb{U}$. Then $Q = $ SELECT $v_1, \ldots, v_n$ D WHERE P is a SELECT query. We call P the WHERE pattern of Q and $v_1, \ldots, v_n$ the select variables of Q.*

Example 7.4.8 *The following query extracts a list of visitor attractions in Berlin from DBpedia (everything that is in a subcategory of visitor attraction). The location is restricted using a bounding box on the geographic coordinates of the visitor attraction.*

```
1 SELECT ?attraction
2 FROM <http://dbpedia.org>
3 WHERE {
4      { ?attraction skos:subject ?category.
5        ?category skos:broader dbpedia:Category:Visitor_attractions.
6        ?attraction geo:lat ?lat.
7        ?attraction geo:long ?long }
8    FILTER (?lat > "52.3"^^xsd:float &&
9            ?lat < "52.7"^^xsd:float &&
10           ?long > "13.1"^^xsd:float &&
11           ?long < "13.5"^^xsd:float)
12 }
```

Definition 7.4.9 (CONSTRUCT Query) *Let D be a dataset declaration and P be a group graph pattern. Let C be a BGP. Then $Q = $ CONSTRUCT {C} D WHERE P is a CONSTRUCT query. We call P the WHERE pattern of Q and C the CONSTRUCT pattern of Q. We call a statement pattern in C a CONSTRUCT statement pattern of Q.*

Example 7.4.10 *The following CONSTRUCT query matches multiple properties used in DBpedia for places of birth and returns a new graph which uses only one such property.*

```
1 CONSTRUCT {?person a foaf:Person. ?person placeOfBirth ?city}
2 WHERE {
3     {   ?person a foaf:Person }
4     OPTIONAL {
5             { {?person dbpprop:birthPlace ?city} UNION
6               {?person dbpprop:placeOfBirth ?city}
7             } UNION {?person dbpprop:cityOfBirth ?city}
8     }
9 }
```

Definition 7.4.11 (ASK Query) *Let D be a dataset declaration and P be a group graph pattern. Then Q = ASK D P is an ASK query.*

Example 7.4.12 *The following query checks whether we have a German label for* dbpedia:Berlin.

```
1 ASK
2 FROM <http://dbpedia.org>
3 {   {dbpedia:Berlin rdfs:label ?label}
4     FILTER (LANGMATCHES(LANG(?label),"DE"))
5 }
```

7.4.2 Semantics

It remains to define the semantics of a valid SPARQL query. Prud'homme and Seaborne (2007) define an algebraic semantics which we summarize here. SPARQL employs a multi-set-based semantics, that is to say, a query result may contain duplicate solutions, in order to ease implementation. For ease of specification we use a set-based semantics. We also slightly restrict the syntax, such that the FILTER operator can be formalized more simply compared to Prud'homme and Seaborne (2007), who allow FILTER expressions to occur anywhere inside a BGP, which requires a more complex definition of the semantics of FILTER.

BGP matching forms the basis for the SPARQL semantics.

Definition 7.4.13 (BGP matching) *Let P be a BGP. Let $\mathbb{V}_P$ be the set of variables used in P and $\mathbb{B}_P$ be the set of blank nodes used in P. Let A be an RDF instance mapping as defined in Definition 6.2.12 and T be a* term mapping *of the form $T : \mathbb{V} \rightarrow \mathbb{U} \cup \mathbb{L} \cup \mathbb{B}$. We say that P matches an RDF graph G if there exist A, T such that $P/(A \cup T)$ is a subgraph of G. We call $A \cup T$ a pattern instance mapping. A solution μ of P with respect to G is the restriction of $A \cup T$ to the variables in P. The only solution of the empty BGP is the empty solution μ_0.*

Obviously, μ is a mapping of the form $\mathbb{V} \to \mathbb{U} \cup \mathbb{L} \cup \mathbb{B}$. Note that SPARQL uses the subgraph relationship instead of entailment for matching. This avoids infinite redundant mappings from blank nodes to other blank nodes. Instead, there is one solution per pattern instance mapping, and the number of pattern instance mappings is bound by the number of terms used in P and G.

SPARQL graph pattern matching takes place on the extension of an RDF graph, that is, the closure of all statements, which can be inferred from the knowledge base. Of course, an implementation does not need to materialize this extension. The exact semantics for which this extension is computed is defined in an *entailment regime*. The standard entailment regime is basic RDF graph pattern matching. In this case, the above required match would not be retrieved. However, if the RDF repository supports lightweight OWL reasoning (in particular, transitivity of properties), the graph pattern would be matched against the RDFS closure of the repository. In this case, the statement `ex:Felix rdf:type ex:Cat` could be inferred and hence the pattern would match. Analogously, entailment regimes can be defined for RDFS and more expressive fragments of OWL. One of the most expressive entailment regimes proposed is SPARQL-DL (Sirin and Parsia 2007), which supports querying OWL ontologies and instances. In contrast, many OWL reasoners only support conjunctive queries. An OWL fragment supported by many repositories, which allow for lightweight OWL reasoning, is OWL Horst (ter Horst 2005) – the fragment of OWL expressible using forward chaining rules.

Operators such as OPTIONAL and UNION combine solutions.

Definition 7.4.14 (Compatibility and Merge of Solutions) *Let μ_1, μ_2 be solutions. We say that μ_1 and μ_2 are compatible, if $(x, y) \in \mu_1 \wedge (x, z) \in \mu_2 \Rightarrow y = z$, that is, μ_1 and μ_2 map shared variables to the same terms. If μ_1 and μ_2 are compatible, then* merge$(\mu_1, \mu_2) = \mu_1 \cup \mu_2$.

Obviously, μ_0 is trivially compatible with every solution and merge$(\mu, \mu_0) = \mu$.

Filter expressions in SPARQL are functions returning a literal or the constant `error`. The definitions of the logical connectives `&&`, `||` and `!` are defined on `error` and the effective boolean values of their parameters. This means that, if some filter function produces an error, the entire solution is *silently* discarded. Here we give a only very brief definition of filter functions and refer the reader to Prud'homme and Seaborne (2007) for the complete definition.

A *filter function* is a mapping of the form $f : n^{\mathbb{V} \cup \mathbb{U} \cup \mathbb{B} \cup \{error\}} \to \mathbb{U} \cup \mathbb{B} \cup \{error\}$. If in a filter expression a parameter again is a filter expression, it is evaluated first and replaced by its result, before evaluation continues. Parameters are cast into suitable types, if a suitable casting function is defined by Berglund et al. (2007). If a filter expression is not defined (e.g. because its parameters are outside its domain, or variables are not bound), it evaluates to `error`. Every filter function returns `error`, if one of its parameters is `error`. The special filter function `BOUND` returns `true` if its parameter is a variable which is not bound, `false` if its parameter is a variable which is bound, and error otherwise. It can be used to model negation by failure. The logical connectives of filter expressions work on a three-valued logic, including the `error`. The definitions of `&&(a,b)`, `||(a,b)` and `!a` are as follows (ebv is a function returning the effective boolean value of a variable):

ebv(a)	ebv(b)	$\|$(a,b)	&&(a,b)
true	true	true	true
true	false	true	false
true	error	true	error
false	true	true	false
false	false	false	false
false	error	error	error
error	true	true	error
error	false	error	error
error	error	error	error

ebv(a)	!a
false	true
true	false
error	error

Next the semantics of graph pattern matching is defined, before we finish with the three types of queries. Our semantics is based on sets, instead of multisets as in Prud'homme and Seaborne (2007). The latter authors use multisets to ease implementation, even though this introduces duplicates in query results. Additional operators (DISTINCT and REDUCED) are then introduced to remove duplicates. For this work, set semantics makes no significant difference. In order to keep the specification as short as possible, we therefore use sets and hence do not need to track cardinalities. We leave out all additional solution modifiers, such as ORDER BY, LIMIT and OFFSET, which do not increase expressiveness, but can be added easily.

Definition 7.4.15 (Semantics of SPARQL Operators) *Let B be a BGP and P, Q be graph patterns. Let G be an RDF graph called the* active graph. *Let μ_0 be the empty solution and Ω_0 be the set containing exactly μ_0. eval is the SPARQL evaluation function, which defines the semantics of SPARQL graph patterns with respect to a dataset D. eval is recursively defined as follows:*

expression	semantics
eval(B, G, D)	$\{\mu \mid \mu$ *is a solution for B with respect to G*$\}$
eval({P}, D)	*eval(P, G)*
eval(P. Q)	$\{\mu \mid \mu = \text{merge}(\mu_1, \mu_2) : \mu_1 \in eval(P, G, D),$ $\mu_2 \in eval(Q, G, D)$ *and* μ_1, μ_2 *are compatible*
eval(P FILTER(F), G, D)	$\{\mu \mid \mu \in eval(P, G, D)$ *and* $ebv(F(\mu)) = true)$
eval(P OPTIONAL Q, G, D)	$\{\mu \mid \begin{cases} \mu \in eval(P, G, D), if \nexists \mu_2 \in eval(Q, G, D), \\ \quad such\ that\ \mu, \mu_2\ are\ compatible \\ \mu = \text{merge}(\mu_1, \mu_2) : \mu_1 \in eval(P, G, D), \\ \quad \mu_2 \in eval(Q, G, D)\ and \\ \quad \mu_1, \mu_2\ are\ compatible, \\ otherwise. \end{cases}$

expression	semantics
eval(P UNION Q, G, D)	eval(P, G, D) ∪ eval(Q, G, D)
eval(GRAPH g (P), G, D)	$\begin{cases} eval(P, g, D) \text{ if } g \in \mathbb{U} \wedge g \in D_N; \\ eval(\{GRAPH\, n_1\, P\}UNION \\ \quad \cdots \\ UNION\{GRAPH\, n_n\, P\}, G, D)| \\ \{n_1, \ldots, n_n\} = D_N; \\ \Omega_0 \; otherwise. \end{cases}$

Both SELECT and CONSTRUCT queries project out some variables used in the graph pattern to the result.

Definition 7.4.16 (Projection) *Let V be a set of variables and μ be a solution. Then the projection π of μ to V is*

$$\pi(V, \mu) = \left\{ (x, y)| \begin{cases} x \in V \wedge \nexists(x, z) \in \mu \wedge y = null \\ x \in V \wedge (x, y) \in \mu \end{cases} \right\}.$$

Definition 7.4.17 (Semantics of a SELECT Query) *Let $Q = SELECT\; v_1, \ldots, v_n\; D$ WHERE P be a SELECT query. A result of Q is $\pi(\{v_1, \ldots, v_n\}, eval(P, D_G, D))$.*

Example 7.4.18 *Consider the query Q from Example 7.4.8. The dataset is D with $D_N = \{\}$, $D_D = \{< http://dbpedia.org >\}$. Let BGP =*

```
{ ?attraction skos:subject ?category.
  ?category skos:broader
                   dbpedia:Category:Visitor_attractions.
  ?attraction geo:lat ?lat.
  ?attraction geo:long ?long }
```

Then $\mu = \{$(?attraction, dbpedia:Brandenburg_Gate), (?lat, "52.516272"^^xsd:double), (?long, "13.377722"^^xsd:double)$\}$ *is a solution for eval(BGP, D_G, D). Let F =*

```
?lat  > "52.3"hathatxsd:float &&
?lat  < "52.7"hathatxsd:float &&
?long > "13.1"hathatxsd:float &&
?long < "13.5"hathatxsd:float
```

μ is also a solution for eval(\{BGP\} FILTER F, D_G, D), because F/μ evaluates to true. *Hence, $v = \{$*(?attraction, dbpedia:Brandenburg_Gate)*$\}$ is a solution for Q, because π(?attraction, eval(\{\{BGP\} FILTER F\}, D_G, D)) = v.*

Definition 7.4.19 (Semantics of a CONSTRUCT Query) *Let* $Q = CONSTRUCT \{C\} D$
WHERE P be a CONSTRUCT query. The result of Q is an RDF graph $G = \{(s, p, o) | \exists \mu \in$
$eval(P, D_G, D) : (s, p, o) \in C/\mu$ *and* (s, p, o) *is a valid RDF statement*$\}$.

Intuitively speaking, we evaluate P and instantiate C with every solution. From the
result we only take the valid RDF statements, that is, only partially instantiated statements
are discarded, as are those with blank nodes or literals in predicate positions.

Example 7.4.20 *DBpedia contains the following statements:*

---

1 dbpedia:Richard_Feynman **a** foaf:Person ;
2   dbpprop:placeOfBirth dbpedia:Queens,_New_York .
3 dbpedia:Theodor_Mommsen **a** foaf:Person ;
4   dbpprop:birthplace dbpedia:Schleswig,
5   dbpedia:Garding .

Hence, the following are solutions for the graph pattern of the query in Example 7.4.10:

{ { (?person, dbpedia:Richard_Feynman),
    (?city, dbpedia:Queens,_New_York)},
 { (?person, dbpedia:Theodor_Mommsen),
    (?city, dbpedia:Schleswig)},
 { (?person, dbpedia:Theodor_Mommsen),
    (?city, dbpedia:Garding)}
}

The result of the query therefore contains the following statements:

---

1 *# Include this information without modifications*
2 dbpedia:Richard_Feynman a foaf:Person ;
3   dbpprop:placeOfBirth dbpedia:Queens,_New_York .
4 *# The predicate has changed here*
5 dbpedia:Theodor_Mommsen a foaf:Person ;
6   dbpprop:placeOfBirth dbpedia:Schleswig,
7   dbpedia:Garding .

Definition 7.4.21 (Semantics of an ASK Query) *Let* $Q = ASK\ DP$ *be an ASK query.*
The solution of Q is true *if* $\exists \mu : \mu \in eval(P, D_G, S)$. *It is* false *otherwise.*

Example 7.4.22 *The query from Example 7.4.12 evaluates to* true. *As the reader can*
verify by visiting http://dbpedia.org/sparql?query = SELECT+*+FROM+\%3C
http://dbpedia.org\%3E+WHERE+\%7B\%7B\%3Chttp://dbpedia.org/
resource/Berlin\%3E+rdfs:label+?label\%7D+FILTER+(LANGMATCHES
(LANG(?label),\%22DE\%22))+\%7D, *there is at least one solution for the graph*
pattern of the query [as of 2010-07-21].

7.4.3   Default Negation in SPARQL

As SPARQL queries are evaluated on the extension of the knowledge base, it locally closes the world for the specified dataset. The underlying knowledge representation languages RDF and OWL, in contrast, assume an open world. Based on this locally closed world, SPARQL can be used to model negation by failure using a combination of OPTIONAL graph patterns and the BOUND filter, which checks whether a variable has been bound during graph pattern matching. We will call such a construct in a query *bound negation*. The semantics of such a construct intuitively is: 'Try to find all solutions which I want to exclude from the result. Introduce a new (and possibly redundant) variable, which is bound only if such a solution is found. Then throw away all solutions, where this variable is bound.'

In general, bound negation looks as follows. Let ?x be a variable occurring in GP2, ?y a new variable which does not occur anywhere else in the query, and s p ?x a statement pattern in GP2.

---

1 {GP1} **OPTIONAL** {{GP2. s p ?y} **FILTER** (?x = ?y)} **FILTER** (!**BOUND**(?y))

Obviously, using negation in SPARQL puts a heavy burden on the user. As a consequence, the next version of SPARQL (Harris and (eds) 2010), which at the time of writing is undergoing standardization, will provide an explicit operator for negation.

Example 7.4.23 *The following query selects all geographic entities from DBpedia, which have a mapping to an entry from GeoNames, but of which DBpedia does not contain geocoordinates. These could then be imported from GeoNames. The query selects all places from DBpedia, which have some mapping (lines 7,8). For each mapping, we check, whether it is a mapping to a place in GeoNames (line 11). Then we try to find geocoordinates for this place in DBpedia (lines 13–18). If we find them, the place is discarded from the result (line 19), leaving only those without geocoordinates.*

```
 1 PREFIX gno <http://www.geonames.org/ontology#>
 2 SELECT ?geoent
 3 FROM NAMED <http://dbpedia.org>
 4 FROM NAMED <http://sws.geonames.org>
 5 WHERE {
 6     {  { GRAPH <http://dbpedia.org> {
 7             ?geoent a <http://dbpedia.org/ontology/Place>;
 8                     owl:sameAs ?gnentry
 9         } .
10         GRAPH <http://sws-geonames.org> {
11                 ?gnentry gno:featureClass gno:P
12         }
13     } OPTIONAL {
14         GRAPH <http://dbpedia.org> {
15             ?geoent geo:lat ?lat ;
16                     geo:lon ?lon
17         }
```

```
18          }
19      }  FILTER (!BOUND(?lat))
20 }
```

7.4.4 Well-Formed Queries

OPTIONAL patterns can lead to strange behavior, if queries are not well formed. Queries
are not well formed if OPTIONALs are nested, such that the inner OPTIONAL uses some
variable from the surrounding graph pattern, which is not used in the outer OPTIONAL
(Pérez et al. 2006a).

Definition 7.4.24 (Well-Formed Query) *We say that a query is* well formed *if, for all
nested* OPTIONAL *patterns* GP1 { GP2 OPTIONAL GP3 }, $vars(P3) \cap vars(P1) \subseteq vars(P2) \cap vars(P1)$.

To illustrate the problem, consider the following simple query, which is not well formed:

```
1 SELECT ?x
2 WHERE {
3      {?x ex:b ?y} . {{?x ex:c ?z} OPTIONAL {?z ex:b ?y}}
4 }
```

When evaluated against the graph

```
1 ex:a ex:b ex:c .
2 ex:a ex:c ex:d .
```

there is exactly one solution {(?x, ex:z)}. Now if we add a single statement

```
1 ex:d ex:b ex:a .
```

the solution is empty.

7.4.5 Querying for Multimedia Metadata

Queries on semantically enriched media assets vary from navigating the decomposition
of a video into shots and keyframes to retrieving all documents annotated with a given
concept. Sophisticated queries may even take background knowledge into account or ask
for complete scene descriptions represented in RDF. Hence, RDF and SPARQL allow for
much richer but still semantically well-defined annotations than most of the state of the
art (cf. Section 7.4.7). In the scenario presented in Section 2.1, we might be interested,
for example, in all images showing the heads of the United States and the Soviet Union
together. To answer this query, we need to take decompositions of images, semantic
annotations, and domain-specific knowledge into account in order to determine whether
the persons depicted are heads of the USA or USSR.

In order to process and answer such queries, we are faced with various challenges with
respect to the potential size of the dataset, complexity of queries, recursiveness of queries,
and interactive access to media asset annotations. These challenges are elaborated below.

As SPARQL alone is not sufficient to address all of them, we discuss proposed extensions where appropriate.

Large Datasets

The queried datasets may become extremely large. We estimate annotations in COMM of 1 million triples for one hour of video, which is decomposed into keyframes and annotated region based. If basic inferencing is done to compute subclass and instance relations, this may easily result in an increase by a large constant factor. On the other hand, most state-of-the-art RDF repositories scale to tens or hundreds of millions of statements.[10] Since about 2008, when the billion triples mark was exceeded (Harth et al. 2007; Kiryakov et al. 2005b; Schenk et al. 2009), performance of RDF stores has increased dramatically. Today, multi-billion statement repositories are possible (see also Section 6.6.1). However, such repositories usually require powerful hardware or even clusters of repositories. Compared to this scale, typical datasets of background world knowledge, like DBpedia,[11] can almost be considered small.

Complex Queries

Queries can become extremely complex. A typical instantiation of a pattern in the COMM, which will be described in Chapter 9, results in up to 20 statements. This complexity is not COMM-specific, but typical of multimedia annotation, in order to capture the necessary expressivity (Troncy etal. 2007a). In turn, this results in a query with 20 statement patterns and 19 joins. Given the size of the datasets, this is a challenge that also most existing relational databases fail to meet. In order to avoid errors, it is desirable to hide these complex queries from application developers. In the case of COMM, COMM-API provides an abstraction layer for developers, which allows COMM items to be accessed as Java objects without writing SPARQL queries.

As an example of a COMM-based SPARQL query, suppose that we wish to retrieve all images showing a cat which is at least 3 years old. Additionally we would like to know what type of cat we are faced with. We would like to select results 5 to 10. Then we could use the following query:

```
1 SELECT DISTINCT ?URI
2 FROM <http://example.org/annotationGraph>
3 WHERE {
4 ?image              rdf:type          core:image-data;
5                      core:realized-by  ?URI;
6                      core:plays        ?annotated-data-role.
7 ?annotated-data-role rdf:type          core:annotated-data-role.
8 ?annotation         rdf:type          core:semantic-annotation;
9                      core:setting-for  ?image;
10                     core:setting-for  ?label;
11                     core:satifies     [
12                           rdf:type core:method;
```

---

[10] See http://esw.w3.org/topic/RdfStoreBenchmarking for an overview of RDF benchmarks.
[11] http://wiki.dbpedia.org/Datasets

```
13                              core:defines ?annotated-data-role;
14                              core:defines ?semantic-label-role].
15 ?cat              core:plays      ?semantic-label-role.
16 ?semantic-label-role rdf:type    core:semantic-label-role.
17 ?cat              rdf:type        ex:cat.
18 ?cat              ex:age          ?age.
19 FILTER (?age > "3"^^xsd:integer).
20 } LIMIT 5 OFFSET 10
```

Complex Recursive Queries

Annotations to multimedia items can be done on a variety of levels of decomposition. For example, a whole image can be annotated with a concept but also only a segment showing the concept or a compound Open Document Format (ODF) document containing the image. Hence, retrieval queries need to recursively follow the decompositions. Extensions to SPARQL that add support for regular path expressions have been proposed (Alkhateeb et al. 2009; Kochut and Janik 2007; Pérez et al. 2010; Polleres et al. 2007), featuring regular graph expressions. However, such regular expressions are not expressive enough to capture all patterns used in COMM to annotate media assets. For this reason, a metadata repository must additionally support a specialized set of rules allowing decompositions (or other patterns) to be (recursively) followed during retrieval. Such rules can be expressed, for example, in the Semantic Web Rule Language (Horrocks et al. 2004), N3 (Berners-Lee 2001) or – using SPARQL syntax – networked graphs (Schenk 2008). For less complex (in particular non-cyclic) rules, also DL-save rules (Motik et al. 2005) or regular path expressions (Alkhateeb et al. 2009; Kochut and Janik 2007; Pérez et al. 2010) for SPARQL could be used.

7.4.6 Partitioning Datasets

In contrast to many sources of world knowledge, multimedia metadata can easily be split horizontally. This means that annotations of two media assets are to a very large degree independent of each other. The links between them are usually indirect, specified through world knowledge. For example, two images could show the same scenery from different angles. However, the scenery is not part of the actual multimedia annotation but world knowledge. As a result, one possible approach to scaling querying of multimedia metadata is to distinguish between multimedia annotation and world knowledge and to split the datasets and queries accordingly. This allows us to come up with easier problems due to shorter queries and a smaller dataset. On the other hand, new challenges arise when splitting queries and datasets, such as determining relevant fragments for answering (a part of) a query or joining query results like efficiently handling distributed joins. Even though many of these challenges are well known from distributed and federated relational databases, they are more problematic for RDF as schema information is not reflected in the structure of data and an extremely high number of joins have to be handled compared to relational databases. For illustration, remember that in relational databases the table structure implicitly reflects the schema of the data. In contrast, in RDF we have triples as the only structure and schema information is expressed explicitly using special predicates.

Systems supporting federated SPARQL evaluation include those of Quilitz and Leser (2008) and Schenk et al. (2009). Prud'hommeaux (2007) specifies a SPARQL extension to natively support federation in the language. Federation is a proposed optional feature of SPARQL1.1 (Prud'hommeaux and Seaborne 2010). This extension only supports semi-joins and the marking up of subqueries, which should be evaluated remotely. The problem of query optimization for distributed queries is not addressed. Several state of the art systems support clustering, for example 4 store (Harris et al. 2009) and Virtuoso.[12]

7.4.7 Related Work

Related work for querying semantic multimedia includes languages which are similarly generic to SPARQL, but also specialized multimedia query languages.

Other RDF Query Languages

Several RDF query languages have been designed, which implement parts of the SPARQL extensions proposed above, including RQL (Karvounarakis et al. 2002), SeRQL (Broekstra and Kampman 2003), TRIPLE (Sintek and Decker 2002), RDQL (Seaborne 2004) and N3 (Berners-Lee 2001). A comparison of supported query operators and capabilities in these languages is given by Haase et al. (2004), analyzing support for graph operators such as simple and optional path expressions, relational operators such as union and set difference, quantification, aggregation operators such as count, sum and grouping, recursive operators in addition to schema-level recursion in transitive predicates, abstract operators to query reification and collections, operators to handle languages and datatypes in literals, and support for RDF(S) entailment in query answering.

The semantics and computational behavior of SPARQL has been studied (de Bruijn et al. 2005; Muñoz et al. 2007), revealing for example some computationally and semantically problematic corner cases (Pérez et al. 2006b). Also, approaches have been developed to translate SPARQL queries into relational algebra (Cyganiak 2005; Harris and Shadbolt 2005), enabling SPARQL query answering using relational databases, and into Datalog (Polleres 2007; Schenk 2007), allowing engines to combine SPARQL queries with ontological rules.

At the time of writing, the SPARQL Working Group[13] is specifying SPARQL 1.1, which will bring the following additions to SPARQL (Harris and (eds) 2010). Select expressions will allow results to be returned which are computed from bindings of variables. For example, a price could be selected in euros together with an exchange rate for the dollar and the dollar price could be returned. Subselects allow for the specification of subqueries, which can use the full range of result modifiers, including for example select expressions and LIMIT. Property paths allow regular expressions to be specified for predicates. For example, one could write a query which follows a transitive property. With the current version of SPARQL, only paths of a fixed length can be matched. Optional features include entailment regimes for the popular languages RDFS, all OWL profiles and the Rule Interchange Format (RIF). Moreover, an update language and basic support for query federation are being added.

---

[12] http://virtuoso.openlinksw.com/

[13] http://www.w3.org/2009/sparql/wiki/Main_Page

Multimedia Query Languages

Besides the approach described above for querying media assets by the use of a semantic database, there are also other approaches and solutions to query semantic multimedia. While these languages are tailored to particular needs of multimedia querying, they lack the ability to query Semantic Web background knowledge.

For example, the commercial database Oracle, with its Oracle Multimedia[14] feature, provides for retrieving images, audio, and video. The multimedia package is an extension of the relational Oracle database. It supports the extraction of metadata from media assets and allows querying for media assets by specific indices.

The Digital Memory Engineering group at the Research Studios in Austria developed with the multimedia database METIS a sophisticated storage and management solution for structured multimedia content and its semantics (King et al. 2004; Ross et al. 2004). The METIS database provides a flexible concept for the definition and management of arbitrary media elements and their semantics. It is adaptable and extensible to the requirements of a concrete application domain by integrating application-specific plugins and defining domain-specific (complex) media types. In METIS, the semantic relationship of specific media elements and their semantics can be described to form new, independent multimedia data types. Those domain-specific media types can be bundled up and distributed in form of so-called *semantic packs*.

The multimedia presentation algebra (MPA) by Adalí et al. (2000) extends the relational model of data and allows for the dynamic creation of new presentations from (parts of) existing presentations. With the MPA, a page-oriented view on multimedia content is given. A multimedia presentation is considered as an interactive presentation that consists of a tree, which is stored in a database. Each node of this tree represents a non-interactive presentation, for example a sequence of slides, a video element, or an HTML page. The branches of the tree reflect different possible playback variants of a set of presentations. A transition from a parent node to a child node in this tree corresponds to an interaction. The proposed MPA allows for specifying a query on the database based on the contents of individual nodes as well as querying based on the presentation's tree structure. It provides extensions and generalizations of the `select` and `project` operations in the relational algebra. However, it also allows the authoring of new presentations based on the nodes and tree structure stored in the database. The MPA defines operations such as `merge`, `join`, `path-union`, `path-intersection`, and `path-difference`. These extend the algebraic join operation to tree structures and allow the authoring of new presentations by combining existing presentations and parts of presentations.

The main advantage of multimedia-specific approaches based on algebras is that the requested multimedia content is specified as a query in a formal language. However, a great deal of effort is typically necessary to learn the algebra and operators and it is very difficult to apply such a formal approach. Consequently, the algebras presented remain purely academic so far.

---

[14] http://www.oracle.com/technology/products/intermedia/index.html

8

Multimedia Metadata Standards

Peter Schallauer,[1] Werner Bailer,[1] Raphaël Troncy[2] and Florian Kaiser[3]

[1]*JOANNEUM RESEARCH – DIGITAL, Graz, Austria*
[2]*Centrum Wiskunde & Informatica, Amsterdam, The Netherlands*
[3]*Technische Universität Berlin, Institut für Telekommunkationssysteme, Fachgebiet Nachrichtenübertragung, Berlin, Germany*

The term 'multimedia object' encompasses a wide variety of media items (both analog and digital), with different modalities and in a broad range of applications. Also each multimedia object passes through a number of processes throughout its life cycle that use, modify or amend it (see Chapter 3). Multimedia metadata needs to capture this diversity in order to describe the multimedia object itself and its context. A large number of multimedia metadata standards exist, coming from different application areas, focusing on different processes, supporting different types of metadata (cf. Section 8.2) and providing description at different levels of granularity and abstraction.

This chapter discusses a number of commonly used multimedia metadata standards. We also present a list of criteria that can be used to assess metadata standards. The standards are then compared with respect to these criteria. The standards and criteria discussed here are those that reflect the use cases discussed in Chapter 2, and that were found to be relevant across a number of overviews on multimedia metadata standards coming from different application domains. Note that this chapter does not cover multimedia ontologies – these are covered in Chapter 9.

van Ossenbruggen et al. (2004) discuss requirements for semantic content description and review the capabilities of standards coming from the W3C and MPEG communities. In the second part of their article (Nack et al. 2005) they focus on the formal semantic definition of these standards which determines the expressiveness of semantic content description and enables mapping between descriptions. The report in Eleftherohorinou et al. (2006) surveys multimedia ontologies and related standards and defines requirements for a multimedia ontology, many of which are also relevant for multimedia metadata standards. A comprehensive overview of multimedia metadata standards and formats has

Multimedia Semantics: Metadata, Analysis and Interaction, First Edition.
Edited by Raphaël Troncy, Benoit Huet and Simon Schenk.
© 2011 John Wiley & Sons, Ltd. Published 2011 by John Wiley & Sons, Ltd.

been prepared by the W3C Multimedia Semantics XG (Hausenblas et al. 2007), which takes into consideration the photo and music use cases discussed in Chapter 2, among others. The third use case of Chapter 2 is covered by a chapter in Granitzer et al. (2008), and this book addresses also other application areas and multimedia retrieval. A recent review in Pereira et al. (2008) focuses on the standards from MPEG but also discusses interoperability issues between standards in the context of general multimedia metadata application scenarios.

8.1 Selected Standards

8.1.1 MPEG-7

MPEG-7, formally named *Multimedia Content Description Interface* (MPEG-7 2001), is an ISO/IEC standard developed by the Moving Picture Experts Group. The standard targets the description of multimedia content in a wide range of applications. MPEG-7 defines a set of description tools, called description schemes and descriptors. Descriptors represent single properties of the content, while description schemes are containers for descriptors and other description schemes. It is important to highlight the fact that MPEG-7 standardizes the description format. Neither the description extraction nor its consumption falls within the scope of the standard.

Figure 8.1 shows an overview of the parts of the standard. The definition of description schemes and descriptors is based on the *Description Definition Language* (DDL, part 2), which extends XML Schema by types such as vectors and matrices. Though attractive regarding syntactic interoperability issues, defining the DDL using XML Schema does not allow semantics to be defined in a formal way as Semantic Web Languages can do (see Chapters 6 and 7). Persistent representations for MPEG-7 descriptions are either XML (textual format, TeM) or a binary format (binary format, BiM) as specified by the standard. The actual content description tools are specified in *Multimedia Description Schemes* (MDS, part 5), *Visual* (part 3) and *Audio* (part 4).

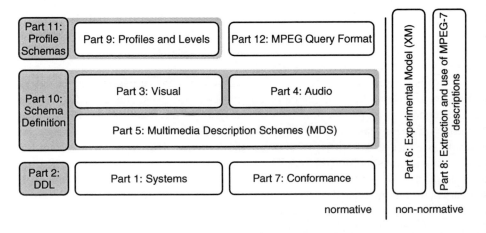

Figure 8.1   Parts of the MPEG-7 standard.

MDS contains tools for the description of media information, creation and production information, content structure, usage of content, semantics, navigation and access, content organization and user interaction. The structuring tools in particular are very flexible and allow the description of content at different levels of granularity. Part 3 defines visual low- and mid-level descriptors.

Part 4 defines a set of audio low-level descriptors (Kim et al. 2005), that describe temporal and spectral components of sounds (cf. Chapter 4). One can easily use this low-level signal descriptions to measure similarity between two audio files. Fusing low-level descriptors with the mean from statistical analysis or machine learning methods (cf. Chapter 5), one can extract structural and semantic information about the audio. MPEG-7 Audio already provides the following description schemes: timbre description, melody contour description, spoken content description, sound classification and similarity, audio signature. With regard to the music use case (cf. Chapter 2), such descriptions allow for browsing audio collections in different ways: search and filtering through perceptual features like timbre, query-by-humming, keyword spotting, etc. See Kim et al. (2005) and Quackenbush and Lindsay (2001) for in depth description of MPEG-7 Audio capabilities and application scenarios.

The *Systems* part of MPEG-7 deals with streaming of descriptions and description fragments, while the *Conformance* part specifies the rules for conformance checking of documents. Further non-normative parts describe the use of the MPEG-7 description tools and provide examples and reference implementations.

As the complexity of MPEG-7 is considerable and the standard has also been defined for a wide range of applications, the potential complexity and interoperability problems were quickly recognized. Profiles have thus been proposed as a possible solution. The specification of a profile consists of three parts MPEG Requirements Group (2001), namely: (a) description tool selection, that is, the definition of the subset of description tools to be included in the profile; (b) description tool constraints, that is, definition of constraints on the description tools such as restrictions on the cardinality of elements or on the use of attributes; and (c) semantic constraints that further describe the use of the description tools in the context of the profile.

The first two parts of a profile specification are used to address the complexity problem, that is, the complexity of a description that can be measured by its size or the number of descriptors used. Limiting the number of descriptors and description schemes (either by excluding elements or constraining their cardinality) reduces this complexity. The third part of a profile specification tackles the interoperability problem, with semantic constraints being expressed in natural language.

Three MPEG-7 profiles have been defined in part 9 of the standard MPEG-7 (2005). The *Simple Metadata Profile* (SMP) allows the description of single instances or simple collections of multimedia content. The motivation for this profile is to support simple metadata tagging similar to ID3 (cf. Section 8.1.11) for music and EXIF (cf. Section 8.1.9) for images, and to support mobile applications such as the 3rd Generation Partnership Project.[1] A partial mapping from these formats to SMP has been specified.

The *User Description Profile* contains tools for describing the personal preferences and usage patterns of users of multimedia content in order to enable automatic discovery,

---

[1] http://www.3gpp.org

selection, personalization and recommendation of multimedia content. The tools in this profile were adopted by the TV-Anytime standard (cf. Section 8.1.5).

The *Core Description Profile* allows image, audio, video and audiovisual content as well as collections of multimedia content to be described. Tools for the description of relationships between content, media information, creation information, usage information and semantic information are included.

None of these profiles includes the visual and audio description tools defined in parts 3 and 4 of MPEG-7. Nor do they make use of the third step in a profile definition, namely of specifying semantic constraints on the included description tools. The *Detailed Audiovisual Profile* (Bailer and Schallauer 2006) has been proposed to cover many of the requirements of the audiovisual media production process and also contains a set of semantic constraints (i.e. the third step in a profile definition) on the included metadata elements, mainly concerning structuring of the description. In addition, a number of applications and research projects have defined their subset and way of using MPEG-7, although they have not yet been formalized as profiles. Recently, the EC-M SCAIE working group[2] of the European Broadcast Union (EBU) has started to harmonise DAVP and efforts of the Japanese Broadcasting Corporation (NHK) and to develop a profile called Audiovisual Description Profile (AVDP), which is expected to become an ISO MPEG standard in 2011.

Parts 10 and 11 provide the XML Schema of the description tools described in parts 3–5 and 9 respectively. The *MPEG Query Format* is the latest part of the standard and does not deal with content description, but defines a query interface for multimedia metadata supporting various types of queries.

8.1.2 EBU P_Meta

The European Broadcasting Union (EBU) has defined P_Meta (PME 2007) as a metadata vocabulary for program exchange in the professional broadcast industry. It is not intended as an internal representation of a broadcaster's system but as an exchange format for program-related information in a business-to-business (B2B) use cases (e.g. between content creators, distributors, broadcasters, archives). The core entities of the P_Meta data model are *programs* (which can be organized in *program groups*) consisting of *Items* (constituent parts of a program), *item groups* and *media objects* (temporally continuous media components). The standard covers identification, technical metadata, program description and classification, creation and production information, rights and contract information and publication information.

In version 2.0 P_Meta has been significantly revised and defined based on XML Schema. The metadata sets of earlier versions have thus been replaced by a set of metadata tools defined by XML Schema constructs and application specifications for certain application areas. Classification schemes are used for controlled values for attributes whenever possible, including classification schemes defined by the EBU, but also from MPEG-7 or TV-Anytime. In addition there exists a descriptive metadata scheme (DMS) definition of P_Meta that allows P_Meta descriptions to be embedded in MXF files.

8.1.3   SMPTE Metadata Standards

The Society of Motion Picture and Television Engineers (SMPTE) has issued standards for technical and descriptive metadata of audiovisual content, including a container format, metadata scheme and structured list of metadata elements. Like P_Meta, the SMPTE standards are rather geared toward business-to-business use cases.

The Material Exchange Format (MXF 2004) specifies a file format for wrapping and transporting essence and metadata in a single container targeted at the interchange of captured, ingested, finished or 'almost finished' audiovisual content. Support for technical metadata is directly integrated in MXF; descriptive metadata can be added using a flexible plugin mechanism called descriptive metadata schemes. The 'native' metadata scheme is the Descriptive Metadata Scheme 1 (DMS-1). Like all SMPTE file formats, MXF files are encoded using KLV,[3] including the metadata elements. The MXF standard has a concept called operational patterns, which is a kind of profiling for the essence container and multiplexing functionality of the standard, but this concept does not exist for the descriptive metadata scheme or the metadata dictionary.

DMS-1 (2004), formerly known as the Geneva Scheme, uses metadata sets defined in the SMPTE Metadata Dictionary. Metadata sets are organized into descriptive metadata (DM) frameworks and DMS-1 defines three such frameworks, corresponding to different description granularities: *production*, describing the entire media item; *clip*, describing a continuous temporal segment of the audiovisual content; and *scene*, containing the metadata of a narratively or dramatically coherent unit. DMS-1 descriptions can be serialized in XML (based on an XML Schema definition) or in KLV. The latter is mainly used when embedding DMS-1 descriptions in MXF files.

The SMPTE Metadata Dictionary (SMP 2004) is a structured list of 1500 metadata elements in six distinct classes: identification, administration, interpretation, parametric, process, relational and spatio-temporal. The elements can be represented in KLV format and the standard defines the key, value format and semantics for each of the elements. It is used for all metadata embedded in MXF files, but the elements defined in the dictionary are also used outside MXF. The context of the elements and their contextual semantics are not specified.

8.1.4   Dublin Core

The Dublin Core Metadata Initiative (DCMI)[4] is an open organization engaged in the development of interoperable online metadata standards. The Dublin Core metadata standard (DCMI 2003) targets the high-level description of networked electronic resources such as text documents, images, video and audio. Focusing on simplicity, the standard contains 15 elements belonging to three groups (content, version and intellectual property). The elements can be seen as a set of key–value pairs, with the values being free text. Dublin Core descriptions can be represented using XML.

Qualified Dublin Core introduces three additional elements (audience, provenance and rights holder) as well as qualifiers that allow for semantic refinement of the simple DC

---

[3] KLV (Key-Length-Value) encodes items into key–length–value triplets, where key identifies the data or metadata item, length specifies the length of the data block, and value is the data itself (SMP 2001).

[4] http://dublincore.org

elements (e.g. they allow it to be specified that a certain description element contains an abstract). For some of the elements the qualifiers also allow the encoding of the element's content to be specified.

Due to the very limited and generic definition of the Dublin Core elements, each user has to define how to apply the standard to a certain application. Thus application profiles have been proposed (Heery and Patel 2004) for this purpose. This proposal has led to the definition of guidelines for the specification of Dublin Core Application Profiles (DCAPs); see Baker et al. (2005). In contrast to profiles of other standards, a DCAP is not only a restriction of Dublin Core, but may also specify extensions. A DCAP describes the terms and attributes used for a certain application (including URIs, namespaces) as well as their semantics in textual form. Examples are the Dublin Core Library Application Profile (DC-Lib) and the Collections Application Profile.

8.1.5 TV-Anytime

TV-Anytime (TVA 2005a,b) targets business-to-consumer (B2C) metadata applications in the broadcast domain, such as electronic program guides, interactive TV and video on demand. It allows consumers to find, navigate and manage content from a variety of internal (PVR, videos downloaded to a PC) and external (broadcast, IPTV, Web) sources. Metadata is generated during the process of content creation and content delivery and typically derived from the in-house metadata model of the content provider. The standard defines three basic types of metadata: content description, instance description, and consumer metadata. In addition, the standard defines segmentation metadata and metadata origination. The information used by a consumer or software agent to decide whether or not to acquire a particular piece of content is called an attractor.

Content description metadata includes basic descriptive metadata of individual programs and program groups, reviews and purchase information. Instance description metadata targets the discrimination of different instances of the same program, for example, broadcast by different broadcasters or at different time.

The consumer metadata usage history and user preferences description schemes are defined by the MPEG-7 User Description Profile (MPEG-7 2005), and other MPEG-7 descriptors and description schemes are also used in TV-Anytime.

8.1.6 METS and VRA

The Metadata Encoding and Transmission Standard (METS 2008) is intended for the management of digital objects (including multimedia objects) in repositories and the exchange of such objects. The standard is a registered NISO standard and defined based on XML Schema. It is organized into sections: a *descriptive metadata section*; an *administrative metadata section*, including technical, rights and access, source and digitization metadata; a *behavior section*, describing the object's behavior in the context of a certain repository architecture, and *file groups* and *structural map sections* that define the files associated with an object and how to navigate them (e.g. scanned pages of a book).

METS supports the concept of profiles in order to ensure conformance of METS documents for a certain application area or institution. A METS profile is expressed as an XML document based on a schema. A list of 13 components, including a unique identifier, the structural definition of the profile, rules that must be met my documents conforming to the profile, and tools and validators, must be included in a profile definition (METS 2006). A number of profiles dealing with different kinds of multimedia objects exist.

By extension different standards can be used for the different sections of the METS container structure, whether they are represented as XML or not. The standards that can be used for descriptive metadata include Dublin Core (cf. Section 8.1.4), Metadata Object Description Schema (MODS), MARC 21 XML and VRA Core. MODS and MARC are not specifically designed for multimedia objects but general digital library objects.

VRA Core (VRA 2007) is a standard for representing visual objects (such as paintings) in the cultural heritage domain. The schema consists of a set of 19 elements that are represented as XML. There is a restricted flavor that defines value lists for certain attributes.

8.1.7 MPEG-21

MPEG-21 (2001) is a suite of standards defining a normative open framework for multimedia delivery and consumption for use by all stakeholders in the delivery and consumption chain. The main concepts in MPEG-21 are *digital items*, that is, any single multimedia content item or group of items, and the *users* interacting with digital items (e.g. creating, modifying, annotating, consuming). The standard is still under development, and currently includes 20 parts, covering many aspects of declaring, identifying and adapting digital items along the distribution chain, including file formats and binary representations. MPEG-21 is complementary to other standards belong to the MPEG family, for example, MPEG-7 can be used for the description of digital items.

From a metadata perspective, the parts of MPEG-21 describing rights and licensing of digital items are most intersting. Part 5, Rights Expression Language (REL), provides a machine-readable XML-based standard format for licenses and rights representation, and the possibility of defining new profiles and extensions to target specific requirements. Three REL profiles, addressing business-to-consumer use cases, have been defined so far. The terms to be used in rights expressions are defined in Part 6, Rights Data Dictionary.

A recent addition is Part 19, the Media Value Chain Ontology (MVCO), an ontology for formalizing the value chain of a digital item. It represents intellectual entities, actions and user roles as an OWL ontology.

8.1.8 XMP, IPTC in XMP

The main goal of XMP[5] is to attach more powerful metadata to media assets in order to enable better management of multimedia content, and better ways to search and retrieve content in order to improve consumption of these multimedia assets. Furthermore, XMP aims to enhance reuse and re-purposing of content and to improve interoperability between different vendors and systems.

The Adobe XMP specification standardizes the definition, creation, and processing of metadata by providing a data model, storage model (serialization of the metadata as a stream of XML), and formal schema definitions (predefined sets of metadata property

---

[5] http://partners.adobe.com/public/developer/en/xmp/sdk/XMPspecification.pdf

definitions that are relevant for a wide range of applications). XMP makes use of RDF in order to represent the metadata properties associated with a document.

With XMP, Adobe provides a method and format for expressing and embedding metadata in various multimedia file formats. It provides a basic data model as well as metadata schemas for storing metadata in RDF, and provides a storage mechanism and a basic set of schemas for managing multimedia content such as versioning support. The most important components of the specification are the data model and the predefined (and extensible) schemas:

- XMP Data Model is derived from RDF and is a subset of the RDF data model. Metadata properties have values, which can be structured (structured properties) or simple types or arrays. Properties may also have properties (property qualifiers) which may provide additional information about the property value.
- XMP Schemas consist of predefined sets of metadata property definitions. Schemas are essentially collections of statements about resources which are expressed using RDF. It is possible to define new external schemas, to extend the existing ones or to add some if necessary. There are some predefined schemas included in the specification, for example, a Dublin Core Schema, a basic rights schema or a media management schema.

There are a growing number of commercial applications that already support XMP. For example, the International Press and Telecommunications Council (IPTC) has integrated XMP in its Image Metadata specifications and almost every Adobe application such as Photoshop or In-Design supports XMP. IPTC Metadata for XMP can be considered as a multimedia metadata format for describing still images and could soon be the most popular format.

8.1.9 EXIF

One of today's commonly used metadata formats for digital images is the Exchangeable Image File Format (EXIF).[6] The standard specifies the formats to be used for images and sounds, as well as tags in digital still cameras and for other systems handling the image and sound files recorded by digital cameras. The so-called EXIF header carries the metadata for the captured image or sound.

The metadata tags which the EXIF standard provides cover metadata related to the capture of the image and the context situation of the capturing. This includes metadata related to the image data structure (e.g. height, width, orientation), capturing information (e.g. rotation, exposure time, flash), recording offset (e.g. image data location, bytes per compressed strip), image data characteristics (e.g. transfer function, color space transformation), geolocation, as well as general tags (e.g. image title, copyright holder, manufacturer). Lastly, we observe that metadata elements pertaining to the image are stored in the image file header and are identified by unique tags, which serve as an element identifier.

Recently, there have been efforts to represent the EXIF metadata tags in an RDFS ontology. The two approaches presented here are semantically very similar, but are both described for completeness:

---

[6] http://www.digicamsoft.com/exif22/exif22/html/exif22_1.htm

- The Kanzaki EXIF RDF Schema[7] provides a serialization of the basic EXIF metadata tags in an RDF Schema. We also note here that relevant domains and ranges are used as well. In addition, Kanzaki provides an EXIF conversion service which extracts EXIF metadata from images and automatically maps them to their RDF representation.
- The Norm Walsh EXIF RDF Schema[8] provides another encoding of the basic EXIF metadata tags in RDF Schema. In addition, Walsh provides JPEGRDF, which is a Java application that provides an API to read and manipulate EXIF metadata stored in JPEG images. Currently, JPEGRDF can extract, query, and augment the EXIF/RDF data stored in the file headers. In particular, we note that the API can be used to convert existing EXIF metadata in file headers to the schema defined by Walsh.

8.1.10   DIG35

The DIG35 specification[9] includes a standard set of metadata for digital images which promotes interoperability and extensibility, as well as a uniform underlying construct to support interoperability of metadata between various digital imaging devices.

The metadata properties are encoded within an XML Schema and cover:

- basic image parameter (a general-purpose metadata standard);
- image creation (e.g. the camera and lens information);
- content description (who, what, when and where);
- history (partial information about how the image got to the present state);
- intellectual property rights;
- fundamental metadata types and fields (define the format of the field defined in all metadata blocks).

A DIG35 ontology[10] has been developed by the IBBT Multimedia Lab (University of Ghent). This ontology provides an OWL-Full schema covering the entire specification. For the formal representation of DIG35, no other ontologies have been used. However, relations with other ontologies such as EXIF and FOAF have been created to give the DIG35 ontology a broader semantic range.

8.1.11   ID3/MP3

ID3 is a metadata container for audio. Mainly targeted at files encoded with MP3, it rapidly became a *de facto* standard as the size of digital audio libraries considerably increased in the late 1990s. It is now in its second version (ID3v2),[11] and covers the metadata categories *Editorial* and *Cultural* (see Chapter 2). Supported by most personal media players, ID3 targets end user applications such as search and retrieval within digital audio libraries.

ID3 tags are constructed as an association of information frames, each frame encoding specific metadata. The huge set of predefined frames includes identification, technical metadata, rights and licensing, lyrics, comments, pictures, URL links, etc. To be flexible

---

[7] http://www.kanzaki.com/ns/exif
[8] http://sourceforge.net/projects/jpegrdf
[9] http://xml.coverpages.org/FU-Berlin-DIG35-v10-Sept00.pdf
[10] http://multimedialab.elis.ugent.be/users/chpoppe/Ontologies/DIG35.zip
[11] http://www.id3.org/

and extensible, users are allowed to create new frame types. Also, any software supporting ID3 will ignore frames it does not support or recognize. An ID3v2 tag can contain up to 256 MB, the size of each frame being limited to 16MB. The characters are coded using the Unicode standard.

8.1.12   NewsML G2 and rNews

To facilitate the exchange of news, the International Press Telecommunications Council (IPTC) has developed the News Architecture for G2-Standards[12] whose goal is to provide a single generic model for exchanging all kinds of newsworthy information, thus providing a framework for a future family of IPTC news exchange standards. This family includes NewsML-G2, SportsML-G2, EventsML-G2, ProgramGuideML-G2 and a future WeatherML. All are XML-based languages used for describing not only the news content (traditional metadata), but also their management, packaging, or matters related to the exchange itself (transportation, routing).

As part of this architecture, specific controlled vocabularies, such as the IPTC News Codes, are used to categorize news items together with other industry-standard thesauri. While news is still mainly in the form of text-based stories, these are often illustrated with graphics, images and videos. Media-specific metadata formats, such as EXIF, DIG35 and XMP, are used to describe the media. The use of different metadata formats in a single production process leads to interoperability problems within the news production chain itself. It also excludes linking to existing web knowledge resources and impedes the construction of uniform end-user interfaces for searching and browsing news content.

In order to allow these different metadata standards to interoperate within a single information environment, Troncy has designed an OWL ontology for the IPTC News Architecture, linked with other multimedia metadata standards (Troncy 2008). Furthermore, Troncy has also converted the IPTC NewsCodes into SKOS thesauri and demonstrated how news metadata can be enriched using natural language processing and multimedia analysis and integrated with existing knowledge already formalized on the Semantic Web. The IPTC has recently announced the development of the rNews format[13] which aims to be a standard for using RDFa to annotate news-specific metadata in HTML documents.

8.1.13   W3C Ontology for Media Resources

The aim of the W3C Media Annotations Working Group (MAWG)[14] is to define a core set of descriptive and technical metadata properties for the deployment of multimedia content on the Web. The proposed approach is to provide an interlingua ontology and an API designed to facilitate cross-community data integration of information related to media resources on the Web, such as video, audio, and images.

The set of core properties that constitute the Ontology for Media Resources 1.0 (Lee et al. 2010b) is based on a list of the most commonly used annotation properties from media metadata schemas currently in use. This set is derived from the work of the W3C

---

[12] http://www.iptc.org/NAR/
[13] http://dev.iptc.org/rNews
[14] http://www.w3.org/2008/WebVideo/Annotations/

Incubator Group Report on Multimedia Vocabularies on the Semantic Web and a list of use cases (Lee et al. 2010a), compiled after a public call. The use cases involve heterogeneous media metadata schemas used in different communities (interactive TV, cultural heritage institutions, etc.).

The set of core properties defined in the ontology consists of 20 descriptive and eight technical metadata properties. The technical properties, specific to certain media types, apply to a certain instantiation of the content, while the descriptive properties can also be used in a more abstract context, for example, to describe a work. For many of the descriptive properties, subtypes that optionally further qualify the property have been defined, for example, qualify a title as main or secondary. The descriptive properties contain identification metadata such as identifiers, titles, languages and the locator of the media resource. Other properties describe the creation of the content (the creation date, creation location, the different kinds of creators and contributors, etc.), the content description as free text, the genre, a rating of the content by users or organizations and a set of keywords. There are also properties to describe the collections the described resource belongs to, and to express relations to other media resources, for example, source and derived works, thumbnails or trailers. Digital rights management is considered out of scope, thus the set of properties only contains a copyright statement and a reference to a policy (e.g. Creative Commons or MPEG-21 licenses). The distribution-related metadata includes the description of the publisher and the target audience in terms of regions and age classification. Annotation properties can be attached to the whole media or to part of it, for example using the Media Fragments URI (Troncy et al. 2010) specification for identifying multimedia fragments. The set of technical properties has been limited to the frame size of images and video, the duration, the audio sampling rate and frame rate, the format (specified as MIME type), the compression type, the number of tracks and the average bit rate. These were the only properties that were needed for the different use cases listed by the group.

This core set of annotation properties often has correspondences with existing metadata standards. The working group has therefore further specified a mapping table that defines *one-way* mappings between the ontology's core properties and the metadata fields from 24 other standards (Lee et al. 2010b). The proposed ontology has been formalized using an OWL representation.

8.1.14   EBUCore

The aim of the EBU Core Metadata set (EBUCore) (EC-M 2010) is to define a minimum set of attributes for video and audio content for a wide range of applications in the broadcast domain. It is based on Dublin Core to facilitate interoperability beyond the broadcast and audiovisual media domain, for example, with the digital library world.

EBUCore adds a number of attributes that are crucial for audiovisual media, as well as qualifiers, to the Dublin Core elements. Wherever possible, the use of controlled vocabularies is supported. An important advantage over Dublin Core is also the support for parts (e.g. fragments, segments) of media resources, which enables annotation of time-based metadata. An OWL representation of EBUCore has been published.

8.2   Comparison

In this section we compare the standards discussed above by a set of criteria. The list of criteria is compiled from experience of building multimedia metadata applications in different domains including those that turned out to be crucial for the successful use of metadata standards. Of course such a selection of criteria is necessarily incomplete. Table 8.1 assesses the standards discussed in the previous section w.r.t. the following criteria.

Content Types
A basic property of a multimedia metadata standard is the type of media that are supported, such as video, audio and still images. Some standards are specifically designed for one media type, while others are more versatile.

Types of Metadata
There are many ways to classify the huge number of metadata elements that exist across standards. Depending on the intended application areas and use cases these elements are grouped differently in the existing standards. A basic distinction is between *technical* (sometimes called *structural*) metadata (MXF 2004) and *descriptive* metadata. The classification used in the following is similar to the one in Bailer and Schallauer (2008) and has been compiled from the structure of various metadata standards.

Identification information usually contains IDs as well as the titles related to the content (working titles, titles used for publishing, etc.).

Production information describes metadata related to the creation of the content, such as location and time of capture as well as the persons and organizations contributing to the production.

Rights. Information about the rights holders of the content as well as permitted use, modification, distribution, etc. This class of metadata elements ranges from a simple reference to a license (e.g. Creative Commons) to a very detailed description (e.g. for each shot in a newscast that may come from a different source) of the conditions and permitted use of each of the segments.

Publication information describes previous use of content (e.g. broadcasting, syndication) and related information (e.g. contracts, revenues).

Process-related This is a separate group in some standards describing the history of production and post-production steps that occurred, for example, information about capture, digitization, encoding and editing steps in the workflow.

Content-related metadata is descriptive metadata in the narrowest sense, describing the structure of the content (e.g. shots, scenes) and related information.

Context-related. The semantics of multimedia content is in many cases strongly influenced by the context in which it is used and consumed.

Relational/enrichment information describes links between the content and external data sources, such as other multimedia content or related textual sources.

Extensibility
It is hardly possible to foresee the requirements of all possible applications or application domains of a specific metadata standard. Thus extensibility is an important issue.

While some standards have strict definitions that can only be changed through a formal standardization process, other standards provide hooks for extensions and define rules about conformance to the standard.

Controlled Vocabulary

Many metadata elements contain one out of a set of possible values or reference entities such as persons, organizations or geographical entities. In this case the use of controlled vocabulary is required to ensure consistent annotations, especially in large organizations. Multimedia metadata standards differ in their support for controlled vocabulary: some provide no support at all, some allow only the use of specific vocabularies defined by the standards, others also references to external vocabularies (e.g. a broadcaster's thesaurus).

Structuring

For annotation of time-based metadata, as required for example in the professional annotation use case described in Chapter 2, structuring capabilities are crucial. Some standards allow only global metadata elements per media item, while others provide predefined structures or allow arbitrary spatial, temporal or spatiotemporal segments to be defined and support annotation on these segments.

Modularity

There are often applications that only need specific parts of a metadata standard. The question is whether a standard is modular enough to support this requirement or whether it causes overhead for certain applications. Another aspect of modularity is whether or not description elements referring to different features of the multimedia item (e.g. different modalities, different abstraction levels) have interdependencies.

Complexity

There are two aspects of complexity that are relevant in this context. The *conceptual complexity* specifies how difficult it is to understand the model and the concepts of the standards – the learning curve of someone using it. The *implementation complexity* is related to the resources needed by an implementation of the standard in terms of memory, processing power, etc. and constrains the range of target systems on which it can be used (e.g. mobile devices, set-top boxes).

Serialization

Many applications need to be persistently stores or serialized for transmission over certain network protocols. The persistent representation is in many cases a tradeoff between efficient storage (binary) and human-readable formats that are easy to parse or for which standard tools are available (text, XML).

Formalization

The definition of multimedia metadata standards differs in terms of formality. Some standards provide only a textual description of the elements, many provide a formal specification of the syntactic aspects (e.g. using DTD, XML Schema, ASN.1) and a

Table 8.1 Comparison of selected multimedia metadata standards (+ ... well supported/applies, o ... average/partly applies, – ... weakly or not supported/does not apply)

Criteria	MPEG-7	EBU P_Meta	SMPTE	Dublin Core	TV-Anytime	METS and VRA	MPEG-21	XMP, IPTC in XMP	EXIF	DIG35	ID3/MP3	NewsML G2	W3C MAWG	EBUCore
Content types														
Still image	+	–	–	+	–	+	+	+	+	+	–		+	–
Video	+	+	+	+	+	o	+	+	–	–	–		+	+
Audio	+	+	+	+	–	o	+	+	–	–	+		+	+
Metadata types														
Identification	+	+	+	o	+	o	+	+	+	+	+		+	+
Production	+	+	+	–	o	–	o	+	+	+	+		o	+
Rights	o	+	–	o	+	o	+	+	–	+	o		–	o
Publication	o	+	–	–	+	o	o	+	+	o	o		–	o
Process-related	–	–	+	–	–	+	o	–	+	+	–		–	–
Content-related	+	o	o	o	o	o	–	+	–	o	–		o	o
Context-related	–	–	–	–	–	o	–	+	o	o	–		–	–
Relational	o	o	–	–	o	–	–	–	–	o	o		–	o
Extensibility	+	o	+	o	–	+	+	+	–	–	o		o	o
Controlled voc.	+	+	–	o	o	+	+	+	–	–	o		+	+
Modularity	o	o	+	+	o	+	+	+	o	o	+		+	+
Complexity														
Conceptual	+	o	o	–	o	o	+	+	o	o	–		–	–
Implementation	+	o	o	–	o	–	o	+	+	–	–		–	–
Serialisation														
Text	–	–	–	+	–	–	–	–	o	–	–		–	–
XML	+	+	o	+	+	+	+	+	o	+	–		+	+
Binary	+	+	+	–	–	–	+	–	+	–	+		–	–
Formalization														
Text	+	+	+	+	+	+	+	+	+	o	+		+	+
DTD	–	–	–	+	–	–	–	o	–	–	–		–	–
XML Schema	+	+	o	+	+	+	+	+	–	+	–		–	+
Formal semantics	–	–	–	–	–	–	o	+	+	o	–		+	+
Process support														
Premeditate	–	o	o	–	–	o	o	o	o	o	–		–	–
Create	o	o	+	–	–	–	o	+	+	+	–		o	o
Annotate	+	+	o	+	–	o	–	+	+	+	+		+	+
Package	o	+	o	–	+	o	o	+	–	o	–		–	–
Query	+	o	–	o	–	o	–	+	+	–	+		+	+
Construct message	–	–	–	–	–	–	–	–	–	–	–		–	–
Organize	+	–	–	–	o	o	–	o	o	o	–		o	o
Publish	o	+	o	o	+	o	+	+	+	o	–		o	o
Distribute	o	o	o	o	+	–	+	+	+	o	o		o	o

(continued overleaf)

Table 8.1 (*continued*)

Criteria	MPEG-7	EBU P_Meta	SMPTE	Dublin Core	TV-Anytime	METS and VRA	MPEG-21	XMP, IPTC in XMP	EXIF	DIG35	ID3/MP3	NewsML G2	W3C MAWG	EBUCore
Target applications														
Professional	+	+	+	+	−	o	+	+	+	+	−		+	+
End user	o	−	−	+	+	−	+	+	+	−	+		+	+
Education	+	−	−	+	−	+	+	+	+	−	−		−	−

textual description of the semantics of the elements, and some standards provide a formal definition of both syntax and semantics (e.g. using RDF).

Process Support
The canonical processes of media production (cf. Chapter 3) define basic process steps that can be used to model different processes related to multimedia production. Some metadata standards can only provide support in specific processes while others cover several steps of the production process.

Target Applications
Finally, metadata standards differ in the target applications. While some standards are geared toward professional users and business-to-business scenarios, others target metadata delivery to consumers or educational applications.

8.3 Conclusion

We have reviewed some commonly used metadata standards and compared them by a set of criteria that take different application areas and use cases into account. The analysis of the different metadata standards shows that many of them target certain types of application, stages of the media production process or types of metadata. There is no single standard or format that satisfactorily covers all aspects of audiovisual content descriptions; the ideal choice depends on type of application, process and required complexity.

This observation has two consequences. First, it means that in practice systems have to deal with several multimedia metadata standards and map between them. Second, such mappings are not trivial, as none of the standards can be used as the central hub or 'Rosetta Stone', but the standards involved overlap only partially. As conversion and mapping between standards is inevitable, it is necessary to define interoperable subsets for certain applications. For example, the W3C Media Annotations Working Group[15] defines a

---

[15] http://www.w3.org/2008/WebVideo/Annotations/

core set of descriptive and technical metadata properties for the deployment of multimedia content on the Web, including mappings of these properties to relevant standards.

Definition of such mappings is a tedious manual task that needs to be done for each relevant pair of standards. In order to facilitate this task, formal definitions of metadata standards are necessary, complementing the textual description of the semantics of elements of the standards by formal semantics. Troncy et al. (2006) have shown how such an approach can be used for semantic validation of MPEG-7 documents with respect to profiles. Formalized definitions of metadata standards could be used to automate mapping and the validation of mapping results.

9

The Core Ontology for Multimedia

Thomas Franz,[1] Raphaël Troncy[2] and Miroslav Vacura[3]

[1]ISWeb – Information Systems and Semantic Web, University of Koblenz-Landau, Koblenz, Germany
[2]Centrum Wiskunde & Informatica, Amsterdam, The Netherlands
[3]University of Economics, Prague, Czech Republic

In this chapter, we present the Core Ontology for Multimedia (COMM), which provides a formal semantics for multimedia annotations to enable interoperability of multimedia metadata among media tools. COMM maps the core functionalities of the MPEG-7 standard to a formal ontology, following an ontology design approach that utilizes the foundational DOLCE ontology to safeguard conceptual clarity and soundness as well as extensibility towards new annotation requirements. We analyze the requirements underlying the semantic representation of media objects, explain why the requirements are not fulfilled by most semantic multimedia ontologies and present our solutions as implemented by COMM.

9.1 Introduction

Multimedia assets obtained from various sources (Web, personal camera, TV, surveillance systems) require the addition of machine-readable semantic information prior to any advanced knowledge processing. Such processing can sometimes take the form of complex reasoning employing a number of sophisticated logic tools and techniques (see Chapter 11). Sometimes we may simply require the ability to directly search and retrieve the multimedia objects using text keywords or strings. An individual multimedia asset becomes much more useful when semantic information about what it represents or provenance metadata about how or when it was created is included. In some cases

Multimedia Semantics: Metadata, Analysis and Interaction, First Edition.
Edited by Raphaël Troncy, Benoit Huet and Simon Schenk.
© 2011 John Wiley & Sons, Ltd. Published 2011 by John Wiley & Sons, Ltd.

even machine-readable metadata providing information regarding low-level multimedia features is necessary.

A number of tools and services have been developed by the multimedia research community to make it possible to extract low-level features of multimedia assets, track provenance metadata and annotate multimedia with semantic information. Interoperability among such media tools is commonly enabled by human negotiation on the syntax of language constructs used to represent media annotations. As a result, agreed-upon conventions need to be implemented in supporting media tools that are confined to the constructs used and the agreements negotiated.

MPEG-7 is an international standard that tries to provide a common syntax for attaching complex descriptions to various parts of multimedia objects (MPEG-7 2001; Nack and Lindsay 1999). MPEG-7 can be used to describe individual parts or segments of a multimedia object and covers a wide range of characteristics from low-level features to semantic labels. However, the standard has a number of drawbacks, recently identified by Arndt et al. (2008): (i) because it is based on XML Schema it is hard to directly machine process semantic descriptions; (ii) its use of URNs is cumbersome for the Web; (iii) it is not open to Web standards, which represent knowledge on the Internet (such as RDF, OWL) and make use of existing controlled vocabularies; (iv) it is debatable whether annotations by means of simple text string labels can be considered semantic. Some of these drawbacks will later be considered in detail.

That being said, we do not wish to propose abandoning the MPEG-7 standard altogether. Rather we suggest combining the descriptive abilities and experience of MPEG-7 with up-to-date Semantic Web-based technology. We present the Core Ontology for Multimedia as a result.

The rest of the chapter is divided to several sections. First we present a use case based on a scenario introduced in previous chapters. Then we discuss related work and requirements for designing a multimedia ontology that have been defined in the literature. Section 9.5 presents a representation of MPEG-7 by means of formal Semantic Web languages. Section 9.6 returns to the scenario and shows how problems described in the use case can be solved by employing the COMM ontology. The chapter concludes with lessons learned and some final thoughts.

9.2   A Multimedia Presentation for Granddad

Let us recall the scenario described in Chapter 2 where Ellen Scott and her family wish to create a multimedia presentation for granddad. We observed that Ellen's family took many photos and recorded videos using their digital cameras. Ellen uses an MPEG-7 compliant authoring tool to create the presentation for granddad.

As a first step, Ellen wishes to create a selection of photos that show family members. Her son, Mike, has already annotated many videos and photos with MPEG-7 compliant annotation tools. Using those tools he has associated segments of photos and sequences of videos with descriptive keywords indicating what or who is depicted and what is happening. As shown in Figure 9.1, Mike has marked a (still) region (SR2) of an image and associated it with the text *Leaning Tower of Pisa*. Ellen wants to import such descriptions into her authoring tool to find interesting photos and videos based on the keywords entered by her son. Next to such manually created descriptions, Ellen wants to import

```
...
<StillRegion id="SR1">
 <Semantic>
  <Label><Name>Ellen</Name></Label>
 </Semantic>
</StillRegion>
...
<StillRegion id="SR2">
 <TextAnnotation><!--TextAnnotationType-->
  <KeywordAnnotation>
   <Keyword>Pisa Tower</Keyword>
  </KeywordAnnotation>
 </TextAnnotation>
</StillRegion>
...
<StillRegion id="SR3">
 <Semantic>
  <Definition><!--Also TextAnnotationType-->
   <StructuredAnnotation><Who><Name>Mike</Name></Who></StructuredAnnotation>
  </Definition>
 </Semantic>
</StillRegion>
...
```

Figure 9.1   Family portrait near Pisa Cathedral and the Leaning Tower.

further photo descriptions created by a face recognition service that she has found on the Web. She wants to run the service on all photos and import the extraction results into her authoring tool to find photos showing specific family members or photos depicting the whole family. Region SR1 in Figure 9.1, for example, was detected by the face recognizer and automatically associated with the keyword Ellen. As Ellen selects suitable photos, she sometimes spots photos that lack descriptions, for example when the face recognizer has not correctly identified a family member. SR3 in Figure 9.1 is an example of a region that was described neither by Mike nor by the face recognizer. In such cases, Ellen uses the annotation feature of her authoring tool to add missing descriptions.

Next to photos of the family, Ellen wants to add media about interesting sights, such as the Tower of Pisa or Etruscan and Roman remains. Knowing that granddad is always interested in detailed facts, she wants to annotate photos and videos with resources from the web that contain detailed descriptions and stories about the buildings and ancient remains depicted. For instance, she wants to link SR2, which depicts the Tower of Pisa, with the corresponding Wikipedia page[1] which lists interesting details such as the height of the tower, when it was built and so on.

When Ellen has finalized the presentation, she plans to deliver it in an ODF document embedding the chosen images and videos. Using existing solutions, however, Ellen faces several problems:

Fragment identification. Particular regions of images need to be localized – anchor value in Halasz and Schwartz (1994). However, the current web architecture does not provide a means for uniquely identifying sub-parts of multimedia assets, in the same way that the fragment identifier in the URI can refer to part of an HTML or XML document. Indeed, for almost any other media type, the semantics of the fragment identifier has

---
[1] http://en.wikipedia.org/wiki/Leaning_Tower_of_Pisa

not been defined or is not commonly accepted. Providing an agreed way to localize sub-parts of multimedia objects (e.g. sub-regions of images, temporal sequences of videos or tracking moving objects in space and time) is fundamental (Geurts et al. 2005).[2] For images, one can use either MPEG-7 or SVG snippet code to define the bounding box coordinates of specific regions. For temporal locations, one can use MPEG-7 code or the TemporalURI Internet Draft[3]. MPEG-21 specifies a normative syntax to be used in URIs for addressing parts of any resource but whose media type is restricted to MPEG (MPEG-21 2006). The MPEG-7 approach requires an indirection: an annotation is *about* a fragment of an XML document that *refers* to a multimedia document, whereas the MPEG-21 approach does not have this limitation (Troncy et al. 2007b).

Semantic annotation. MPEG-7 is a natural candidate for representing the extraction results of multimedia analysis software such as a face recognition Web service. The language, standardized in 2001, specifies a rich vocabulary of multimedia descriptors, which can be represented in either XML or a binary format. While it is possible to specify very detailed annotations using these descriptors, it is not possible to guarantee that MPEG-7 metadata generated by different agents will be mutually understood due to the lack of formal semantics of this language (Hunter 2001; Troncy 2003). The XML code of Figure 9.1 illustrates the inherent interoperability problems of MPEG-7: several descriptors, semantically equivalent and representing the same information while using different syntax, can coexist (Troncy et al. 2006). For instance, Ellen wants to use face recognition Web services and annotations Mike has created with two MPEG-7 compli-ant annotation tools. The resulting statements for the regions SR1, SR2 and SR3 differ from each other even though they are all syntactically correct. While the Web service uses the MPEG-7 SemanticType for assigning the <Label> *Ellen* to still region SR1, Mike's MPEG-7 tool uses a <KeywordAnnotation> for associating the text *Leaning Tower of Pisa* to still region SR2. Ellen's tool expresses the fact that still region SR3 is labeled *Mike* by a <StructuredAnnotation> (which can be used within the SemanticType). Consequently, alternative ways of annotating the still regions render almost impossible the retrieval of the different results within the authoring tool since the corresponding XPath query has to deal with these syntactic variations. As a result, Ellen cannot easily scan through images depicting family members as her tool does not expect semantic labels of still regions as part of the <KeywordAnnotation> element.

Web interoperability. Ellen would like to link the multimedia presentation to information that is already available on the web. She has also found semantic metadata about the relationships between some of the depicted remains that could improve the automatic generation of the multimedia presentation. However, she realizes that MPEG-7 cannot be combined with these concepts defined in domain-specific ontologies because of its relationship to the web. As this example demonstrates, although MPEG-7 provides ways of associating semantics with (parts of) non-textual media assets, it is incompatible with (Semantic) Web technologies and has no formal description of the semantics encapsulated implicitly in the standard.

---

[2] See also the W3C Media Fragments Working Group, http://www.w3.org/2008/01/media-fragments-wg.html
[3] http://www.annodex.net/TR/URI_fragments.html

Embedding into compound documents. Finally, Ellen needs to compile the semantic annotations of the images, videos and textual stories into a semantically annotated compound document. However, the current state of the art does not provide a framework which allows the semantic annotation of compound documents. MPEG-7 solves the problem only partially as it is restricted to the description of audiovisual compound documents. Bearing the growing number of multimedia office documents in mind, this limitation is a serious drawback.

9.3 Related Work

In the field of semantic image understanding, using a multimedia ontology infrastructure is regarded as the first step in closing the so-called semantic gap between low-level signal processing results and explicit semantic descriptions of the concepts depicted in images. Furthermore, multimedia ontologies have the potential to increase the interoperability of applications producing and consuming multimedia annotations. The application of multimedia reasoning techniques on top of semantic multimedia annotations is also currently under investigation (Neumann and Möller 2006). A number of drawbacks of MPEG-7 have been reported (Nack et al. 2005; van Ossenbruggen et al. 2004). As a solution, multimedia ontologies based on MPEG-7 have been proposed.

Hunter (2001) provided the first attempt to model parts of MPEG-7 in RDFS, later integrated with the ABC model. Tsinaraki et al. (2004) start from the core of this ontology and extend it to cover the full MDS part of MPEG-7 (see 8.1.1), in an OWL DL ontology. A complementary approach was explored by Isaac and Troncy (2004), who proposed a core audiovisual ontology inspired by several terminologies such as MPEG-7, TV-Anytime and ProgramGuideML. Garcia and Celma (2005) produced the first complete MPEG-7 ontology, automatically generated using a generic mapping from XSD to OWL. Finally, Bloehdorn et al. (2005) proposed an OWL DL Visual Descriptor Ontology[4] based on the Visual part of MPEG-7 and used for image and video analysis.

These ontologies have recently been compared with COMM according to three criteria: (i) the way the multimedia ontology is linked with domain semantics; (ii) the MPEG-7 coverage of the multimedia ontology; and (iii) the scalability and modeling rationale of the conceptualization (Troncy et al. 2007a). Unlike COMM, all the other ontologies perform a one-to-one translation of MPEG-7 types into OWL concepts and properties. This translation does not, however, guarantee that the intended semantics of MPEG-7 is fully captured and formalized. On the contrary, the syntactic interoperability and conceptual ambiguity problems illustrated in Section 9.2 remain. Although COMM is based on a foundational ontology, the annotations proved to be no more verbose than those in MPEG-7.

Finally, general models for annotations of non-multimedia content have been proposed by librarians. The Functional Requirements for Bibliographic Records[5] model specifies the conventions for bibliographic description of traditional books. The CIDOC Conceptual Reference Model (CRM)[6] defines the formal structure for describing the concepts and

---

[4] http://image.ece.ntua.gr/~gstoil/VDO
[5] http://www.ifla.org/VII/s13/frbr/
[6] http://cidoc.ics.forth.gr/

relationships used in cultural heritage documentation. Hunter (2002) has described how an MPEG-7 ontology could specialize CIDOC-CRM for describing multimedia objects in museums. Interoperability with such models is an issue but, interestingly, the design rationale used in these models is often comparable and complementary to a foundational ontologies approach.

9.4   Requirements for Designing a Multimedia Ontology

Requirements for designing a multimedia ontology have been gathered together and reported in the literature (Hunter and Armstrong 1999). Here, we compile these and use our scenario to present a list of requirements for a Web-compliant multimedia ontology.

MPEG-7 compliance. MPEG-7 is an existing international standard, used in both the signal processing and broadcasting communities. It contains a wealth of accumulated experience that needs to be included in a Web-based ontology. In addition, existing annotations in MPEG-7 should be easily expressible in our ontology.

Semantic interoperability. Annotations are only reusable when the captured semantics can be shared among multiple systems and applications. Obtaining similar results from reasoning processes about terms in different environments can only be guaranteed if the semantics is sufficiently explicitly described. A multimedia ontology has to ensure that the intended meaning of the captured semantics can be shared among different systems.

Syntactic interoperability. Systems are only able to share the semantics of annotations if there is a means of conveying this in some agreed syntax. Given that the (Semantic) Web is an important repository of both media assets and annotations, a semantic description of the multimedia ontology should be expressible in a Web language (e.g. OWL, RDF/XML or RDFa).

Separation of concerns. Clear separation of subject matter (i.e. knowledge about depicted entities, such as the person Ellen Scott) from knowledge that is related to the administrative management or the structure and the features of multimedia documents (e.g. Ellen's face is to the left of Mike's face) is required. Reusability of multimedia annotations can only be achieved if the connection between both ontologies is clearly specified by the multimedia ontology.

Modularity. A complete multimedia ontology can be very large, as demonstrated by MPEG-7. The design of a multimedia ontology should thus be made modular, to minimize the execution overhead when used for multimedia annotation. Modularity is also a good engineering principle.

Extensibility. While we intend to construct a comprehensive multimedia ontology, this can never be complete, as demonstrated by ontology development methodologies. New concepts will always need to be added to the ontology. This requires a design that can always be extended, without changing the underlying model and assumptions and without affecting legacy annotations.

9.5   A Formal Representation for MPEG-7

MPEG-7 specifies the connection between semantic annotations and parts of media assets. We take it as a knowledge base that needs to be expressible in our ontology. Therefore, we re-engineer MPEG-7 according to the intended semantics of the written standard. We satisfy our semantic interoperability not by aligning our ontology to the XML Schema definition of MPEG-7, but by providing a formal semantics for MPEG-7. We use a methodology based on a foundational, or top-level, ontology as a basis for designing COMM (Sure et al. 2009). This provides a domain-independent vocabulary that explicitly includes formal definitions of foundational categories, such as processes or physical objects, and eases the linkage of domain-specific ontologies because of the shared definitions of top-level concepts. We briefly introduce our chosen foundational ontology in Section 9.5.1, and then present our multimedia ontology, COMM, in Sections 9.5.2 and 9.5.3.[7] Finally, we discuss why our ontology satisfies all our stated requirements in Section 9.5.4.

9.5.1 DOLCE as Modeling Basis

Using the review in Oberle et al. (2006), we select the Descriptive Ontology for Linguistic and Cognitive Engineering (DOLCE) (Borgo and Masolo 2009) as a modeling basis. Our choice is influenced by two of the main design patterns: *Descriptions & Situations* (D&S) and *Ontology of Information Objects* (OIO); see Gangemi et al. (2004). The former can be used to formalize contextual knowledge, while the latter implements a semiotics model of communication theory. We suppose that the annotation process is a *situation* (i.e. a reified context) that needs to be described.

9.5.2 Multimedia Patterns

The patterns for D&S and OIO need to be extended for the purpose of representing MPEG-7 concepts since they are not sufficiently specialized to the domain of multimedia annotation. This section introduces these extended multimedia design patterns, while Section 9.5.3 details two central concepts underlying these patterns: digital data and algorithms (Gangemi and Presutti 2009). In order to define design patterns, one has to identify repetitive structures and describe them at an abstract level. The two most important functionalities provided by MPEG-7 are: the *decomposition* of media assets and the (semantic) *annotation* of their parts, which we include in our multimedia ontology.

Decomposition. MPEG-7 provides descriptors for spatial, temporal, spatio-temporal and media source decompositions of multimedia content into segments. A segment is the most general abstract concept in MPEG-7 and can refer to a region of an image, a piece of text, a temporal scene of a video or even to a moving object tracked over a period of time.

Annotation. MPEG-7 defines a very large collection of descriptors that can be used to annotate a segment. These descriptors can be low-level visual features, audio features or more abstract concepts. They allow the annotation of the content of multimedia documents or the media asset itself.

In the following, we first introduce the notion of multimedia data and then present the patterns that formalize the decomposition of multimedia content into segments, or allow the annotation of these segments. The decomposition pattern handles the structure of a multimedia document, while the media annotation pattern, the content annotation pattern

---

[7] COMM is available at http://multimedia.semanticweb.org/COMM/

and the semantic annotation pattern are useful for annotating the media, the features and the semantic content of the multimedia document, respectively.

Multimedia Data

This encapsulates the MPEG-7 notion of multimedia content and is a subconcept of digital-data[8] (introduced in more detail in Section 9.5.3). multimedia-data is an abstract concept that has to be further specialized for concrete multimedia content types (e.g. image-data corresponds to the pixel matrix of an image). According to the OIO pattern, multimedia-data is realized by some physical media (e.g. an image). This concept is needed for annotating the physical realization of multimedia content.

Decomposition Pattern

Following the D&S pattern, we suppose that a decomposition of a multimedia-data entity is a situation[9] (a segment-decomposition) that satisfies a description, such as a segmentation-algorithm or a method (e.g. a user drawing a bounding box around a depicted face), which has been applied to perform the decomposition (see Figure 9.2B). Of particular importance are the roles that are defined by a segmentation-algorithm or a method. output-segment-roles express the fact that some multimedia-data entities are segments of a multimedia-data entity that plays the role of an input segment (input-segment-role). These data entities have as setting a segment-decomposition situation that satisfies the roles of the applied segmentation-algorithm or method. output-segment-roles as well as segment-decompositions are then specialized according to the segment and decomposition hierarchies of MPEG-7 (MPEG-7 (2001), part 5, section 11). In terms of MPEG-7, unsegmented (complete) multimedia content also corresponds to a segment. Consequently, annotations of complete multimedia content start with a root segment. In order to designate multimedia-data instances that correspond to these root segments the decomposition pattern provides the root-segment-role concept. Note that root-segment-roles are not defined by methods which describe segment-decompositions. They are rather defined by methods which cause the production of multimedia content. These methods as well as annotation modes which allow the description of the production process (e.g. MPEG-7 (2001), part 5, section 9) are currently not covered by our ontology. Nevertheless, the prerequisite for enhancing the COMM into this direction is already given.

The decomposition pattern also reflects the need for localizing segments within the input segment of a decomposition as each output-segment-role requires a mask-role. Such a role has to be played by one or more digital-data entities which express a localization-descriptor. An example of such a descriptor is an ontological representation of the MPEG-7 RegionLocatorType[10] for localizing regions in an image (see Figure 9.2C). Hence, the mask-role concept corresponds to the notion of a mask in MPEG-7.

The specialization of the pattern for describing image decompositions is shown in Figure 9.2F. According to MPEG-7, an image or an image segment (image-data) can be composed into still regions. Following this modeling, the concepts output-segment-role and root-segment-role are specialized by the concepts still-region-role

---

[8] Sans serif font indicates ontology concepts or instances.
[9] Cf. Borgo and Masolo (2009).
[10] Typewriter font indicates MPEG-7 language descriptors.

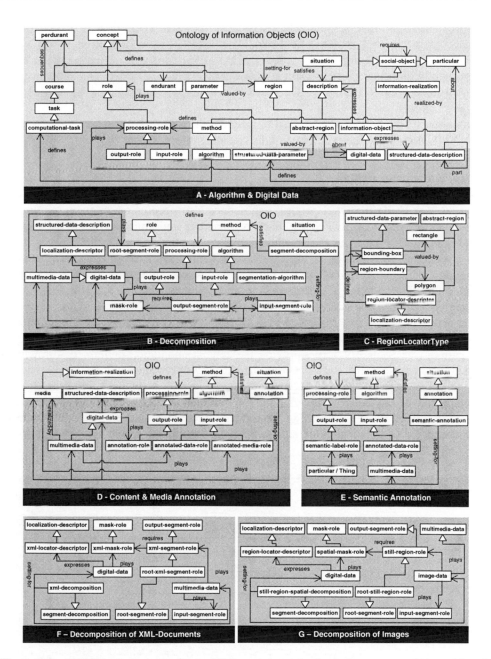

Figure 9.2 COMM: design patterns in UML notation – basic design patterns (A), multimedia patterns (B, D, E) and modeling examples (C, F).

and root-still-region-role respectively. Note that root-still-region-role is a subconcept of still-region-role *and* root-segment-role. The MPEG-7 decomposition mode which can be applied to still regions is called StillRegionSpatialDecompositionType. Consequently, the concept still-region-spatial-decomposition is added as a subconcept of segment-decomposition. Finally, the mask-role concept is specialized by the concept spatial-mask-role.

Analogously, the pattern can be used to describe the decomposition of a video asset or of an XML document, for example an ODF document (see Figure 9.3).

Content Annotation Pattern

This formalizes the attachment of metadata (i.e. annotations) to multimedia-data (Figure 9.2D). Using the D&S pattern, annotations also become situations that represent the state of affairs of all related digital-data (metadata and annotated multimedia-data). digital-data entities represent the attached metadata by playing an annotation-role. These roles are defined by methods or algorithms. The former are used to express manual (or semi-automatic) annotation while the latter serve as an explanation for the attachment of automatically computed features, such as the dominant colors of a still region. It is mandatory that the multimedia-data entity being annotated plays an annotated-data-role.

The actual metadata that is carried by a digital-data entity depends on the structured-data-description that is expressed by it. These descriptions are formalized using the digital data pattern (see Section 9.5.3). Applying the content annotation pattern for formalizing a specific annotation, for example a dominant-color-annotation which corresponds to the connection of a MPEG-7 DominantColorType with a segment, requires only the specialization of the concept annotation, for example dominant-color-annotation. This concept is defined by being a setting for a digital-data entity that expresses a dominant-color-descriptor (a subconcept of structured-data-description which corresponds to the DominantColorType).

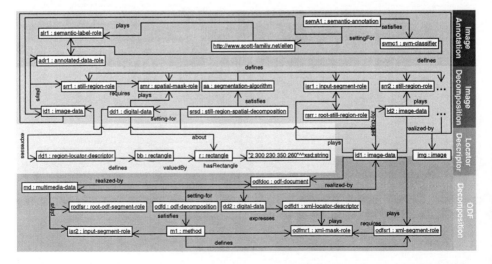

Figure 9.3 Annotation of the image from Figure 9.1 and its embedding into the multimedia presentation for granddad.

Media Annotation Pattern

This forms the basis for describing the physical instances of multimedia content (Figure 9.2D). It differs from the content annotation pattern in only one respect: it is the media that is being annotated and therefore plays an annotated-media-role.

One can thus represent the fact that some visual content (e.g. a digital camera picture) is realized by a JPEG image, 462 848 bytes in size, using the MPEG-7 `MediaFormatType`. Using the media annotation pattern, the metadata is attached by connecting a digital-data entity with the image. The digital-data plays an annotation-role while the Image plays an annotated-media-role. An ontological representation of the `MediaFormatType`, namely an instance of the structured-data-description subconcept media-format-descriptor, is expressed by the digital-data entity. The tuple formed with the scalar '462848' and the string 'JPEG' is the value of the two instances of the concepts file-size and file-format respectively. Both concepts are subconcepts of structured-data-parameter.

Semantic Annotation Pattern

Even though MPEG-7 provides some general concepts (see MPEG-7 (2001), part 5, section 12) that can be used to describe the perceivable content of a multimedia segment, independent development of domain-specific ontologies is more appropriate for describing possible interpretations of multimedia – it is useful to create an ontology specific to multimedia, but it is not useful to try to model the real world within this. An ontology-based multimedia annotation framework should rely on domain-specific ontologies for the representation of the real-world entities that might be depicted in multimedia content. Consequently, this pattern specializes the content annotation pattern to allow the connection of multimedia descriptions with domain descriptions provided by independent world ontologies (Figure 9.2E).

An OWL Thing or a DOLCE particular (belonging to a domain-specific ontology) that is depicted by some multimedia content is not directly connected to it but rather through the way the annotation is obtained. Actually, a manual annotation method or its subconcept algorithm, such as a classification algorithm, has to be applied to determine this connection. It is embodied through a semantic-annotation that satisfies the method applied. This description specifies that the annotated multimedia-data has to play an annotated-data-role and the depicted Thing/particular has to play a semantic-label-role. The pattern also allows the integration of features which might be evaluated in the context of a classification algorithm. In that case, digital-data entities that represent these features would play an input-role.

9.5.3 Basic Patterns

Specializing the D&S and OIO patterns for the purpose of defining multimedia design patterns is enabled through the definition of basic design patterns, which formalize the notion of digital data and algorithm.

Digital Data Pattern

Within the domain of multimedia annotation, the notion of digital data is central – both the multimedia content being annotated and the annotations themselves are expressed as digital data. We consider digital-data entities of arbitrary size to be information-objects, which are used for communication between machines. The OIO design pattern states that descriptions are expressed by information-objects, which have to be about facts (represented by particulars). These facts are settings for situations that have to satisfy the descriptions that are expressed by information-objects. This chain of constraints allows the modeling of complex data structures to store digital information. Our approach is as follows (see Figure 9.2A): digital-data entities express descriptions, namely structured-data-descriptions, which define meaningful labels for the information contained by digital-data. This information is represented by numerical entities such as scalars, matrices, strings, rectangles or polygons. In DOLCE terms, these entities are abstract-regions. In the context of a description, these regions are described by parameters. structured-data-descriptions thus define structured-data-parameters, for which abstract-regions carried by digital-data entities assign values.

The digital data pattern can be used to formalize complex MPEG-7 low-level descriptors. Figure 9.2C shows the application of this pattern by formalizing the MPEG-7 `RegionLocatorType`, which mainly consists of two elements: a `Box` and a `Polygon`. The concept region-locator-descriptor corresponds to the `RegionLocatorType`. The element `Box` is represented by the structured-data-parameter subconcept BoundingBox while the element `Polygon` is represented by the region-boundary concept.

The MPEG-7 code example given in Figure 9.1 highlights the fact that the formalization of data structures so far is not sufficient – complex MPEG-7 types can include nested types that again have to be represented by structured-data-descriptions. In our example, the MPEG-7 `SemanticType` contains the element `Definition` which is of complex type `TextAnnotationType`. The digital data pattern covers such cases by allowing a digital-data instance dd1 to be about a digital-data instance dd2 which expresses a structured-data-description that corresponds to a nested type (see Figure 9.2A). In this case the structured-data-description of instance dd2 would be a part of the one expressed by dd1.

Algorithm Pattern

The production of multimedia annotation can involve the execution of algorithms or the application of computer-assisted methods which are used to produce or manipulate digital-data. The recognition of a face in an image region is an example of the former, while manual annotation of the characters is an example of the latter.

We consider algorithms to be methods that are applied to solve a computational problem (see Figure 9.2A). The associated (DOLCE) situations represent the work that is being done by algorithms. Such situations encompass digital-data[11] involved in the computation, regions that represent the values of parameters of an algorithm, and perdurants[12] that act as computational-tasks (i.e. the processing steps of an algorithm). An algorithm defines roles which are played by digital-data. These roles encode the meaning of data. In order to solve a problem, an algorithm has to process input data and return some output data. Thus, every algorithm defines at least one input-role and one output-role which both have to be played by digital-data.

---

[11] digital-data entities are DOLCE endurants, that is, entities which exist in time and space.
[12] Events, processes or phenomena are examples of perdurants. endurants participate in perdurants.

9.5.4   Comparison with Requirements

We now discuss whether the requirements stated in Section 9.4 are satisfied by our proposed modeling of the multimedia ontology.

The ontology is **MPEG-7 compliant** since the patterns have been designed with the aim of translating the standard into DOLCE. It covers the most important part of MPEG-7 that is commonly used for describing the structure and content of multimedia documents. Our current investigation shows that parts of MPEG-7 that have not yet been considered (e.g. navigation & access) can be formalized analogously to the other descriptors through the definition of further patterns. The technical realization of the basic MPEG-7 data types (e.g. matrices and vectors) is not within the scope of the multimedia ontology. They are represented as ontological concepts, because the about relationship which connects digital-data with numerical entities is only defined between concepts. Thus, the definition of OWL data type properties is required to connect instances of data type concepts (subconcepts of the DOLCE abstract-region) with the actual numeric information (e.g. xsd:string). Currently, simple string representation formats are used for serializing data type concepts (e.g. rectangle) that are currently not covered by W3C standards. Future work includes the integration of the extended data types of OWL 1.1.

Syntactic and semantic interoperability of our multimedia ontology is achieved by an OWL DL formalization.[13] Similar to DOLCE, we provide a rich axiomatization of each pattern using first-order logic. Our ontology can be linked to any Web-based domain-specific ontology through the semantic annotation pattern.

A clear **separation of concerns** is ensured through the use of the multimedia patterns: the decomposition pattern for handling the structure and the annotation pattern for dealing with the metadata. These patterns form the core of the **modular** architecture of the multimedia ontology. We follow the various parts of MPEG-7 and organize the multimedia ontology into modules which cover (i) the descriptors related to a specific media type (e.g. visual, audio or text) and (ii) the descriptors that are generic to a particular media (e.g. media descriptors). We also design a separate module for data types in order to abstract from their technical realization.

Through the use of multimedia design patterns, our ontology is also **extensible**, allowing the inclusion of further media types and descriptors (e.g. new low-level features) using the same patterns. As our patterns are grounded in the D&S pattern, it is straightforward to include further contextual knowledge (e.g. about provenance) by adding roles or parameters. Such extensions will not change the patterns, so that legacy annotations will remain valid.

9.6   Granddad's Presentation Explained by COMM

The interoperability problems Ellen faced in Section 9.2 can be solved by employing the COMM ontology to represent the metadata of all relevant multimedia objects. The COMM ontology provides a programming interface, the COMM API, to embed it into multimedia tools such as authoring tools and analysis Web services. Thus users are not confronted with details of the multimedia ontology.

---

[13] Examples of the axiomatization are available on the COMM website.

Utilizing COMM, the application of the face recognizer for Ellen results in an annotation RDF graph that is depicted in the upper part of Figure 9.3 (visualized by an UML object diagram).[14] The decomposition of Figure 9.1, whose content is represented by id0, into one still region (the bounding box of Ellen's face) is represented by the lighter middle part of the UML diagram. The segment is represented by the image-data instance id1 which plays the still-region-role srr1. It is located by the digital-data instance dd1 which expresses the region-locator-descriptor rld1 (lower part of the diagram). Using the semantic annotation pattern, the still region around Ellen's face can be annotated by connecting it with the URI http://www.family-scott.net/ellen. This URI can denote any resource, for example it can be an instance of a concept defined by some arbitrary domain ontology that defines concepts for persons, families and their relationships.

Further annotations of the Leaning Tower of Pisa (SR2) and Ellen's son Mike (SR3), which may be done manually or by automatic recognizers, will extend the decomposition further by two still regions, that is, the image-data instances id2 and id3 as well as the corresponding still-region-roles, spatial-mask-roles and digital-data instances expressing two more region-locator-descriptors (indicated at the right border of Figure 9.3).

The domain ontologies which provide the instances for annotating id2 and id3 with the semantic annotation pattern do not have to be identical to the one that contains Ellen. In fact, Web pages or ontologies describing Web pages could also be used. For instance, the region depicting the Leaning Tower of Pisa could be annotated with the URI http://en.wikipedia.org/wiki/Leaning_Tower_of_Pisa. If several domain ontologies are used, Ellen can use the OWL sameAs and equivalentClass constructs to align the labelings used in the different annotations to the domain ontology that is best suited for enhancing the automatic generation of the multimedia presentation.

Decomposition of ODF documents is formalized analogously to image segmentation (see Figure 9.2F). Therefore, embedding the image annotation into an ODF document annotation is straightforward. The lower part of Figure 9.3 shows the decomposition of a compound ODF document into image content. This decomposition description could result from copying an image from the desktop and pasting it into an ODF editor such as OpenOffice. A plugin of this program could produce COMM metadata of the document in the background while it is produced by the user. The media-independent design patterns of COMM allow the implementation of a generic mechanism for inserting metadata of arbitrary media assets into already existing metadata of an ODF document. In the case of Figure 9.3, the instance id0 (which represents the whole content of the image shown in Figure 9.1) needs to be connected with three instances of the ODF annotation: (i) the odf-decomposition instance odfd which is a setting-for all top-level segments of the odf-document; (ii) the xml-segment-role instance odfsr1 which identifies id0 as a part of the whole ODF content md (a multimedia-data instance); and (iii) the instance odfdoc, as the image is now also realized-by the odf-document.

The segment id1 is located within md by one digital-data instance (dd2) which expresses a corresponding odf-locator-descriptor instance. The complete instantiation of the odf-locator-descriptor is not shown in Figure 9.3. The modeling of the region-locator-descriptor, which is completely instantiated in Figure 9.3, is shown in Figure 9.2C. The technical details of the odf-locator-descriptor are not presented. However, it is possible

---

[14] The scheme used in Figure 9.3 is instance:Concept, the usual UML notation.

to locate segments in ODF documents by storing an XPath which points to the beginning and end of an ODF segment. Thus, the modeling of the odf-locator-descriptor can be carried out analogously to the region-locator-descriptor.

In order to ease the creation of multimedia annotations with our ontology, we have developed a Java API[15] which provides an MPEG-7 class interface for the construction of metadata at runtime. Annotations which are generated in memory can be exported to Java-based RDF triple stores such as Sesame. For that purpose, the API translates the objects of the MPEG-7 classes into instances of the COMM concepts. The API also facilitates the implementation of multimedia retrieval tools as it is capable of loading RDF annotation graphs (e.g. the complete annotation of an image including the annotation of arbitrary regions) from a store and converting them back to the MPEG-7 class interface. Using this API, the face recognition Web service will automatically create the annotation which is depicted in the upper part of Figure 9.3 by executing the following code:

```
Image img0 = new Image();
StillRegion isr0 = new StillRegion();
img0.setImage(isr0);
StillRegionSpatialDecomposition srsd1 = new
  StillRegionSpatialDecomposition();
isr0.addSpatialDecomposition(srsd1);
srsd1.setDescription(new SegmentationAlgorithm());
StillRegion srr1 = new StillRegion();
srsd1.addStillRegion(srr1);
SpatialMask smr1 = new SpatialMask();
srr1.setSpatialMask(smr1);
RegionLocatorDescriptor rld1 = new RegionLocatorDescriptor();
smr1.addSubRegion(rld1);
rld1.setBox(new Rectangle(300, 230, 50, 30));
Semantic s1 = new Semantic();
s1.addLabel("http://www.family-scott.net/ellen");
s1.setDescription(new SVMClassifier());
srr1.addSemantic(s1);
```

9.7 Lessons Learned

At the time of writing, the COMM has been already employed by multiple multimedia tools, such as the K-Space annotation tool (KAT) described in Chapter 14 and information extraction tools developed within the EU integrated project X-Media.[16]

Throughout the life cycle of COMM, it has been extended and modified in order to adapt to new requirements evolving over time. By the pattern-oriented design of COMM, it has been well prepared to handle such new requirements and extensions. For example, within the X-Media project, new requirements for media annotation have been raised. While the COMM ontology was initially designed to support the description of image and video media, the scenarios in X-Media additionally needed to describe annotations

---

[15] The Java API is available at http://multimedia.semanticweb.org/COMM/api/
[16] http://www.x-media-project.org

of textual and numerical data, for example textual data given in the open document format. The annotation of portions of the data, for example, a text passage or a subset of numerical data, also needed to describe the decomposition of the data into segments. The extension of COMM towards these media types has also resulted in a new ontology module for knowledge acquisition (OAK); see Iria (2009). OAK was designed by specializing concepts defined in COMM in order to adapt our core ontology to the requirements on the integration of knowledge acquisition tools. To describe the decomposition of the newly added media types a number of generic COMM concepts were specialized. Analogous to the concepts Image and Video, the concept Text was introduced. Like Image and Video, it is a subclass of Media that realizes TextData (a specialization of the COMM MultimediaData). Furthermore, new locators for textual data were defined, analogous to the locators already existing for image and video data. For the description of the annotation and decomposition of the newly added media types, existing COMM patterns were reused.

Throughout the utilization of COMM, we have also gathered experience related to the ontology design and the design of the application programming interface (API). APIs are essential to make practical use of core ontologies in concrete applications such as the projects in which we employ COMM. Our experience with such APIs is that a core ontology should not have a fixed programming interface. Although the design of a core ontology should be very stable with respect to its use in different domains, the API should be redesigned based on the requirements of the concrete domain in which it is used. In this way, a much more efficient access to the knowledge modeled and represented with the core ontology can be enabled. However, such APIs that are specifically redesigned for each concrete domain in which they are applied require a flexible and efficient means for API creation. Here, an approach for the automatic generation of APIs for pattern-oriented ontologies is required, such as the a gogo approach (Parreiras et al. 2009). It provides a platform-independent, domain-specific language to automatically generate the mappings that are required between the instances of the ontology design patterns and the classes of the programming language.

9.8 Conclusion

We have presented COMM, an MPEG-7 based multimedia ontology, well founded and composed of multimedia patterns. It satisfies the requirements, as they are described by the multimedia community itself, for a multimedia ontology framework. The ontology is completely formalized in OWL DL and a stable version is available with its API at http://multimedia.semanticweb.org/COMM/. It has been used in projects such as K-Space and X-Media.

The ontology already covers the main parts of the standard, and we are confident that the remaining parts can be covered by following our method for extracting more design patterns. Our modeling approach confirms that the ontology offers even more possibilities for multimedia annotation than MPEG-7 since it is interoperable with existing web ontologies. The explicit representation of algorithms in the multimedia patterns describes the multimedia analysis steps, something that is not possible in MPEG-7. The need to provide this kind of annotation is demonstrated in the algorithm use case of the W3C Multimedia Semantics Incubator Group.[17] The intensive use of the D&S reification mechanism results

---

[17] http://www.w3.org/2005/Incubator/mmsem/XGR-interoperability/

in RDF annotation graphs, which are generated according to our ontology, being quite large compared to those of more straightforwardly designed multimedia ontologies. This presents a challenge for current RDF and OWL stores, but we think it is a challenge worth deep consideration as it is utterly necessary to overcome the isolation of current multimedia annotations and to achieve full interoperability for (nearly) arbitrary multimedia tools and applications.

10

Knowledge-Driven Segmentation and Classification

Thanos Athanasiadis,[1] Phivos Mylonas,[1] Georgios Th. Papadopoulos,[2] Vasileios Mezaris,[2] Yannis Avrithis,[1] Ioannis Kompatsiaris[2] and Michael G. Strintzis[2]

[1]*National Technical University of Athens, Zographou, Greece*
[2]*Informatics and Telematics Institute, Centre for Research and Technology Hellas, Thermi-Thessaloniki, Greece*

In this chapter a first attempt will be made to examine how the coupling of multimedia processing and knowledge representation techniques, presented separately in previous chapters, can improve analysis. No formal reasoning techniques will be introduced at this stage; our exploration of how multimedia analysis and knowledge can be combined will start by revisiting the image and video segmentation problem. Semantic segmentation, presented in Section 10.2, starts with an elementary segmentation and region classification and refines it using similarity measures and merging criteria defined at the semantic level. Our discussion will continue in Sections 10.3 and 10.4 with knowledge-driven classification approaches, which exploit knowledge in the form of contextual information for refining elementary classification results obtained via machine learning. Two relevant approaches will be presented. The first deals with visual context and treats it as interaction between global classification and local region labels. The second deals with spatial context and formulates its exploitation as a global optimization problem. All approaches presented in this chapter are geared toward the photo use case scenario defined in Chapter 2, although their use as part of a semi-automatic annotation process in a professional media production and archiving setting is also feasible.

Multimedia Semantics: Metadata, Analysis and Interaction, First Edition.
Edited by Raphaël Troncy, Benoit Huet and Simon Schenk.
© 2011 John Wiley & Sons, Ltd. Published 2011 by John Wiley & Sons, Ltd.

10.1 Related Work

Starting from an initial image segmentation and the classification of each resulting segment into one of a number of possible semantic categories, using one of the various segmentation algorithms available and one or more of the classifiers discussed in Chapter 5, there are two broad categories of possible analysis errors that one may encounter: segmentation errors and classifications errors. Segmentation errors occur as either under-segmentation or over-segmentation; in both cases, the result is the formation of one or more spatial regions, each of which does not accurately correspond to a single semantic object depicted in the image. Classification errors, on the other hand, occur as a result of the insufficiency of the employed combination of classification technique and feature vector to effectively distinguish between different classes of objects. Clearly, these two categories of analysis errors are not independent of each other: a segmentation error, such as the formation of regions that correspond to only a small part of a semantic object, can render useless any shape features used and thus lead to erroneous classification; there are several similar examples of how segmentation affects classification performance.

Various methods have been proposed to minimize segmentation error. Some focus on the use of a more complete set of visual cues for performing segmentation, for example the combined use of color, texture and position features (Mezaris et al. 2004), and the introduction of new algorithms to exploit these features. Others use post-processing procedures that perform region-merging or even region-splitting operations upon an appropriately generated initial segmentation (Adamek et al. 2005). The knowledge-driven segmentation procedures that will be introduced in this chapter belong to the latter category. A comprehensive review of image segmentation methods that do not exploit semantic information is, however, beyond the scope of this chapter; see Mezaris et al. (2006) for a review on this topic.

Using combinations of feature vectors and classification techniques is not always sufficient to distinguish between different classes of objects. In many cases, using contextual information can help create a more reliable classification of the regions. Contextual information, for the purpose of semantic image analysis, can refer to spatial information, concept co-occurrence information and other kinds of prior knowledge that can contribute to the disambiguation of the semantics of a spatial region based on the semantics of its peers (Luo et al. 2006).

Spatial information in particular has been shown to be effective for discriminating between objects exhibiting similar visual characteristics, since objects tend to be present in a scene within a particular spatial context (Papadopoulos et al. 2007). For example, sky tends to appear over sea in a beach scene. Several approaches have been proposed that use spatial information to overcome the ambiguities and limitations inherent in the visual medium. Hollink et al. (2004) discuss the issue of semi-automatically adding spatial information to image annotations. Directional spatial relations, which denote the relative positions of objects in space, are among the most common spatial context representations used for semantic image analysis. There are two main types of these relations: angle-based approaches and projection-based approaches. Angle-based approaches include Wang et al. (2004), where a pair of fuzzy k-NN classifiers are trained to differentiate between the *Above/Below* and *Left/Right* relations, and Millet et al. (2005), where a fuzzy membership function is defined for every relation and directly applied to the estimated angle histogram. Projection-based approaches include Hollink et al. (2004), where qualitative directional relations are defined in terms of the center and sides of the corresponding object's minimum bounding rectangles.

10.2 Semantic Image Segmentation

10.2.1 Graph Representation of an Image

An image can be described as a structured set of individual objects, allowing a straight-forward mapping to a graph structure. As such, many image analysis problems can be considered as graph theory problems, inheriting a solid theoretical foundation. The attributed relation graph (ARG) is a type of graph often used in computer vision and image analysis for the representation of structured objects (Berretti et al. 2001).

Formally, an attributed relation graph ARG is defined by spatial entities represented as a set of vertices V, and binary spatial relationships represented as a set of edges $E : ARG \equiv \langle V, E \rangle$. Letting G be the set of all connected, non-overlapping regions/segments of an image, then a region $a \in G$ of the image can be represented in the graph by a vertex $v_a \in V$, where $v_a \equiv \langle a, D_a, L_a \rangle$. D_a and L_a are the set of attributes of each graph vertex and in general can be any property that characterizes the respective region. In our case, D_a is the ordered set of MPEG-7 visual descriptors characterizing the region in terms of low-level features, while $L_a = \sum_{i=1}^{|C|} c_i/\mu_a(c_i)$ is the fuzzy set of candidate labels for the region (extracted using a process that will be described in the next section). The adjacency relation between two neighboring regions $a, b \in G$ of the image is represented by a graph edge $e_{ab} = \langle (v_a, v_b), s_{ab} \rangle \in E$, where s_{ab} is a similarity value for the two adjacent regions represented by the pair (v_a, v_b). This value is calculated based on the semantic similarity of the two regions as described by the two fuzzy sets L_a and L_b:

$$s_{ab} = \sup_{c \in C}(t_{\text{norm}}(\mu_a(c), \mu_b(c))), \quad a, b \in G. \tag{10.1}$$

This formula states that the similarity of two regions is the supremum (sup) over all common concepts of the fuzzy intersection (t_{norm}) of the degrees of membership $\mu_a(c)$ and $\mu_b(c)$ for the specific concept of the regions a and b.

Finally, we consider two regions $a, b \in G$ to be connected when at least one pixel of region a is 4-connected to one pixel of region b. In an ARG, a neighborhood N_a of a vertex $v_a \in V$ is the set of vertices whose corresponding regions are connected to a: $N_a = \{v_b : e_{ab} \neq \emptyset\}$, $a, b \in G$. It is clear now that the subset of ARG edges that are incident to region a can be defined as $E_a = \{e_{ab} : b \in N_a\} \subseteq E$.

In the following section we shall focus on the use of the ARG model and provide the guidelines for the fundamental initial region labeling of an image.

10.2.2 Image Graph Initialization

Our intention in this work is to operate on a semantic level where regions are linked to possible labels, rather than just their visual features. As a result, the ARG is used to store both the low-level and the semantic information for each region. Two MPEG-7 visual descriptors, namely dominant color and homogeneous texture (Manjunath et al. 2001), are used to represent each region in the low-level feature space, while fuzzy sets of candidate

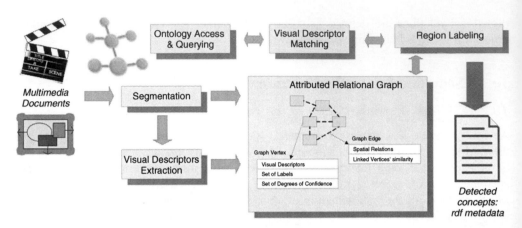

Figure 10.1 Initial region labeling based on attributed relation graph and visual descriptor matching.

concepts are used to model high-level information. To extract these candidate concepts, we use the Knowledge Assisted Analysis (KAA) algorithm developed in Athanasiadis et al. (2005). The general architecture scheme is depicted in Figure 10.1, with the ARG at its center interacting with the rest of the processes.

The ARG is constructed based on an initial recursive shortest spanning tree-like segmentation (Adamek et al. 2005), producing a few tens of regions (approximately 30–40 in our experiments). For every region, dominant color and homogeneous texture are extracted (i.e. for region a: $D_a = [DC_a HT_a]$) and stored in the corresponding graph vertex. The formal definition of dominant color (Manjunath et al. 2001) is $DC \equiv \left[\{c_i, v_i, p_i\}, s\right]$, $i = 1, \ldots, N$, where c_i is the ith dominant color, v_i the color's variance, p_i the color's percentage value, s the spatial coherency and N can be up to 8. The distance function for two descriptors DC_1, DC_2 is

$$d_{DC}(DC_1, DC_2) = \sqrt{\sum_{i=1}^{N_1} p_{1i}^2 + \sum_{j=1}^{N_2} p_{2j}^2 - \sum_{i=1}^{N_1}\sum_{j=1}^{N_2} 2a_{1i,2j} p_{1i} p_{2j}}, \qquad (10.2)$$

where $a_{1i,2j}$ is a similarity coefficient between two colors. Similarly, for homogeneous texture we have $HT \equiv [avg, std, e_1, \ldots, e_{30}, d_1, \ldots, d_{30}]$, where avg is the average intensity of the region, std is the standard deviation of the region's intensity, e_i and d_i are the energy and the deviation for 30 ($i \in [1, \ldots, 30]$) prescribed frequency channels. A distance function is also defined:

$$d_{HT}(HT_1, HT_2) = \sum_{i=1}^{N_{HT}=62} \left| \frac{HT_1(i) - HT_2(i)}{\sigma_i} \right|, \qquad (10.3)$$

where σ_i is a normalization value for each frequency channel. For the sake of simplicity and readability, we will use the following two distance notations equivalently: $d_{DC}(DC_a, DC_b) \equiv d_{DC}(a, b)$ (and similarly for d_{HT}). This is justified as we do not deal with abstract vectors but with image regions a and b represented by their visual descriptors.

Region labeling is based on matching the visual descriptors for each vertex of the ARG with the visual descriptors of all concepts $c \in C$, stored as prototype instances $P(c)$ in the ontological knowledge base. Matching a region $a \in G$ with a prototype instance $p \in P(c)$ for a concept $c \in C$ is done by combining the individual distances of the two descriptors:

$$d(a, p) = d([DC_a HT_a], [DC_p HT_p])$$

$$= w_{DC}(c) \cdot n_{DC}(d_{DC}(a, p)) + w_{HT}(c) \cdot n_{HT}(d_{HT}(a, p)), \qquad (10.4)$$

where d_{DC} and d_{HT} are given in equations (10.2) and (10.3), w_{DC} and w_{HT} are weights depending on each concept c and $w_{DC}(c) + w_{HT}(c) = 1$, for all $c \in C$. n_{DC} and n_{HT} are normalization functions selected to be linear:

$$n(x) = \frac{x - d_{\min}}{d_{\max} - d_{\min}}, \quad n : [d_{\min} \; d_{\max}] \rightarrow [0 \; 1], \qquad (10.5)$$

where $d_{\min}$ and $d_{\max}$ are the minimum and maximum of the two distance functions d_{DC} and d_{HT}.

After exhaustive matching between regions and all prototype instances has been performed, the last step of the algorithm is to populate the fuzzy set L_a for every graph vertex. The degree of membership of each concept c in the fuzzy set L_a is calculated as

$$\mu_a(c) = 1 - \min_{p \in P(c)} d(a, p), \qquad (10.6)$$

where $d(a, p)$ is given in (10.4). This process results in an initial fuzzy labeling of all regions with concepts from the knowledge base, or more formally in a set $L = \{L_a\}$, $a \in G$, whose elements are the fuzzy sets of all regions in the image.

This is obviously a complex task and its efficiency is highly dependent on the domain where it is applied, as well as on the quality of the knowledge base. The primary limitations of this approach are its dependency on the initial segmentation, and the creation of representative prototype instances of the concepts. The latter is easier to manage; we deal with the former presently by suggesting an extension to the algorithm based on region merging and segmentation on a semantic level.

10.2.3 Semantic Region Growing

Overview

The major aim of this work is to improve both image segmentation and labeling of materials and simple objects at the same time, with obvious benefits for problems in the area of image understanding. As mentioned in the introduction, we blend well-established segmentation techniques with mid-level features, such as the fuzzy sets of labels we defined in Section 10.2.1.

In order to emphasize that this approach is independent of the segmentation algorithm, we examine two traditional segmentation techniques, belonging to the general category of region growing algorithms. The first is the watershed segmentation algorithm (Beucher and Meyer 1993), and the second is the recursive shortest spanning tree (RSST) algorithm (Morris et al. 1986). We modify these techniques to operate on the fuzzy sets stored in the ARG in a similar way as if they worked on low-level features (color, texture, etc.). Both variations follow in principle the algorithmic definition of their traditional counterparts, though several adjustments are necessary. We call the overall approach 'semantic region growing'.

Semantic Watershed

The watershed algorithm (Beucher and Meyer 1993) owes its name to the way in which regions are segmented into catchment basins. A catchment basin is the set of points that is the local minimum of a height function (usually the gradient magnitude of the image). After locating these minima, the surrounding regions are incrementally flooded and the places where flood regions touch are the boundaries of the regions. Unfortunately, this strategy leads to over-segmentation of the image; therefore a marker-controlled segmentation approach is usually applied. Markers constrain the flooding process inside their own catchment basins; hence, the final number of regions is equal to the number of markers.

In our semantic approach of watershed segmentation, called the 'semantic watershed', certain regions play the role of markers/seeds. During the construction of the ARG, every region $a \in G$ has been linked to a graph vertex $v_a \in V$ that contains a fuzzy set of labels L_a. A subset of all regions G are selected to be used as seeds for the initialization of the algorithm and form an initial set $S \subseteq G$. The criteria for selecting a region $s \in S$ to be a seed are as follows:

1. The height of its fuzzy set L_a, the largest degree of membership obtained by any element of L_a (Klir and Yuan 1995), should be above a threshold $h(L_a) > T_{seed}$. The threshold T_{seed} is different for every image and its value depends on the distribution of all degrees of membership over all regions of the particular image. The value of T_{seed} discriminates the top p percent of all degrees and this percentage p (calculated only once) is the optimal value derived from a training set of images.
2. The specific region has only one dominant concept, that is, the remaining concepts should have low degrees of membership compared to that of the dominant concept:

$$h(L_a) > \sum_{c \in \{C - c*\}} \mu_a(c), \quad \text{where } c^* : \mu_a(c*) = h(L_a). \tag{10.7}$$

These constraints were designed to ensure that the specific region has been correctly selected as seed for the particular concept c^*.

An iterative process begins checking every initial region seed, $s \in S$, for all its direct neighbors N_s. Let $r \in N_s$ be a neighbor region of s, or in other words, s is the propagator region of r: $s = p(r)$. A propagator region is considered the parent region during the iterative merging process, starting with the region seeds. By comparing the fuzzy sets of those two regions $L_{p(r)}$ and L_r element-wise for every concept they have in common, we can measure their semantic similarity; in so doing we calculate the degree of membership of region r for the concept c, $\mu_r(c)$. If the membership degree is above a merging threshold $\mu_r(c) > K^n \cdot T_{merge}$, then it is assumed that region r is semantically similar to its propagator and was incorrectly segmented, and therefore we merge those two regions. The parameter K is a constant slightly above 1, which increases the threshold at each iteration in a non-linear way with respect to the distance from the initial regions seeds.

Additionally, region r is added to a new set of regions M_s^n (n denotes the iteration step, with $M_s^0 \equiv s$, $M_s^1 \equiv N_s$, etc.), from which the new seeds will be selected for the next iteration. After merging, the algorithm re-evaluates the degrees of membership for each concept of L_r:

$$\mu_{\hat{r}}(c) = \min(\mu_{p(r)}(c), \mu_r(c)), \tag{10.8}$$

where $p(r)$ is the propagator region of r.

The above procedure is repeated until the termination criterion of the algorithm is met, that is, all sets of region seeds in step n are empty: $M_s^n = \emptyset$. At this point, we should underline that when neighbors of a region are examined, previously accessed regions are excluded, that is, each region is reached only once and that is by the closest region seed, as defined in the ARG.

After running this algorithm on an image, some regions will be merged with one of the seeds, while others will remain unaffected. In order to handle these regions, we repeatedly run our algorithm on new ARGs, each consisting of the regions that remained intact after all previous iterations. This hierarchical strategy needs no additional parameters, since new region seeds are be created automatically each time based on a new threshold T_{seed} (apparently with smaller value than before). Obviously, the regions created in the first pass of the algorithm have stronger confidence for their boundaries and their assigned concept than those created in a later pass. This is not a drawback of the algorithm; quite the contrary, we consider this fuzzy outcome an advantage as we retain all the available information.

Semantic RSST

Traditional RSST (Morris et al. 1986) is a bottom-up segmentation algorithm that begins at the pixel level and iteratively merges similar neighbor regions until certain termination criteria are satisfied. RSST uses an internal graph representation of image regions, similar to the ARG described in Section 10.2.1. The algorithm proceeds as follows. First, all edges of the graph are sorted according to a criterion, for example color dissimilarity of the two connected regions determined using the Euclidean distance between the color components. The edge with the least weight is found and the two regions connected by that edge are merged. After each step, the merged region's attributes (e.g. mean color) are recalculated. Traditional RSST will also recalculate weights of related edges as well and re-sort them, so that at every step the edge with the least weight will be selected. This process repeats until some termination criteria are met. The latter criteria may vary, but usually they are either the number of regions or a threshold on the distance.

Following the conventions and notation used so far, we introduce here a modified version of RSST, called 'semantic RSST'. In contrast to the approach described in the previous section, in this case no initial seeds are necessary, but instead we need to define (dis)similarity and termination criteria. The criterion for ordering the edges is based on the similarity measure defined earlier in Section 10.2.1. We define the weight of an edge e_{ab} between two adjacent regions a and b as

$$w(e_{ab}) = 1 - s_{ab}. \tag{10.9}$$

Equation (10.9) can be expanded by substituting s_{ab} from equation (10.1). The edge weight should represent the degree of dissimilarity between the two joined regions; therefore we subtract the estimated value from 1. All graph edges are sorted by this weight. Commutativity and associativity axioms of all fuzzy set operations (thus including default fuzzy union and default fuzzy intersection) ensure that the ordering of the arguments is indifferent.

Let us now examine in detail one iteration of the semantic RSST algorithm. First, the edge with the least weight is selected: $e_{ab}^{*} = \arg\min_{e_{ab} \in E} (w(e_{ab}))$. Then regions a and b are merged to form a new region $\hat{a}$. Region b is removed completely from the ARG and region a is updated. This update procedure consists of the following two actions:

1. Update the fuzzy set L_a by re-evaluating all degrees of membership in a weighted average fashion:

$$\mu_{\hat{a}}(c) = \frac{A(a) \cdot \mu_a(c) + A(b) \cdot \mu_b(c)}{A(a) + A(b)}, \quad \forall c \in C. \qquad (10.10)$$

The quantity $A(a)$ is a measure of the size (area) of region a and is the number of pixels belonging to this region.
2. Readjust the ARG's edges:
   (a) Remove edge e_{ab}.
   (b) Re-evaluate the weight of all affected edges e: the union of those incident to region a and those incident to region b: $e \in E_a \bigcup E_b$.

This procedure continues until the edge e^* with the least weight in the ARG is above a threshold: $w(e^*) > T_w$. This threshold is calculated in a preprocessing stage (as in traditional RSST), based on the cumulative histogram of the weights of all graph edges.

Figure 10.2 illustrates an example derived from the beach domain. In order to make segmentation results comparable, we predefined the final number of regions produced by traditional RSST to be equal to that produced by the semantic watershed. An obvious observation is that RSST segmentation performance in Figure 10.2b is rather poor; persons are merged with sand, whereas the sea on the left and in the middle under the big cliff is divided into several regions and adjacent regions of the same cliff are classified as different ones. The results of the application of the semantic watershed algorithm are shown in Figure 10.2c and are considerably better. More specifically, we observe that both parts of the sea are merged together, and the rocks on the right are represented by only two large regions, despite their original variations in texture and color information. Moreover, the persons lying on the sand are identified as separate regions. Semantic RSST (Figure 10.2d) is shown to perform similarly well.

10.3   Using Contextual Knowledge to Aid Visual Analysis

10.3.1   Contextual Knowledge Formulation

Ontologies (Staab and Studer 2004) present a number of advantages over other knowledge representation strategies. In the context of this work, ontologies are suitable for expressing multimedia content semantics in a formal machine-processable representation that allows manual or automatic analysis and further processing of the extracted semantic descriptions. As an ontology is a formal specification of a shared understanding of a domain, this formal

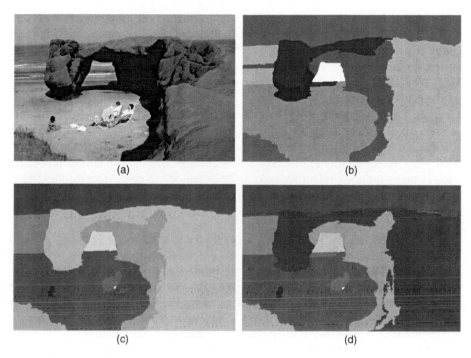

Figure 10.2 Experimental results for an image from the beach domain: (a) input image; (b) RSST segmentation; (c) semantic watershed; (d) semantic RSST.

specification is usually carried out using a subclass hierarchy with relationships among the classes, where one can define complex class descriptions – for example, in DL (Baader et al. 2003) or OWL (Bechhofer et al. 2004). One possible way to describe ontologies can be formalized as

$$O = \{C, \{R_{pq}\}\} \quad \text{where } R_{pq} : C \times C \to \{0, 1\}, \tag{10.11}$$

where O is an ontology, C is the set of concepts described by the ontology, $p, q \in C$ are two concepts, and R_{pq} is the semantic relation amongst these concepts. The above knowledge model encapsulates a set of concepts and the relations between them, forming the basic elements toward semantic interpretation. In general, semantic relations describe specific kinds of links or relationships between any two concepts. In the crisp (non-fuzzy) case, a semantic relation either relates ($R_{pq} = 1$) or does not relate ($R_{pq} = 0$) a pair of concepts p, q with each other. Although almost any type of relation may be included to construct such knowledge representation, the two categories commonly used are *taxonomic* (i.e. ordering) and *compatibility* (i.e. symmetric) relations. However, as extensively discussed in (Akrivas et al. 2004), compatibility relations fail to assist in the determination of the context and the use of ordering relations is necessary for such tasks. Thus, the first challenge is to use information from the taxonomic relations meaningfully to exploit context for semantic image segmentation and object labeling.

For a knowledge model to be highly descriptive, it must contain many distinct and diverse relations among its concepts. A side-effect of using many diverse relations is that available information will then be scattered across them, making any individual relation inadequate for describing context in a meaningful way. Consequently, relations need to be combined to provide a view of the knowledge that suffices for context definition and estimation. In this work we use three types of relations, whose semantics are defined in the MPEG-7 (2001) standard, namely the *specialization* relation Sp, the *part of* relation P and the *property* relation Pr.

One more important point must be considered when designing a knowledge model: real-life data is often considerably different from research data. Real-life information is, in principle, governed by uncertainty and fuzziness. It can therefore be more accurately modeled using fuzzy relations. The commonly encountered crisp relations above can be modeled as fuzzy ordering relations, and can be combined to generate a meaningful fuzzy taxonomic relation. To tackle such complex types of relations we propose a 'fuzzification' of the previous ontology definition:

$$O_F = \{C, \{r_{pq}\}\}, \quad \text{where } r_{pq} = F(R_{pq}) : C \times C \rightarrow [0, 1], \qquad (10.12)$$

in which O_F defines a fuzzy ontology, C is again the set of all possible concepts it describes and r_{pq} denotes a fuzzy relation among the two concepts $p, q \in C$. In the fuzzy case, a fuzzy semantic relation relates a pair of concepts p, q with each other to a given degree of membership, that is, the value of r_{pq} lies in the interval $[0, 1]$. More specifically, given a universe U, a crisp set C is described by a membership function $\mu_C : U \rightarrow \{0, 1\}$ (as already observed in the crisp case for R_{pq}), whereas according to Klir and Yuan (1995), a *fuzzy set* F on C is described by a membership function $\mu_F : C \rightarrow [0, 1]$. We may describe the fuzzy set F using the widely applied sum notation (Miyamoto 1990):

$$F = \sum_{i=1}^{n} c_i/w_i = \{c_1/w_1, c_2/w_2, \ldots, c_n/w_n\},$$

where $n = |C|$ is the cardinality of set C and concept $c_i \in C$. The membership degree w_i describes the membership function $\mu_F(c_i)$, that is, $w_i = \mu_F(c_i)$, or, for the sake of simplicity, $w_i = F(c_i)$. As in Klir and Yuan (1995), a *fuzzy relation* on C is a function $r_{pq} : C \times C \rightarrow [0, 1]$ and its *inverse* relation is defined as $r_{pq}^{-1} = r_{qp}$. Based on the relations r_{pq} and for the purposes of image analysis here, we construct a relation T using the transitive closure of the fuzzy taxonomic relations Sp, P and Pr:

$$T = Tr^t(Sp \cup P^{-1} \cup Pr^{-1}). \qquad (10.13)$$

In these relations, fuzziness has the following meaning. High values of $Sp(p, q)$ imply that the meaning of q approaches the meaning of p, in the sense that when an image is semantically related to q, then it is likely related to p as well. On the other hand, as $Sp(p, q)$ decreases, the meaning of q becomes 'narrower' than the meaning of p, in the sense that an image's relation to q will not imply a relation to p as well with a high probability, or to a high degree. Likewise, the degrees of the other two relations can also be interpreted as conditional probabilities or degrees of implied relevance. MPEG-7 (2001) contains several types of semantic relations meaningful to multimedia analysis, defined together with their inverses. Sometimes, the semantic interpretation of a relation is not meaningful whereas the inverse is. In our case, the relation *part* $P(p, q)$ is defined as:

p *part* q if and only if q is *part of* p. For example, let p be New York and q Manhattan. It is obvious that the inverse relation *part of* P^{-1} is semantically meaningful, since Manhattan is *part of* New York. There is, similarly, a meaningful inverse for the *property* relation Pr. On the other hand, following the definition of the *specialization* relation $Sp(p, q)$, p is a *specialization* of q if and only if q is a specialization in meaning of p. For example, let p be mammal and q dog; $Sp(p, q)$ means that dog is a *specialization* of a mammal, which is exactly the semantic interpretation we wish to use (and not its inverse). Based on these roles and semantic interpretations of Sp, P and Pr, it is easy to see that (10.13) combines them in a straightforward and meaningful way, utilizing inverse functionality where it is semantically appropriate, that is, where the meaning of one relation is semantically contradictory to the meaning of the rest on the same set of concepts. The transitive closure Tr^t is required in order for T to be taxonomic, since the union of transitive relations is not necessarily transitive, as discussed in Akrivas et al. (2002).

Representation of our concept-centric contextual knowledge model follows the Resource Description Framework (RDF) standard (Bechhofer et al. 2004). RDF is the framework in which Semantic Web metadata statements are expressed and usually represented as graphs. The RDF model is based upon the idea of making statements about resources in the form of a subject–predicate–object expression. Predicates are traits or aspects of a resource that express a relationship between the subject and the object. The relation T can be visualized as a graph in which every node represents a concept and each edge constitutes a contextual relation between these concepts. Additionally each edge has an associated membership degree, which represents the fuzziness within the context model. Representing the graph in RDF is straightforward, sincethe RDF structure itself is based on a similar graph model.

To represent fuzzy relations, we use reification (Brickley and Guha RV (eds) 2004) – a method for making statements about other statements in RDF. In our model, the reified statements capture the degree of membership for the relations. This method of representing fuzziness is a novel but acceptable one, since the reified statement should not be asserted automatically. For instance, having a statement such as '*Sky PartOf BeachScene*' and a membership degree of 0.75 for this statement does not imply that the sky is always a part of a beach scene.

A small clarifying example is provided in Figure 10.3 for an instance of the specialization relation Sp. As discussed, $Sp(x, y) > 0$ implies that the meaning of x 'includes' the meaning of y; the most common forms of specialization are sub-classing (i.e. x is a generalization of y) and thematic categorization (i.e. x is the thematic category of y). In the example, the RDF subject *wrc* (World Rally Championship) has *specializationOf* as an RDF predicate and *rally* forms the RDF object. The reification process introduces a statement about the *specializationOf* predicate, stating that the membership degree for the relation is 0.90.

10.3.2 Contextual Relevance

The idea behind the use of visual context information responds to the fact that not all human acts are relevant in all situations, and this also holds when dealing with image analysis problems. Since visual context is a difficult notion to grasp and capture

```
<rdf:Description rdf:about="#s1">
    <rdf:subject rdf:resource="&dom;wrc"/>
    <rdf:predicate rdf:resource="&dom;specializationOf"/>
    <rdf:object>rdf:resource="&dom;rally"</rdf:object>
    <rdf:type rdf:resource="http://www.w3.org/1999/02/
    22-rdf-syntax-ns#Statement"/>
    <context:specializationOfrdf:datatype="http://www.w3.org/
    2001/XMLSchema#float">0.90</context:specializationOf>
</rdf:Description>
```

Figure 10.3   Fuzzy relation representation: RDF reification.

(Mylonas and Avrithis 2005), we restrict it here to the notion of ontological context, defined in terms of the fuzzy ontologies presented in Section 10.3.1. Here the problems to be addressed include how to meaningfully readjust the membership degrees of segmented (and possibly merged) image regions, and how to use visual context to improve the performance of knowledge-assisted image analysis. Based on the mathematical background introduced in the previous subsections, we develop an algorithm used to readjust the degree of membership $\mu_a(c)$ of each concept c in the fuzzy set L_a associated with a region $a \in G$ in a scene. Each concept $k \in C$ in the application domain's ontology is stored together with its degrees of relationship r_{kl} to every other related concept $l \in C$. To tackle cases in which one concept is related to multiple concepts, we use the term *context relevance*, denoted $cr_{dm}(k)$, which refers to the overall relevance of concept k to the *root element* characterizing each domain dm. For instance, the *root element* of *beach* and *motorsports* domains are concepts *beach* and *motorsports*. All possible routes in the graph are taken into consideration, forming an exhaustive approach to the domain, with respect to the fact that all routes between concepts are reciprocal.

An estimation of each concept's value is derived from the direct and indirect relationships between the concept and other concepts, using a *compatibility indicator* or distance metric. The ideal distance metric for two concepts is one that quantifies their semantic correlation. Depending on the nature of the domains under consideration, the best indicator could be either the max or the min operator. For the problem at hand, the *beach* and *motorsports* domains, the max value is a meaningful measure of correlation for both. Figure 10.4 presents a simple example. The concepts are: *motorsports* (the *root element*, denoted by m), *asphalt* (a), *grass* (g), and *car* (c). Their relationships can be

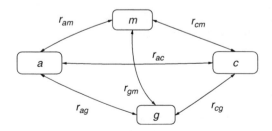

Figure 10.4   Graph representation example: compatibility indicator estimation.

summarized as follows: let concept a be related to concepts m, g and c directly with r_{am}, r_{ag} and r_{ac}, while concept g is related to concept m with r_{gm} and concept c is related to concept m with r_{cm}. Additionally, c is related to g with r_{cg}. The context relevance for concept a is given by

$$cr_{dm}(a) = \max\{r_{am}, r_{ag}r_{gm}, r_{ac}r_{cm}, r_{ag}\cdot r_{cg}r_{cm}, r_{ac}r_{cg}r_{gm}\} \qquad (10.14)$$

The general structure of the degree of membership re-evaluation algorithm is as follows:

1. Identify an optimal normalization parameter np to use within the algorithm's steps, according to the domain(s) considered. np is also referred to as a domain similarity or dissimilarity measure and $np \in [0, 1]$.
2. For each concept k in the fuzzy set $I_{\cdot a}$ associated with a region $a \in G$ in a scene with a degree of membership $\mu_a(k)$, obtain the contextual information in the form of its relations to all other concepts: $\{r_{kl} : l \in C, l \neq k\}$.
3. Calculate the new degree of membership $\mu_a(k)$ associated with region a, based on np and the context's relevance value. In the case of multiple concept relations in the ontology, relating concept k to more than one concept, rather than relating k solely to the 'root element' r^e, an intermediate aggregation step should be applied for k: $cr_k = \max\{r_{kr^e}, \ldots, r_{km}\}$. We express the calculation of $\mu_a(k)$ with the recursive formula

$$\mu_a^n(k) = \mu_a^{n-1}(k) - np(\mu_a^{n-1}(k) - cr_k), \qquad (10.15)$$

where n denotes the iteration number. Equivalently, for an arbitrary iteration n,

$$\mu_a^n(k) = (1 - np)^n \cdot \mu_a^0(k) + (1 - (1 - np)^n), \cdot cr_k \qquad (10.16)$$

where $\mu_a^0(k)$ represents the original degree of membership.

In practice, typical values for n lie between 3 and 5. Interpretation of both equations (10.15) and (10.16) implies that this contextual approach will favor confident degrees of membership for a region's concept over non-confident or misleading degrees of membership. It amplifies their differences, while diminishing confidence in clearly misleading concepts for each region. Further, based on the supplied ontological knowledge, it will clarify and solve ambiguities in cases of similar concepts or difficult-to-analyze regions.

The key step remaining is the definition of a meaningful normalization parameter np. When re-evaluating the confidence values, the ideal np is always defined with respect to the particular domain of knowledge and is the value that quantifies their semantic correlation to the domain. In our work we conducted a series of experiments on a training set of 120 images for both the beach and the motorsports application domains and selected the np value that resulted in the best overall evaluation score values for each domain. The algorithm presented above readjusts the initial degrees of membership, using semantics in the form of the contextual information residing in the constructed fuzzy ontology.

Figure 10.5 presents indicative results for a beach image. Contextualization, which works on a per region basis, is applied after semantic region growing. In this example we

Figure 10.5 Contextual experimental results for a beach image.

have selected the unified sea region in the upper left part of the image (illustrated by an artificial blue color). The contextualized results are presented in red in the right column at the bottom of the tool. Context favors strongly the fact that the merged region belongs to sea, increasing its degree of membership from 86.15% to 92.00%. The (irrelevant for this region) membership degree for person is extinguished, whereas degrees of membership for the rest of the possible beach concepts are slightly increased, due to the ontological knowledge relations that exist in the knowledge model.

Finally, in the process of evaluating this work and testing its tolerance to imprecision in initial labels, we provide an evaluation overview of the application of the presented methodologies on the dataset of the beach domain; for a more detailed description of the evaluation methodology and the dataset, see (Athanasiadis et al. 2007). Table 10.1 shows detection results for each concept, as well as the overall score derived from the six beach domain concepts. Apparently, the concept sky has the best score among the rest, since its color and texture are relatively invariable. Visual context indeed aids the labeling process, even with the overall marginal improvement of approximately 2%, given in Table 10.1, a fact mainly justified by the diversity and the quality of the image dataset provided. Apart from that, the efficiency of visual context depends also on the particularity of each specific concept; for instance, in Table 10.1 we observe that in the case of the semantic watershed algorithm and for the concepts sea and person the improvement measured over the complete dataset is 5% and 7.2%, respectively. Similarly, in the case of the semantic RSST and for concepts sea, sand and person we see an overall increase significantly above the 2% average, namely 7.2%, 7.3% and 14.6%, respectively. It should be noted that adding visual context to the segmentation algorithms is not an expensive process in terms of computational complexity or timing.

Table 10.1 Comparison of segmentation variants and their combination with visual context, with evaluation scores per concept (SW, semantic watershed; SRSST, semantic RSST)

Concept	Watershed	SW	SW + Context	RSST	SRSST	SRSST + Context
SKY	0.93	0.95	0.94	0.90	0.93	0.93
SEA	0.79	0.80	0.84	0.81	0.83	0.89
SAND	0.72	0.79	0.81	0.72	0.82	0.88
PERSON	0.44	0.56	0.60	0.48	0.48	0.55
ROCK	0.61	0.75	0.77	0.66	0.76	0.78
VEGETATION	0.67	0.70	0.71	0.63	0.68	0.69
Total	**0.69**	**0.77**	**0.79**	**0.70**	**0.77**	**0.78**

10.4 Spatial Context and Optimization

10.4.1 Introduction

In this section, a semantic approach to image analysis using spatial context is presented. To analyze an image, it is first spatially segmented, and support vector machines (SVMs) are used to perform an initial association of each image region with a set of predefined high-level semantic concepts, based solely on visual information. Then, a genetic algorithm (GA) is used to estimate a globally optimal region-concept assignment, taking into account the spatial context, represented by fuzzy directional relations.

10.4.2 Low-Level Visual Information Processing

Segmentation and Feature Extraction

To perform the initial region-concept association procedure, the image has to be segmented into regions, and suitable low-level descriptions have to be extracted for each segment. In the current implementation, an extension of the RSST algorithm is used to segment the image. The output of this segmentation algorithm is a segmentation mask, where each spatial region s_n, $n = 1, \ldots, N$, is likely to represent a meaningful semantic object. For every image segment the following MPEG-7 descriptors, discussed in Chapter 4, are extracted and form a *region feature vector*: scalable color, homogeneous texture, region shape, and edge histogram.

Fuzzy Spatial Relations Extraction

In the present analysis framework, eight fuzzy directional relations are supported, namely North (N), East (E), South (S), West (W), South-East (SE), South-West (SW), North-East (NE) and North-West (NW). Their extraction builds on the principles of projection- and angle-based methodologies. It consists of the following steps. First, a *reduced box* is computed from the minimum bounding rectangle (MBR) of the *ground* region (the reference region, colored dark gray in Figure 10.6), so as to include the region in a more representative way. The computation of this *reduced box* is performed

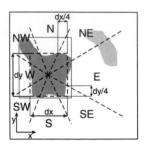

Figure 10.6   Fuzzy directional relations definition.

in terms of the MBR compactness value v, which is defined as the fraction of the region's area to the area of the respective MBR: if the initially computed v is below a threshold T, the ground region's MBR is repeatedly reduced until the threshold is met. Then, eight cone-shaped regions are formed on top of this reduced box, as illustrated in Figure 10.6, each corresponding to one of the defined directional relations. The percentage of the *figure* region (the region whose relative position is to be estimated and is colored light gray in Figure 10.6) pixels that are included in each of the cone-shaped regions determines the degree to which the corresponding directional relation is satisfied. The value of the threshold T was empirically chosen to be 0.85.

10.4.3   Initial Region-Concept Association

SVMs, which were discussed in Chapter 5, have been widely used in semantic image analysis tasks due to their reported generalization ability (Papadopoulos et al. 2007). In this work, SVMs are used to perform an initial association between the image regions and one of the predefined high-level semantic concepts using the estimated region feature vector. An individual SVM is introduced for each concept c_l, $l = 1, \ldots, L$, to detect the corresponding instances, and is trained using the *one-against-all* approach. Each SVM returns, for every segment, a numerical value in the range $[0, 1]$ denoting the degree of confidence, h_{nl}^C, to which the corresponding region is assigned to the concept associated with the particular SVM. The degree of confidence is given by

$$h_{nl}^C = \frac{1}{1 + e^{-p \cdot z_{nl}}}, \qquad (10.17)$$

where z_{nl} is the distance of the input feature vector from the corresponding SVM's separating hyperplane, and p is a slope parameter, chosen empirically. For every region, $\arg\max(h_{nl}^C)$ indicates its concept assignment, whereas $H_n^C = \{h_{nl}^C, l = 1, \ldots, L\}$ constitutes its concept hypothesis set. Note that any other classification algorithm can be adopted during this step, provided that a similar hypothesis set is estimated for every image region. Additional techniques for refining these hypothesis sets can also be applied prior to the use of spatial context, including the semantic segmentation and visual context exploitation algorithms discussed in previous sections of this chapter.

10.4.4   Final Region-Concept Association

Spatial Constraints Estimation

In this section, the procedure for estimating the values of the spatial relations (spatial-related contextual information) between all the defined high-level semantic concepts is described. These values are calculated according to the following learning approach. Let

$$R = \{r_k, k = 1, \ldots, K\} = \{N, NW, NE, S, SW, SE, W, E\}, \tag{10.18}$$

denote the set of the supported spatial relations. Then the degree to which region s_i satisfies relation r_k with respect to region s_j can be denoted by $I_{r_k}(s_i, s_j)$ and is estimated according to the procedure of Section 10.4.2. In order to acquire the contextual information, this function needs to be evaluated over a set of segmented images with ground truth annotations, which serves as a training set. For that purpose, an appropriate image set, $\mathcal{B}_{tr}$, is assembled. Then, using this training set, the mean values, $I_{r_k\text{mean}}$, of I_{r_k} for every k over all pairs of regions assigned to concepts (c_i, c_j), $i \neq j$, are estimated. These constitute the constraints of an optimization problem which will be solved using a genetic algorithm.

Spatial Constraints Verification Factor

The learnt fuzzy spatial relations, obtained as described above, serve as constraints denoting the 'allowed' spatial topology of the supported concepts. Here, the exploitation of these constraints is detailed. Let $I_S(g_{ij}, g_{pq})$ be defined as a function that receives values in the interval $[0, 1]$ and returns the degree to which the spatial constraint between the g_{ij}, g_{pq} concept to region mappings is satisfied. To calculate this value the following procedure is followed. Initially, a normalized Euclidean distance $d(g_{ij}, g_{pq})$ is calculated as

$$d(g_{ij}, g_{pq}) = \frac{\sqrt{\sum_{k=1}^{8}(I_{r_k\text{mean}}(c_j, c_q) - I_{r_k}(s_i, s_p))^2}}{\sqrt{8}}, \tag{10.19}$$

which receives values in the interval $[0,1]$. The function $I_S(g_{ij}, g_{pq})$ is then defined as

$$I_S(g_{ij}, g_{pq}) = 1 - d(g_{ij}, g_{pq}). \tag{10.20}$$

Implementation of Genetic Algorithm

The genetic algorithm employed realizes semantic image analysis as a global optimization problem, while taking into account both visual and spatial-related information. Under this approach, each chromosome represents a possible solution. Consequently, the number of the genes comprising each chromosome equals the number N of the regions s_i produced by the segmentation algorithm and each gene assigns a supported concept to an image segment.

The genetic algorithm is initialized using 200 randomly generated chromosomes. A *fitness function* is used to provide a quantitative measure of each solution's fitness, that is, to determine the degree to which each interpretation is plausible:

$$f(Q) = \lambda \cdot FS_{\text{norm}} + (1 - \lambda) \cdot SC_{\text{norm}}, \tag{10.21}$$

where Q denotes a particular chromosome, FS_{norm} refers to the degree of low-level descriptor matching, SC_{norm} stands for the degree of consistency with respect to the provided spatial knowledge, and variable λ, $\lambda \in [0, 1]$, modulates the impact of FS_{norm} and SC_{norm} on the final outcome. The value of λ is estimated according to an optimization procedure, described in the next subsection.

The values of SC_{norm} and FS_{norm} are computed as follows:

$$FS_{\text{norm}} = \frac{\sum_{i=1}^{N} I_M(g_{ij}) - I_{\min}}{I_{\max} - I_{\min}}, \quad SC_{\text{norm}} = \frac{\sum_{l=1}^{W} I_{S_l}(g_{ij}, g_{pq})}{W}, \tag{10.22}$$

where $I_M(g_{ij}) = h_{ij}^C$, $I_{\min} = \sum_{i=1}^{N} \min_j I_M(g_{ij})$, $I_{\max} = \sum_{i=1}^{N} \max_j I_M(g_{ij})$, and W denotes the number of the constraints examined.

After the genetic algorithm is initialized, new generations are iteratively produced until an optimal solution is reached. Each new generation is computed from the previous one through the application of the following operators:

- Selection: the Tournament Selection Operator (Goldberg and Deb 1991) with replacement is used to select a pair of chromosomes from the previous generation to serve as parents for the generation of a new offspring.
- Crossover: uniform crossover with a probability of 0.7 is used.
- Mutation: every gene of the offspring chromosome processed is likely to be mutated with a probability of 0.008.

To ensure that chromosomes with high fitness will contribute to the next generation, the overlapping populations approach (Mitchel 1996) is used. The above iterative procedure continues until the diversity of the current generation is less than or equal to 0.001, or the number of generations exceeds 50.

An indicative example of region-concept associations obtained using the spatial context exploitation algorithm is presented in Figure 10.7. Depending on a number of factors (including the domain of experimentation, the dataset, the efficiency of the segmentation algorithm, the number of concepts, the accuracy of the initial hypothesis sets, etc.), the use of spatial context has been shown to result in an increase in the region correct classification rate of up to 15% (Papadopoulos et al. 2009).

Parameter Optimization

Since the selection of the value of parameter λ (equation (10.21)) is of crucial importance in the behavior of the overall approach, a GA is also used to estimate an optimal value. The task is to select the value of λ that gives the highest correct concept association rate. For that purpose, *concept accuracy* (*CoA*) is used as a quantitative performance measure, and is defined as the fraction of the number of the correctly assigned concepts to the total number of image regions to be examined.

Input image

Initial region-concept
association

Region-concept association
after application of the GA

Figure 10.7 Indicative region-concept association results.

λ is optimized as follows: each GA's chromosome Q represents a possible solution, that is, a candidate λ value. Under this approach, the number of genes of each chromosome is set to 5. The genes represent the binary coded value of λ assigned to the respective chromosome, according to the equation

$$Q = [q_1 \ q_2 \ldots q_5], \quad \text{where} \ \sum_{i=1}^{5} q_i \cdot 2^{-i} = \lambda, \tag{10.23}$$

where $q_i \in \{0, 1\}$ represents the value of gene i. With respect to the corresponding GA's fitness function, this is defined as equal to the CoA metric already defined, where CoA is calculated over all images that are included in a validation set $\mathcal{B}_{val}$ (similar to the set $\mathcal{B}_{tr}$ described earlier), after applying the GA for spatial context optimization with $\lambda = \sum_{i=1}^{5} q_i \cdot 2^{-i}$.

Regarding the GA's implementation details, an initial population of 100 randomly generated chromosomes is employed. New generations are successively produced based on the same evolution mechanism, used by the GA for spatial context optimization. The differences are that the maximum number of generations is set to 30 and the probabilities of mutation and crossover are empirically set to 0.4 and 0.2, respectively. Note that the large divergence in the value of the probability of the mutation operator, compared to the respective value used by the GA for spatial context optimization, denotes its increased importance in this particular optimization problem.

10.5 Conclusions

In this chapter three multimedia analysis algorithms that incorporate domain knowledge have been presented. The first focused on semantic segmentation: it used similarity measures and merging criteria defined at a semantic level to refine an initial segmentation. The next two algorithms concentrated on the use of contextual information to refine an initial region classification result: the first of these two examined the use of taxonomic relations such as specialization, while the second considered spatial relations between region-concept pairs. While none of the algorithms presented in this chapter fully bridges the 'semantic gap' by itself, all three were experimentally shown to contribute toward improved analysis results by efficiently exploiting domain knowledge. These as well as other knowledge-driven algorithms that improve distinct steps of the multimedia analysis chain can be valuable components of a complete system attempting to address all aspects of multimedia analysis.

11

Reasoning for Multimedia Analysis

Nikolaos Simou,[1] Giorgos Stoilos,[1] Carsten Saathoff,[2] Jan Nemrava,[3] Vojtěch Svátek,[3] Petr Berka[3] and Vassilis Tzouvaras[1]
[1]*National Technical University of Athens, Greece*
[2]*ISWeb – Information Systems and Semantic Web, University of Koblenz-Landau, Koblenz, Germany*
[3]*University of Economics, Prague, Czech Republic*

In this chapter the focus is on demonstrating how different reasoning algorithms can be applied in multimedia analysis. Extracting semantics from images and videos has proved to be a very difficult task. On the other hand, artificial intelligence has made significant progress, especially in the area of knowledge technologies. Knowledge representation and reasoning form a research area that has been chosen by many researchers to enable the interpretation of the content of an image scene or a video shot. The rich theoretical background, the formality and the soundness of reasoning algorithms can provide a very powerful framework for multimedia analysis. Section 11.1 introduces the fuzzy extension of the expressive DL language $\mathcal{SHIN}$, f-$\mathcal{SHIN}$, together with the fuzzy reasoning engine, FiRE, that supports it. By using f-shin and FiRE the fairly rich – though imprecise – information extracted from multimedia analysis algorithms on a multimedia document, that is not usable by a traditional DL system, can be used to extract addition implicit knowledge. Section 11.2 presents a model that uses explicitly represented knowledge about the typical spatial arrangements of objects, so-called spatial constraint templates, which are acquired from labeled examples. *Fuzzy constraint reasoning* is used to represent the problem and to find a solution that provides an optimal labeling with respect to both low-level features and the spatial. Section 11.3 presents the NEST expert system, which is used for estimation of image regions dissimilarity. Finally, section 11.4 presents an approach for reasoning over resources complementary to audiovisual streams using FiRE.

Multimedia Semantics: Metadata, Analysis and Interaction, First Edition.
Edited by Raphaël Troncy, Benoit Huet and Simon Schenk.
© 2011 John Wiley & Sons, Ltd. Published 2011 by John Wiley & Sons, Ltd.

11.1 Fuzzy DL Reasoning

Realizing the content of multimedia documents is an extremely difficult task because of
the multimedia content's wide range and potential complexity. For that purpose the effective management and exploitation of multimedia documents requires highly expressive
representation formalisms, capable of supporting efficient reasoning. Description logics
(DLs) fulfill these requirements since they are based on a strong theoretical background,
providing sound and complete decision procedures. However, despite the fact that multimedia analysis algorithms can produce fairly rich – though imprecise – information on a
multimedia document, traditional DL systems cannot reason using such information. This
section introduces the fuzzy extension of the expressive DL language $\mathcal{SHIN}$, f-$\mathcal{SHIN}$,
together with the fuzzy reasoning engine, FiRE, that supports it.

11.1.1 The Fuzzy DL f-$\mathcal{SHIN}$

The DL f-$\mathcal{SHIN}$ is a fuzzy extension of the DL $\mathcal{SHIN}$ (Horrocks et al. 2000) which
similarly consists of an alphabet of distinct concepts names (**C**), role names (**R**) and individual names (**I**). Using DLs, the construction of new concepts and roles is possible. For
that purpose DLs include a set of constructors to construct concept and role descriptions.
These constructors specify the name of the DL language (Baader et al. 2003) and, in the
case of f-$\mathcal{SHIN}$, these are $\mathcal{ALC}$ constructors (i.e. negation ¬, conjunction ⊓, disjunction
⊔, full existential quantification ∃ and value restriction ∀) extended by transitive roles
($\mathcal{S}$), roles hierarchy ($\mathcal{H}$), inverse roles ($\mathcal{I}$), and number restrictions ($\mathcal{N} \leq, \geq$). Hence, if
R is a role then R^- is also a role, namely the inverse of R, while f-$\mathcal{SHIN}$ concepts are
inductively defined as follows:

1. If $C \in$ **C**, then C is an f-$\mathcal{SHIN}$ concept.
2. If C and D are concepts, R is a role, S is a simple role and $n \in \mathbb{N}$, then $(\neg C)$, $(C \sqcup D)$,
   $(C \sqcap D)$, $(\forall R.C)$, $(\exists R.C)$, $(\geq nS)$ and $(\leq nS)$ are also f-$\mathcal{SHIN}$ concepts.

Unlike crisp DLs, the semantics of fuzzy DLs are given by a fuzzy interpretation
(Straccia 2001). A fuzzy interpretation is a pair $\mathcal{I} = \langle \Delta^{\mathcal{I}}, \cdot^{\mathcal{I}} \rangle$, where $\Delta^{\mathcal{I}}$ is a non-empty
set of objects and $\cdot^{\mathcal{I}}$ is a fuzzy interpretation function which maps an individual name a
to elements of $a^{\mathcal{I}} \in \Delta^{\mathcal{I}}$ and a concept name A (role name R) to a membership function
$A^{\mathcal{I}} : \Delta^{\mathcal{I}} \to [0, 1]$ ($R^{\mathcal{I}} : \Delta^{\mathcal{I}} \times \Delta^{\mathcal{I}} \to [0, 1]$).

By using fuzzy set-theoretic operations the fuzzy interpretation function can be extended
to give semantics to complex concepts, roles and axioms (Klir and Yuan 1995). In this
case the standard fuzzy operators of $1 - x$ for fuzzy complement (c), and max t-conorm
(u), min t-norm (t) are used for fuzzy union and intersection respectively. The complete
set of semantics is presented in Table 11.1.

An f-$\mathcal{SHIN}$ knowledge base Σ is a triple $\langle \mathcal{T}, \mathcal{R}, \mathcal{A} \rangle$, where $\mathcal{T}$ is a fuzzy TBox, $\mathcal{R}$ is
a fuzzy RBox and $\mathcal{A}$ is a fuzzy ABox. A TBox is a finite set of fuzzy concept axioms
which are of the form $C \sqsubseteq D$, called fuzzy concept inclusion axioms, or $C \equiv D$, called
fuzzy concept equivalence axioms, saying that C is equivalent or C is a sub-concept of D,
respectively. Similarly, an RBox is a finite set of fuzzy role axioms of the form Trans(R),
called fuzzy transitive role axioms, and $R \sqsubseteq S$, called fuzzy role inclusion axioms, saying

Table 11.1 Semantics of concepts and roles

Constructor	Syntax	Semantics
top	$\top$	$\top^{\mathcal{I}}(a) = 1$
bottom	$\bot$	$\bot^{\mathcal{I}}(a) = 0$
general negation	$\neg C$	$(\neg C)^{\mathcal{I}}(a) = c(C^{\mathcal{I}}(a))$
conjunction	$C \sqcap D$	$(C \sqcap D)^{\mathcal{I}}(a) = t(C^{\mathcal{I}}(a), D^{\mathcal{I}}(a))$
disjunction	$C \sqcup D$	$(C \sqcup D)^{\mathcal{I}}(a) = u(C^{\mathcal{I}}(a), D^{\mathcal{I}}(a))$
exists restriction	$\exists R.C$	$(\exists R.C)^{\mathcal{I}}(a) = \sup_{b \in \Delta^{\mathcal{I}}}\{t(R^{\mathcal{I}}(a, b), C^{\mathcal{I}}(b))\}$
value restriction	$\forall R.C$	$(\forall R.C)^{\mathcal{I}}(a) = \inf_{b \in \Delta^{\mathcal{I}}}\{\mathcal{J}(R^{\mathcal{I}}(a, b), C^{\mathcal{I}}(b))\}$
at-most	$\leq pR$	$\inf_{b_1,\ldots,b_{p+1} \in \Delta^{\mathcal{I}}} \mathcal{J}(t_{i=1}^{p+1} R^{\mathcal{I}}(a, b_i), u_{i<j}\{b_i = b_j\})$
at-least	$\geq pR$	$\sup_{b_1,\ldots,b_p \in \Delta^{\mathcal{I}}} t(t_{i=1}^p R^{\mathcal{I}}(a, b_i), t_{i<j}\{b_i \neq b_j\})$
inverse role	R^-	$(R^-)^{\mathcal{I}}(b, a) = R^{\mathcal{I}}(a, b)$
equivalence	$C \equiv D$	$\forall a \in \Delta^{\mathcal{I}}.C^{\mathcal{I}}(a) = D^{\mathcal{I}}(a)$
sub-concept	$C \sqsubseteq D$	$\forall a \in \Delta^{\mathcal{I}}.C^{\mathcal{I}}(a) \leq D^{\mathcal{I}}(a)$
transitive role	$\text{Trans}(R)$	$\forall a, b \in \Delta^{\mathcal{I}}.R^{\mathcal{I}}(a, b) \geq \sup_{c \in \Delta^{\mathcal{I}}}\{t(R^{\mathcal{I}}(a, c), R^{\mathcal{I}}(c, b))\}$
sub-role	$R \sqsubseteq S$	$\forall a, b \in \Delta^{\mathcal{I}}.R^{\mathcal{I}}(a, b) \leq S^{\mathcal{I}}(a, b)$
concept assertions	$\langle a : C \bowtie n \rangle$	$C^{\mathcal{I}}(a^{\mathcal{I}}) \bowtie n$
role assertions	$\langle \langle a, b \rangle : R \bowtie n \rangle$	$R^{\mathcal{I}}(a^{\mathcal{I}}, b^{\mathcal{I}}) \bowtie n$

that R is transitive and R is a sub-role of S, respectively. Finally, an ABox is a finite set of fuzzy assertions of the form $\langle a : C \bowtie n \rangle$, $\langle \langle a, b \rangle : R \bowtie n \rangle$, where $\bowtie$ stands for $\geq$, $>$, $\leq$, $<$ and $a \neq b$, for $a, b \in \mathbf{I}$. Fuzzy representation enriches expressiveness, so a fuzzy assertion of the form $\langle a : C \geq n \rangle$ means that a participates in the concept C with a membership degree that is at least equal to n.

11.1.2 The Tableaux Algorithm

Consistency is checked with tableaux algorithms that try to construct a fuzzy tableau for a fuzzy ABox $\mathcal{A}$ (Stoilos et al. 2005), which is an abstraction of a model of $\mathcal{A}$ (Horrocks et al. 2000). The tableau has a forest-like structure with nodes representing the individuals that appear in $\mathcal{A}$, and edges between nodes, which represent the relations that hold between two individuals. Each node is labeled with a set of triples of the form $\langle D, \bowtie, n \rangle$, which denote the concept, the type of inequality and the membership degree with which the individual of the node has been asserted to belong to D. We call such triples 'membership triples'. For triples of a single node, the concepts of conjugated, positive and negative triples can be defined in the obvious way that is, $C > 0.6$ is conjugated with $C < 0.4$. Furthermore, the symbols $\bowtie^-$, $\rhd^-$ and $\lhd^-$ are used to denote their reflections for example, the reflection of $\leq$ is $\geq$ and that of $<$ is $>$. Since the expansion rules decompose the initial concept, the concepts that appear in triples are sub-concepts of the initial concept. Sub-concepts of a concept D are denoted by $sub(D)$. The set of all sub-concepts that appear within an ABox is denoted by $sub(\mathcal{A})$. An f-$\mathcal{SHIN}$ tableau has been defined as follows:

Definition 11.1.1 *Let $\mathcal{A}$ be an f_{KD}-$\mathcal{SHIN}$ ABox, R_A the set of roles occurring in $\mathcal{A}$ together with their inverses, $\mathbf{I}_A$ the set of individuals in $\mathcal{A}$, X the set $\{\geq, >, \leq, <\}$ and $\mathcal{R}$ a fuzzy RBox. A fuzzy tableau T for $\mathcal{A}$ with respect to $\mathcal{R}$ is a quadruple $(S, \mathcal{L}, \mathcal{E}, \mathcal{V})$ such that:*

- *S is a non-empty set of individuals (nodes);*
- *$\mathcal{L} : S \rightarrow 2^{sub(\mathcal{A})} \times X \times [0, 1]$ maps each element of S to membership triples;*
- *$\mathcal{E} : R_A \rightarrow 2^{S \times S} \times X \times [0, 1]$ maps each role to membership triples;*
- *$\mathcal{V} : \mathbf{I}_A \rightarrow S$ maps individuals occurring in $\mathcal{A}$ to elements in S.*

For all $s, t \in S$, $C, E \in sub(\mathcal{A})$, and $R \in R_A$, T satisfies the following:

1. *If $\langle \neg C, \bowtie, n \rangle \in \mathcal{L}(s)$, then $\langle C, \bowtie^-, 1 - n \rangle \in \mathcal{L}(s)$.*
2. *If $\langle C \sqcap E, \triangleright, n \rangle \in \mathcal{L}(s)$, then $\langle C, \triangleright, n \rangle \in \mathcal{L}(s)$ and $\langle E, \triangleright, n \rangle \in \mathcal{L}(s)$.*
3. *If $\langle C \sqcup E, \triangleleft, n \rangle \in \mathcal{L}(s)$, then $\langle C, \triangleleft, n \rangle \in \mathcal{L}(s)$ and $\langle E, \triangleleft, n \rangle \in \mathcal{L}(s)$.*
4. *If $\langle C \sqcup E, \triangleright, n \rangle \in \mathcal{L}(s)$, then $\langle C, \triangleright, n \rangle \in \mathcal{L}(s)$ or $\langle E, \triangleright, n \rangle \in \mathcal{L}(s)$.*
5. *If $\langle C \sqcap E, \triangleleft, n \rangle \in \mathcal{L}(s)$, then $\langle C, \triangleleft, n \rangle \in \mathcal{L}(s)$ or $\langle E, \triangleleft, n \rangle \in \mathcal{L}(s)$.*
6. *If $\langle \forall R.C, \triangleright, n \rangle \in \mathcal{L}(s)$ and $\langle \langle s, t \rangle, \triangleright', n_1 \rangle \in \mathcal{E}(R)$ is conjugated with $\langle \langle s, t \rangle, \triangleright^-, 1 - n \rangle$, then $\langle C, \triangleright, n \rangle \in \mathcal{L}(t)$.*
7. *If $\langle \exists R.C, \triangleleft, n \rangle \in \mathcal{L}(s)$ and $\langle \langle s, t \rangle, \triangleright, n_1 \rangle \in \mathcal{E}(R)$ is conjugated with $\langle \langle s, t \rangle, \triangleleft, n \rangle$, then $\langle C, \triangleleft, n \rangle \in \mathcal{L}(t)$.*
8. *If $\langle \exists R.C, \triangleright, n \rangle \in \mathcal{L}(s)$, then there exists $t \in S$ such that $\langle \langle s, t \rangle, \triangleright, n \rangle \in \mathcal{E}(R)$ and $\langle C, \triangleright, n \rangle \in \mathcal{L}(t)$.*
9. *If $\langle \forall R.C, \triangleleft, n \rangle \in \mathcal{L}(s)$, then there exists $t \in S$ such that $\langle \langle s, t \rangle, \triangleleft^-, 1 - n \rangle \in \mathcal{E}(R)$ and $\langle C, \triangleleft, n \rangle \in \mathcal{L}(t)$.*
10. *If $\langle \exists S.C, \triangleleft, n \rangle \in \mathcal{L}(s)$, and $\langle \langle s, t \rangle, \triangleright, n_1 \rangle \in \mathcal{E}(R)$ is conjugated with $\langle \langle s, t \rangle, \triangleleft, n \rangle$, for some $R \sqsubseteq S$ with $\mathsf{Trans}(R)$, then $\langle \exists R.C, \triangleleft, n \rangle \in \mathcal{L}(t)$.*
11. *If $\langle \forall S.C, \triangleright, n \rangle \in \mathcal{L}(s)$ and $\langle \langle s, t \rangle, \triangleright', n_1 \rangle \in \mathcal{E}(R)$ is conjugated with $\langle \langle s, t \rangle, \triangleright^-, 1 - n \rangle$, for some $R \sqsubseteq S$ with $\mathsf{Trans}(R)$, then $\langle \forall R.C, \triangleright, n \rangle \in \mathcal{L}(t)$.*
12. *$\langle \langle s, t \rangle, \bowtie, n \rangle \in \mathcal{E}(R)$ if and only if $\langle \langle t, s \rangle, \bowtie, n \rangle \in \mathcal{E}(\mathsf{Inv}(R))$.*
13. *If $\langle \langle s, t \rangle, \triangleright, n \rangle \in \mathcal{E}(R)$ and $R \sqsubseteq S$, then $\langle \langle s, t \rangle, \triangleright, n \rangle \in \mathcal{E}(S)$.*
14. *If $\langle \geq pR, \triangleright, n \rangle \in \mathcal{L}(s)$, then $|\{t \in S \mid \langle \langle s, t \rangle, \triangleright, n \rangle \in \mathcal{E}(R)\}| \geq p$.*
15. *If $\langle \leq pR, \triangleleft, n \rangle \in \mathcal{L}(s)$, then $|\{t \in S \mid \langle \langle s, t \rangle, \triangleleft^-, 1 - n \rangle \in \mathcal{E}(R)\}| \geq p + 1$.*
16. *If $\langle \geq pR, \triangleleft, n \rangle \in \mathcal{L}(s)$, then $|\{t \in S \mid \langle \langle s, t \rangle, \triangleright, n_i \rangle \in \mathcal{E}(R)\}| \leq p - 1$, conjugated with $\langle \langle s, t \rangle, \triangleleft, n \rangle$.*
17. *If $\langle \leq pR, \triangleright, n \rangle \in \mathcal{L}(s)$, then $|\{t \in S \mid \langle \langle s, t \rangle, \triangleright', n_i \rangle \in \mathcal{E}(R)\}| \leq p$ conjugated with $\langle \langle s, t \rangle, \triangleright^-, 1 - n \rangle$.*
18. *There do not exist two conjugated triples in any label of any individual $x \in S$.*
19. *If $\langle a : C \bowtie n \rangle \in \mathcal{A}$, then $\langle C, \bowtie, n \rangle \in \mathcal{L}(\mathcal{V}(a))$.*
20. *If $\langle (a, b) : R \bowtie n \rangle \in \mathcal{A}$, then $\langle \langle \mathcal{V}(a), \mathcal{V}(b) \rangle, \bowtie, n \rangle \in \mathcal{E}(R)$.*
21. *If $a \neq b \in \mathcal{A}$, then $\mathcal{V}(a) \neq \mathcal{V}(b)$.*

Nodes in the completion forest are labeled with a set of triples $\mathcal{L}(x)$ (node triples) which contain membership triples. We define $\mathcal{L}(x)\{\langle C, \bowtie, n \rangle\}$, where $C \in sub(\mathcal{A})$ and $n \in [0, 1]$. Furthermore, edges $\langle x, y \rangle$ are labeled with a set $\mathcal{L}(\langle x, y \rangle)$ (edge triples), defined as $\mathcal{L}(\langle x, y \rangle) = \{\langle R, \bowtie, n \rangle\}$, where $R \in R_A$. The algorithm expands the tree either by expanding the set $\mathcal{L}(x)$ of a node x with new triples, or by adding new leaf nodes.

If nodes x and y are connected by an edge $\langle x, y \rangle$, then y is called a successor of x and x is called a predecessor of y; ancestor is the transitive closure of predecessor. A node x is called an S-neighbor of a node x if, for some R with $R \sqsubseteq S$, either y is a successor of x and $\mathcal{L}(\langle x, y \rangle) = \langle R, \bowtie, n \rangle$ or y is a predecessor of x and $\mathcal{L}(\langle y, x \rangle) = \langle \mathsf{Inv}(R), \bowtie, n \rangle$. We then say that the edge triple connects x and y to a degree of n.

A node x is blocked if and only if it is not a root node and it is either directly or indirectly blocked. A node x is directly blocked if and only if none of its ancestors is blocked, and it has ancestors x', y and y' such that: (i) y is not a root node; (ii) x is a successor of x' and y a successor of y'; (iii) $\mathcal{L}(x) = \mathcal{L}(y)$ and $\mathcal{L}(x') = \mathcal{L}(y')$; and (iv) $\mathcal{L}(\langle x', x \rangle) = \mathcal{L}(\langle y', y \rangle)$. In this case we say that y blocks x. A node y is indirectly blocked if and only if none of its ancestors is blocked, or it is a successor of a node x and $\mathcal{L}(\langle x, y \rangle) = \emptyset$.

The algorithm initializes a forest $\mathcal{F}_A$ to contain a root node x_0^i, for each individual $a_i \in \mathbf{I}_A$ occurring in the ABox $\mathcal{A}$ and additionally $\{\langle C_i, \bowtie, n \rangle\} \cup \mathcal{L}(x_0^i)$, for each assertion of the form $\langle a_i : C_i \bowtie n \rangle$ in $\mathcal{A}$, and an edge $\langle x_0^i, x_0^j \rangle$ if $\mathcal{A}$ contains an assertion $\langle (a_i, a_j) : R_i \bowtie n \rangle$, with $\{\langle R_i, \bowtie, n \rangle\} \cup \mathcal{L}(\langle x_0^i, x_0^j \rangle)$ for each assertion of the form $\langle (a_i, a_j) : R_i \bowtie n \rangle$ in $\mathcal{A}$. Finally, we initialize the relation $\neq$ as $x_0^i \neq x_0^j$ if $a_i \neq a_j \in \mathcal{A}$ and the relation $\doteq$ to be empty. $\mathcal{F}_A$ is then expanded by repeatedly applying the rules in Table 11.2. We use the notation R^* to denote either the role R or the role returned by $\mathsf{Inv}(R)$, and the notation $\langle *, \bowtie, n \rangle$ to denote any role that participates in such a triple.

For a node x, $\mathcal{L}(x)$ is said to contain a clash if it contains one of the following: (a) two conjugated pairs of triples; (b) one of the triples $\langle \bot, \geq, n \rangle$, $\langle \top, \leq, n \rangle$, with $n > 0$, $n < 1$, $\langle \bot, >, n \rangle$, $\langle \top, <, n \rangle$ $\langle C, <, 0 \rangle$ or $\langle C, >, 1 \rangle$; (c) some triple $\langle \leq pR, \rhd, n \rangle \in \mathcal{L}(x)$ and x has $p + 1$ R-neighbors $y_0, \dots, y_p$, connected to x with a triple $\langle P^*, \rhd', n_i \rangle$, $P \sqsubseteq R$, which is conjugated with $\langle P^*, \rhd^-, 1 - n \rangle$, and $y_i \neq y_j$, for all $0 \leq i < j \leq p$; or (d) some triple $\langle \geq pR, \lhd, n \rangle \in \mathcal{L}(x)$ and x has p R-neighbors $y_0, \dots, y_{p-1}$, connected to x with a triple $\langle P^*, \rhd, n_i \rangle$, $P \sqsubseteq R$, which is conjugated with $\langle P^*, \lhd, n \rangle$, and $y_i \neq y_j$, for all $0 \leq i < j \leq p$. A completion forest is clash-free if none of its nodes contains a clash, and it is complete if none of the expansion rules is applicable.

11.1.3 The FiRE Fuzzy Reasoning Engine

Interface

FiRE[1] is a Java based fuzzy reasoning engine currently supporting f-$\mathcal{SHIN}$ that can be used either as an API by another application or by means of its graphical user interface. The graphical user interface of FiRE consists of the editor panel, the inference services panel and the output panel (Figure 11.1). Hence, the user can create or edit an existing fuzzy knowledge base using the editor panel, and use the inference services panel to put different kinds of queries to the fuzzy knowledge base. Finally, the output panel consists of four different tabs, each one displaying relative feedback depending on the user operation.

---

[1] FiRE can be found at http://www.image.ece.ntua.gr/~nsimou/FiRE/ together with installation instructions and examples.

Table 11.2 Tableau expansion rules

Rule		Description
$(\neg)$	if 1.	$\langle \neg C, \bowtie, n \rangle \in \mathcal{L}(x)$
	2.	and $\langle C, \bowtie^-, 1 - n \rangle \notin \mathcal{L}(x)$
	then	$\mathcal{L}(x) \rightarrow \mathcal{L}(x) \cup \{\langle C, \bowtie^-, 1 - n \rangle\}$
$(\sqcap_{\triangleright})$	if 1.	$\langle C_1 \sqcap C_2, \triangleright, n \rangle \in \mathcal{L}(x)$, x is not indirectly blocked, and
	2.	$\{\langle C_1, \triangleright, n \rangle, \langle C_2, \triangleright, n \rangle\} \nsubseteq \mathcal{L}(x)$
	then	$\mathcal{L}(x) \rightarrow \mathcal{L}(x) \cup \{\langle C_1, \triangleright, n \rangle, \langle C_2, \triangleright, n \rangle\}$
$(\sqcup_{\triangleleft})$	if 1.	$\langle C_1 \sqcup C_2, \triangleleft, n \rangle \in \mathcal{L}(x)$, x is not indirectly blocked, and
	2.	$\{\langle C_1, \triangleleft, n \rangle, \langle C_2, \triangleleft, n \rangle\} \nsubseteq \mathcal{L}(x)$
	then	$\mathcal{L}(x) \rightarrow \mathcal{L}(x) \cup \{\langle C_1, \triangleleft, n \rangle, \langle C_2, \triangleleft, n \rangle\}$
$(\sqcup_{\triangleright})$	if 1.	$\langle C_1 \sqcup C_2, \triangleright, n \rangle \in \mathcal{L}(x)$, x is not indirectly blocked, and
	2.	$\{\langle C_1, \triangleright, n \rangle, \langle C_2, \triangleright, n \rangle\} \cap \mathcal{L}(x) = \emptyset$
	then	$\mathcal{L}(x) \rightarrow \mathcal{L}(x) \cup \{C\}$ for some $C \in \{\langle C_1, \triangleright, n \rangle, \langle C_2, \triangleright, n \rangle\}$
$(\sqcap_{\triangleleft})$	if 1.	$\langle C_1 \sqcap C_2, \triangleleft, n \rangle \mathcal{L}(x)$, x is not indirectly blocked, and
	2.	$\{\langle C_1, \triangleleft, n \rangle, \langle C_2, \triangleleft, n \rangle\} \cap \mathcal{L}(x) = \emptyset$
	then	$\mathcal{L}(x) \rightarrow \mathcal{L}(x) \cup \{C\}$ for some $C \in \{\langle C_1, \triangleleft, n \rangle, \langle C_2, \triangleleft, n \rangle\}$
$(\exists_{\triangleright})$	if 1.	$\langle \exists R.C, \triangleright, n \rangle \in \mathcal{L}(x)$, x is not blocked,
	2.	x has no R-neighbor y connected with a triple $\langle P^*, \triangleright, n \rangle$, $P \sqsubseteq^* R$ and $\langle C, \triangleright, n \rangle \in \mathcal{L}(y)$
	then	create a new node y with $\mathcal{L}(\langle x, y \rangle) = \{\langle R, \triangleright, n \rangle\}$, $\mathcal{L}(y) = \{\langle C, \triangleright, n \rangle\}$,
$(\forall_{\triangleleft})$	if 1.	$\langle \forall R.C, \triangleleft, n \rangle \in \mathcal{L}(x)$, x is not blocked,
	2.	x has no R-neighbor y connected with a triple $\langle P^*, \triangleleft^-, 1 - n \rangle$, $P \sqsubseteq^* R$ and $\langle C, \triangleleft, n \rangle \in \mathcal{L}(y)$
	then	create a new node y with $\mathcal{L}(\langle x, y \rangle) = \{\langle R, \triangleleft^-, 1 - n \rangle\}$, $\mathcal{L}(y) = \{\langle C, \triangleleft, n \rangle\}$,
$(\forall_{\triangleright})$	if 1.	$\langle \forall R.C, \triangleright, n \rangle \in \mathcal{L}(x)$, x is not indirectly blocked, and
	2.	x has an R-neighbor y with $\langle C, \triangleright, n \rangle \notin \mathcal{L}(y)$ and
	3.	$\langle *, \triangleright^-, 1 - n \rangle$ is conjugated with the positive triple that connects x and y
	then	$\mathcal{L}(y) \rightarrow \mathcal{L}(y) \cup \{\langle C, \triangleright, n \rangle\}$,
$(\exists_{\triangleleft})$	if 1.	$\langle \exists R.C, \triangleleft, n \rangle \in \mathcal{L}(x)$, x is not indirectly blocked, and
	2.	x has an R-neighbor y with $\langle C, \triangleleft, n \rangle \notin \mathcal{L}(y)$, and
	3.	$\langle *, \triangleleft, n \rangle$ is conjugated with the positive triple that connects x and y
	then	$\mathcal{L}(y) \rightarrow \mathcal{L}(y) \cup \{\langle C, \triangleleft, n \rangle\}$,
$(\forall_+)$	if 1.	$\langle \forall R.C, \triangleright, n \rangle \in \mathcal{L}(x)$, x is not indirectly blocked, and
	2.	there is some P, with $\mathsf{Trans}(P)$, and $P \sqsubseteq^* R$, and x has a P-neighbor y with $\langle \forall P.C, \triangleright, n \rangle \notin \mathcal{L}(y)$, and
	3.	$\langle *, \triangleright^-, 1 - n \rangle$ is conjugated with the positive triple that connects x and y
	then	$\mathcal{L}(y) \rightarrow \mathcal{L}(y) \cup \{\langle \forall P.C, \triangleright, n \rangle\}$,
$(\exists_+)$	if 1.	$\langle \exists R.C, \triangleleft, n \rangle \in \mathcal{L}(x)$, x is not indirectly blocked, and
	2.	there is some P, with $\mathsf{Trans}(P)$, and $P \sqsubseteq^* R$, x has a P-neighbor y with, $\langle \exists P.C, \triangleleft, n \rangle \notin \mathcal{L}(y)$, and
	3.	$\langle *, \triangleleft, n \rangle$ is conjugated with the positive triple that connects x and y
	then	$\mathcal{L}(y) \rightarrow \mathcal{L}(y) \cup \{\langle \exists P.C, \triangleleft, n \rangle\}$,

(continued overleaf)

Table 11.2   (*continued*)

Rule		Description
$(\geq_\triangleright)$	if 1.	$\langle \geq pR, \triangleright, n\rangle \in \mathcal{L}(x)$, x is not blocked,
	2.	there are no p R-neighbors $y_1, \ldots, y_p$ connected to x with a triple $\langle P^*, \triangleright, n\rangle$, $P \sqsubseteq_* R$,
	3.	and $y_i \neq y_j$ for $1 \leq i < j \leq p$
	then	create p new nodes $y_1, \ldots, y_p$, with $\mathcal{L}(\langle x, y_i\rangle) = \{\langle R, \triangleright, n\rangle\}$ and $y_i \neq y_j$ for $1 \leq i < j \leq p$
$(\leq_\triangleleft)$	if 1.	$\langle \leq pR, \triangleleft, n\rangle \in \mathcal{L}(x)$, x is not blocked, and
	then	apply the $(\geq_\triangleright)$ rule to the triple $\langle \geq (p+1)R, \triangleleft^-, 1-n\rangle$
$(\leq_\triangleright)$	if 1.	$\langle \leq pR, \triangleright, n\rangle \in \mathcal{L}(x)$, x is not indirectly blocked,
	2.	there are $p+1$ R-neighbors $y_1, \ldots, y_{p+1}$ connected to x with a triple $\langle P^*, \triangleright', n_i\rangle$, $P \sqsubseteq_* R$,
	3.	which is conjugated with $\langle P^*, \triangleright^-, 1-n\rangle$, and there are two of them y, z, with no $y \neq z$ and
	4.	y is neither a root node nor an ancestor of z
	then	1. $\mathcal{L}(z) \rightarrow \mathcal{L}(z) \cup \mathcal{L}(y)$ and
		2. if z is an ancestor of x
		then $\mathcal{L}(\langle z, x\rangle) \longrightarrow \mathcal{L}(\langle z, x\rangle) \cup \mathsf{Inv}(\mathcal{L}(\langle x, y\rangle))$
		else $\mathcal{L}(\langle x, z\rangle) \longrightarrow \mathcal{L}(\langle x, z\rangle) \cup \mathcal{L}(\langle x, y\rangle)$
		3. $\mathcal{L}(\langle x, y\rangle) \longrightarrow \emptyset$
		4. set $u \neq z$ for all u with $u \neq y$
$(\geq_\triangleleft)$	if 1.	$\langle \geq pR, \triangleleft, n\rangle \in \mathcal{L}(x)$, and x is not indirectly blocked,
	then	apply the $(\leq_\triangleright)$ rule to the triple $\langle \leq (p-1)R, \triangleleft^-, 1-n\rangle$
$(\leq_{r\triangleright})$	if 1.	$\langle \leq pR, \triangleright, n\rangle \in \mathcal{L}(x)$, and
	2.	there are $p+1$ R-neighbors $y_1, \ldots, y_{p+1}$ connected to x with a triple $\langle P^*, \triangleright', n_i\rangle$, $P \sqsubseteq_* R$,
	3.	conjugated with $\langle P^*, \triangleright^-, 1-n\rangle$, and there are two of them y, z, both root nodes, with no $y \neq z$
	then	1. $\mathcal{L}(z) \rightarrow \mathcal{L}(z) \cup \mathcal{L}(y)$ and
		2. for all edges $\langle y, w\rangle$:
		i. if the edge $\langle z, w\rangle$ does not exist, create it with $\mathcal{L}(\langle z, w\rangle) = \emptyset$
		ii. $\mathcal{L}(\langle z, w\rangle) \longrightarrow \mathcal{L}(\langle z, w\rangle) \cup \mathcal{L}(\langle y, w\rangle)$
		3. for all edges $\langle w, y\rangle$:
		i. if the edge $\langle w, z\rangle$ does not exist, create it with $\mathcal{L}(\langle w, z\rangle) = \emptyset$
		ii. $\mathcal{L}(\langle w, z\rangle) \longrightarrow \mathcal{L}(\langle w, z\rangle) \cup \mathcal{L}(\langle w, y\rangle)$
		4. set $\mathcal{L}(y) = \emptyset$ and remove all edges to/from y
		5. set $u \neq z$ for all u with $u \neq y$ and set $y \doteq z$
$(\geq_{r\triangleleft})$	if 1.	$\langle \geq pR, \triangleleft, n\rangle \in \mathcal{L}(x)$,
	then	apply the $(\leq_{r\triangleright})$ rule to the triple $\langle \leq (p-1)R, \triangleleft^-, 1-n\rangle$

Syntax

As previously mentioned, a fuzzy knowledge base consists of three components: a TBox, an RBox and an ABox. The TBox and RBox are defined using the Knowledge Representation System Specification proposed by KRSS Group (1993) since they that do not

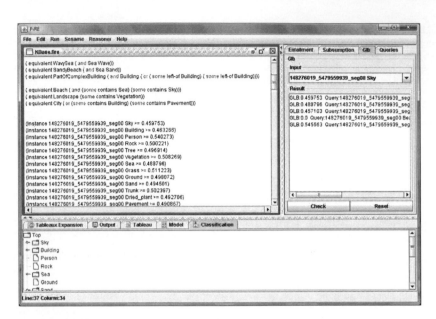

Figure 11.1 The FiRE user interface consists of the editor panel (upper left), the inference services panel (upper right) and the output panel (bottom).

include uncertainty. So transitive roles or the sub-role of another role are defined by using the keywords **transitive** and **parent**, and concept axioms are defined by the keywords **implies** and **equivalent**.

On the other hand, since the assertions are extended to fit imperfect knowledge, the ABox in fuzzy DLs is different. Instances in FiRE are defined using the keyword **instance** followed by the individual, the concept in which the individual participates, the inequality type (one of $<, <=, >, > =$) and the degree of confidence, $degree \in [0, 1]$. Similarly, role assertions are defined by using the keyword **related** followed by subject and object individuals, the inequality type and the degree of confidence. In both cases the inequality type and the degree of confidence are required only for fuzzy assertions; if they are not mentioned then the assertions are assumed as crisp (i.e $> = 1$).

Example **11.1.2** *The syntax of the assertions* $\langle region1 : Sky \rangle$, $\langle region2 : (Sand \sqcap Sea) \geq 0.8 \rangle$, $\langle (region1, region2) : isAboveOf \geq 0.7 \rangle$ *are shown below in FiRE syntax:*

```
(instance region1 Sky)
(instance region2 (and Sea Sand) > 0.8)
(related region1 region2 isAboveOf >= 0.7)
```

Reasoning Services

One of the main advantages of DLs comparared to other formal representation languages is their sound and complete algorithms that can be used for reasoning. The main reasoning services offered by crisp reasoners are checking, subsumption and entailment of concepts and axioms with respect to an ontology. In other words, using these reasoning services, someone is capable of answering questions like 'Can the concept C have any instances in models of the ontology T?' (satisfiability of C), 'Is the concept D more general than the concept C in models of the ontology T?' (subsumption $C \sqsubseteq D$), and 'Does axiom Ψ logically follow from the ontology T?' (entailment of Ψ).

These reasoning services are also available in f-$\mathcal{SHIN}$ together with greatest lower bound queries which are specific to fuzzy assertions. In the case of fuzzy DL, satisfiability questions take the form 'Can the concept C have any instances with degree of participation $\bowtie n$ in models of the ontology T?'. Furthermore, the incorporation of degrees in assertions makes the evaluation of the best lower and upper truth-value bounds of a fuzzy assertion vital. The term of greatest lower bound of a fuzzy assertion with respect to Σ was defined in Straccia (2001). Informally, greatest lower bound are queries like 'What is the greatest degree n that our ontology entails an individual a to participate in a concept C?'.

FiRE uses the tableau algorithm of f-$\mathcal{SHIN}$, presented by Stoilos et al. (2007), in order to decide the key inference problems of a fuzzy ontology. Hence entailment queries that ask whether our knowledge base logically entails the membership of an individual in a specific concept to a certain degree, are specified in the *Entailment* inference tab (see Figure 11.1). Their syntax is the same as that used for the definition of a fuzzy instance. So, for example a statement of the form

```
(instance region1 (and Sea Sky) > 0 8)
```

would ask whether `region1` is `Sea` and `Sky` to a degree greater than or equal to 0.8.

On the other hand, subsumption queries are specified in the *Subsumption* inference tab. Their syntax is of the form

```
(concept1) (concept2)
```

where `concept1` and `concept2` are f-$\mathcal{SHIN}$ concepts. Additionally, the user can perform a global concept classification procedure presenting the concept hierarchy tree in the *Classification* tab of the output panel.

Finally, FiRE permits the user to make greatest lower bound (GLB queries), which are evaluated by FiRE performing entailment queries. During this procedure a set of entailment queries is constructed consisting of an entailment query for every degree contained in the ABox, using the individual and the concept of interest. These queries are performed using the binary search algorithm to reduce the degree search space, resulting in the GLB. The syntax of GLB queries is of the form

```
individual (concept)
```

where `concept` can be either an atomic concept or a result of f-$\mathcal{SHIN}$ constructors. Furthermore, a user can perform a global GLB on a fuzzy knowledge base. The FiRE global GLB service evaluates the greatest lower bound degree of all the concepts of Σ participating in all the individuals of Σ.

11.2 Spatial Features for Image Region Labeling

An important field of research is the identification and labeling of semantically meaningful regions in images, a problem often referred to as image region labeling. Approaches usually involve the segmentation of the image using low-level features, followed by the classification of the resulting regions based on the extracted low-level features (see Chapters 4 and cha:MachineLearning). However, a well-known problem is that low-level features are insufficient to fully describe the semantics of an image, and the same applies to the semantics of a region. Using only low-level information such as color or texture does not suffice to detect the correct concept. Therefore research was initiated toward exploiting additional features, for example spatial relations between regions and domain models regarding the typical spatial arrangements of objects in images (Saathoff and Staab 2008; Yuan et al. 2007).

In this subsection we present a model that uses explicitly represented knowledge about the typical spatial arrangements of objects, so-called spatial constraint templates, which we acquire from labeled examples. We use *fuzzy constraint reasoning* to represent the problem and to find a solution that provides an optimal labeling with respect to both low-level features and the spatial knowledge.

In Section 11.2.1 we introduce fuzzy constraint reasoning as a means to represent and solve systems of variables and constraints under uncertainty assumptions. In Section 11.2.2 we discuss how to label image regions using spatial features, how to acquire spatial background knowledge, how to extract spatial relations, and finally the representation of the overall problem as a fuzzy constraint satisfaction problem.

11.2.1 Fuzzy Constraint Satisfaction Problems

A fuzzy constraint satisfaction problem consists of an ordered set of fuzzy variables $V = \{v_1, \ldots, v_k\}$, each associated with the crisp domain $L = \{l_1, \ldots, l_n\}$ and the membership function $\mu_i : L \to [0, 1]$. The value $\mu_i(l)$, $l \in L$, is called the degree of satisfaction of the variable for the assignment $v_i = l$. Further, we define a set of fuzzy constraints $C = \{c_1, \ldots, c_m\}$. Each constraint c_j is defined on a set of variables $v_1, \ldots, v_q \in V$, and we interpret a constraint as a fuzzy relation $c_j : L^q \to [0, 1]$, which we call the domain of the constraint. The value $c(l_1, \ldots, l_q), v_i = l_i$, is called the degree of satisfaction of the variable assignment $l_1, \ldots, l_q$ for the constraint c. If $c(l_1, \ldots, l_q) = 1$, we say that the constraint is fully satisfied, and if $c(l_1, \ldots, l_q) = 0$ we say that it is fully violated. The purpose of fuzzy constraint reasoning is to obtain a variable assignment that is optimal with respect to the degrees of satisfaction of the variables and constraints. The quality of a solution is measured using a global evaluation function called the *joint degree of satisfaction*.

We first define the joint degree of satisfaction of a variable, which determines to what degree the value assigned to that variable satisfies the problem. Let $P = \{l_1, \ldots, l_k\}$, $k \leq |V|$ be a partial solution of the problem, with $v_i = l_i$. Let $C_i^+ \subseteq C$ be the set of fully instantiated constraints containing the variable v_i; that is, each constraint $c \in C_i^+$ is only defined on variables v_j with $l_j \in P$. Further, let c without explicitly specified labels stand for the degree of satisfaction of c given the current partial solution. Finally, let $C_i^- \subseteq C$ be the set of partially instantiated constraints on v_i; that is, at least one of the variables has no value assigned. We then define the joint degree of satisfaction as

$$\mathrm{dos}(v_i) := \frac{1}{\omega + 1} \left(\frac{1}{|C_i^+ + C_i^-|} \left(\sum_{c \in C_i^+} c + |C_i^-| \right) + \omega \mu_i(l_i) \right),$$

in which ω is a weight used to control the influence of the degree of satisfaction of the variable assignment on the joint degree. In this definition we overestimate the degree of satisfaction of partially instantiated constraints.

We now define the joint degree of satisfaction for a complete fuzzy constraint satisfaction problem. Let $J := \{\mathrm{dos}(v_{i_1}), \ldots, \mathrm{dos}(v_{i_n})\}$ be an ordered multiset of joint degrees of satisfaction for each variable in V, with $\mathrm{dos}(v_{i_k}) \leq \mathrm{dos}(v_{i_l})$ for all $v_{i_k}, v_{i_l} \in V, k < l$. The joint degree of satisfaction of a variable that is not yet assigned a value is overestimated as 1. We can now define a lexicographic order $>_L$ on the multisets. Let $J = \{\gamma_1, \ldots, \gamma_k\}, J' = \{\delta_1, \ldots, \delta_k\}$ be multisets. Then $J >_L J'$ if and only if $\exists i \leq k :$ $\forall j < i : \gamma_j = \delta_j$ and $\gamma_i > \delta_i$. If we have two (partial) solutions P, Q to a fuzzy constraint satisfaction problem with corresponding joint degrees of satisfaction J_P, J_Q, then solution P is better than Q if and only if $J_P >_L J_Q$.

Based on these definitions a fuzzy constraint satisfaction problem can efficiently be solved using algorithms such as branch and bound. Branch and bound creates a search tree and does a depth-first search. Using the evaluation function, it prunes the search tree if it encounters a partial solution that cannot be extended into a full solution that is better than the current best solution. For this pruning the evaluation function has to provide an upper bound, that is, the evaluation score of a partial solution is not allowed to increase if further variables are instantiated. Since we overestimate the degree of satisfaction of uninstantiated variables and partially instantiated constraints, we satisfy this property.

11.2.2 Exploiting Spatial Features Using Fuzzy Constraint Reasoning

Overall Analysis Framework

Figure 11.2 depicts the parts of the overall image processing in our system relevant to region labeling; see Dasiopoulou et al. (2007) for a more detailed system description.

The overall framework consists of two phases. First the background knowledge is created during a constraint acquisition step based on a set of labeled example images. In this step we model a set of spatial prototypes using a semi-automatic approach. A constraint prototype defines the legal arrangements of objects within an image as a set of examples, which are later used to create the constraints within the spatial reasoning step.

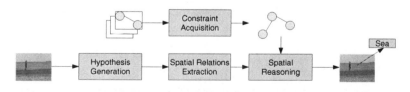

Figure 11.2   The overall analysis chain.

(a) Input image   (b) Hypothesis sets

Figure 11.3   Hypothesis set generation.

The second phase is the image analysis procedure itself. In this phase, the input image is first processed for *hypothesis set generation*, that is, segmenting the input image and classifying each resulting region. For each label a support vector machine was trained based on the characteristic features of each segment. Each SVM provides a confidence score in the interval [0, 1] based on the distance to the separating hyperplane. An example segmentation with simplified hypotheses is shown in Figure 11.3.

In order to integrate the spatial context within our fuzzy constraint reasoning approach, we determine the spatial relations between the regions using the *spatial relations extraction* module. Eventually, we transform the hypothesis sets, spatial relations and spatial prototypes into a fuzzy constraint satisfaction problem. The solution to this problem is a good approximation of an optimal solution to a spatially aware labeling of image segments.

Spatial Relations Extraction

Within our region labeling procedure we consider six relative and two absolute spatial relations to model the spatial arrangements of the regions within an image. The relative spatial relations are either *directional* (*above, below, left, right*) or *topological* (*contains, adjacent*), and the absolute spatial relations are *above-all* and *below-all*.

The directional relations are computed based on the centers of the minimal bounding box containing a region. We illustrate the definition of the directional relations in Figure 11.4a. Based on the angle α, we determine in which area the center of the related region lies, and instantiate the corresponding directional relation. For containment we determine whether the bounding box of one region is fully contained in the bounding box of another region, and instantiate the relation if this is the case. Finally, two regions are adjacent if they share at least one pair of adjacent pixels.

Computing whether a region is *above-all* or *below-all* is again based on the center of the bounding box. We include the regions with the highest center and for which the center lies above a certain threshold, which is given relative to the image size. An example is shown in Figure 11.4b.

Constraint Acquisition

The spatial prototypes constitute the background knowledge in our approach. Manually defining these prototypes is a tedious task, specifically if the number of supported concepts

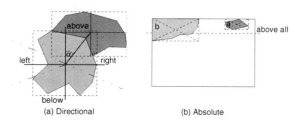

(a) Directional   (b) Absolute

Figure 11.4 Definition of (a) directional and (b) absolute spatial relations.

and spatial relations becomes larger. We derive spatial prototypes representing spatial constraints from example annotations. We mine example annotations using support and confidence as selection criteria. In addition, spatial constraints may be manually refined or deleted.

In order to generate the prototypes we select the concrete spatial relations found in the labeled examples. Each spatial relation relates a number of segments $s_1, \ldots s_n$. Since our images are labeled, each segment s_i has also a related label l_i. Now, we wish to generate the spatial prototypes $p \in P$, which are relations on L^n. We can interpret the relation as a set of tuples, and the purpose of constraint acquisition is to define these tuples.

Let R_{type} be a list (with possibly duplicate entries) of concrete spatial relations found in the labeled examples of type *type*. Since each segment has a related label, we can generate a tuple $(l_1, \ldots, l_n)$ for each spatial relation on segments $(s_1, \ldots, s_n)$. However, using all tuples to generate the spatial prototype would also include a lot of noise for cases where segmentation or spatial relations extraction provided suboptimal results. We therefore use support and confidence for selecting the tuples that can be considered robust. Let $|(l_1, \ldots, l_n)|$ be the number of times a specific tuple was found, and $|(*, l_2, \ldots, L_n)|$ be the number of times a tuple with an arbitrary $l \in L$ in the first position was found. The support is then defined as

$$\sigma((l_1, \ldots, l_n)) = \frac{|(l_1, \ldots, l_n)|}{|R_{type}|},$$

and the confidence as

$$\gamma((l_1, \ldots, l_n)|) = \frac{|(l_1, \ldots, l_n)|}{|(*, l_2, \ldots, l_n)|}.$$

The final set of prototype proposals is then determined based on support and confidence as filter criteria. In a final step, the proposals are manually inspected and refined by adding missing prototypes or removing unsuitable ones.

Representing Image Region Labeling as a Fuzzy Constraint Satisfaction Problem

We now discuss the representation of the regions, hypothesis sets and spatial constraints as a fuzzy constraint satisfaction problem. Let L be the set of labels. Let $S = \{s_1, \ldots, s_m\}$ be the set of regions. Each region s_i is associated with a membership function $\theta_i : L \rightarrow [0, 1]$ which models the hypothesis set, that is, the confidence values obtained during classification. Further, let $R = \{r_1, \ldots, r_k\}$ the set of spatial relations. Each $r \in R$ is defined as $r \in S$ for absolute spatial relations, and $r \in S^2$ for relative ones. Further, each spatial relation is associated with a type $type \in T$, where $type$ refers to the spatial relations extracted in Section 11.2.2.

We must now define the spatial prototypes which represent the background knowledge. For each $type \in T$ a spatial prototype $p \in P$ exists. Each prototype is defined as a fuzzy relation on L, that is, $p : L^n \rightarrow [0, 1]$. Note that a spatial prototype is defined in the same way a constraint is (Section 11.2.1). The difference is that a constraint only exists within the fuzzy constraint satisfaction problem and is associated with a set of variables. The prototype basically defines the domain of the constraint.

A fuzzy constraint satisfaction problem is now created using the following algorithm.

1. For each region $s_i \in S$, create a variable v_i on L with $\mu_i := \theta_i$.
2. For each region $s_i \in S$ and for each spatial relation r of type $type$ defined on s_i and further segments $s_1, \ldots, s_k$, create a constraint c on $v_i, v_1, \ldots, v_k$, with $c := p$, where $p \in P$ is a spatial prototype of type $type$.

11.3   Fuzzy Rule Based Reasoning Engine

Expert systems are typically defined as computer systems that emulate the decision-making ability of a human expert. The power of an expert system is derived from presence of a *knowledge base* filled with expert knowledge, mostly in symbolic form. In addition, there is a generic problem-solving mechanism used as *inference engine* (Durkin 1994). Research in expert systems began in the mid-1970s; classical examples of early systems that influenced other researchers are MYCIN and PROSPECTOR.

Knowledge Representation

NEST uses attributes and propositions, rules, integrity constraints and contexts to express the task-specific (domain) knowledge.

Four types of attributes can be used in the system: binary, single nominal, multiple nominal, and numeric. According to the type of attribute, the derived propositions correspond to:

- values `true` and `false` for a *binary* attribute;
- each value for a *nominal* attribute. The difference between a single and multiple nominal attribute is apparent only when answering the question about the value of the attribute;
- fuzzy intervals for a *numeric* attribute. Each interval is defined using four points: fuzzy lower bound (FL), crisp lower bound (CL), crisp upper bound (CU), fuzzy upper bound (FU). These values need not be distinct; this allows rectangular, trapezoidal and triangular fuzzy intervals to be created.

Rules are defined in the form

$$condition \Rightarrow conclusion(weight), action$$

where *condition* is disjunctive form (disjunction of conjunctions) of literals (propositions or their negations), *conclusion* is a list of literals and *action* is a list of actions (external programs). We distinguish three types of rules:

- *Compositional* rules are such that each literal in *conclusion* has a weight which expresses the uncertainty of *conclusion* if *condition* holds with certainty. The term 'compositional' means that, to evaluate the weight of a proposition, *all* rules with this proposition in *conclusion* are evaluated and combined.
- *Apriori* rules are compositional rules without condition. These rules can be used to assign implicit weights to goals or intermediate propositions.
- *Logical* rules are non-compositional rules without weights; only these rules can infer the conclusion with the weight true or false. *One* activated rule thus fully evaluates the proposition in conclusion.

A list of actions (external programs) can be associated with each rule. These programs are executed if the rule is activated.

Inference Mechanism

During consultation, the system uses rules to compute weights of goals from the weights of questions. This is accomplished by (1) selecting the relevant rule during the current state of the consultation, and (2) applying the selected rule to infer the weight of its conclusion.

1. The selection of the relevant rule can be done using either backward or forward chaining. The actual direction is determined by the user when selecting the consultation mode (see below).
2. For (compositional and apriori) rules with weights, the system combines contributions of rules using the compositional approach described below. For rules without weights, the system uses a non-compositional approach based on (crisp) *modus ponens* – to evaluate the weight of a conclusion – and (crisp) disjunction – to evaluate a set of rules with the same conclusion. The weights are propagated not only toward the actual goal but by using all rules applicable at given moment.

Uncertainty processing in NEST is based on the algebraic theory of Hájek (1985). This theory generalizes the methods of uncertainty processing used in early expert systems such as MYCIN and PROSPECTOR. Algebraic theory assumes that the knowledge base is created by a set of rules in the form

$$condition \Rightarrow conclusion(weight)$$

where *condition* is a conjunction of literals, *conclusion* is a single proposition, and $weight \in [-1, 1]$ expresses the uncertainty of the rule.

During a consultation, all relevant rules are evaluated by combining their weights with the weights of conditions. Weights of questions are obtained from the user, weights of all other propositions are computed by the inference mechanism. Five combination functions are defined to process the uncertainty in such a knowledge base:

(i) $NEG(w)$, to compute the weight of negation of a proposition;
(ii) $CONJ(w_1, w_2, \ldots, w_n)$, to compute the weight of conjunction of literals;

(iii) $DISJ(w_1, w_2, \ldots, w_n)$, to compute the weight of disjunction of literals;

(iv) $CTR(a, w)$, to compute the contribution of the rule to the weight of the conclusion (this is computed from the weight of the rule w and the weight of the condition a);

 (v) $GLOB(w'_1, w'_2, \ldots, w'_n)$, to compose the contributions of more rules with the same conclusion.

Algebraic theory defines a set of axioms, the combination functions must fulfill. Different sets of combination functions can thus be implemented. We call these sets 'inference mechanisms'. The NEST system uses 'standard', 'logical' and 'neural' mechanisms. These differ in the definition of the functions CTR and $GLOB$.

The *standard inference mechanism* is based on 'classical' approach of MYCIN and PROSPECTOR expert systems. The contribution of a rule is computed Mycin-like, that is,

$$w' = CTR(a, w) = a \cdot w, \quad \text{for } a > 0,$$

while the combination of contributions of rules with the same conclusion is computed Prospector-like, that.

$$GLOB(w'_1, w'_2) = \frac{w'_1 + w'_2}{1 + w'_1 \cdot w'_2}.$$

The *logical inference mechanism* is based on an application of the completeness theorem for Lukasiewicz many-valued logic. The task of the inference mechanism is to determine the degree to which each goal logically follows from the set of rules (understood as a fuzzy axiomatic theory) and user's answers during consultation (Berka et al. 1992). This degree can be obtained by using the *fuzzy modus ponens* inference rule,

$$\frac{\alpha, \alpha \Longrightarrow \beta}{\beta} \left(\frac{x, y}{\max(0, x + y - 1)} \right).$$

So the contribution of a rule is computed as

$$w' = CTR(a, w) = \text{sign}(w) \cdot \max(0, a + |w| - 1), \quad \text{for } a > 0.$$

To combine contributions of more rules, the logical inference mechanism uses the *fuzzy disjunction*,

$$GLOB(w'_1, w'_2, \ldots, w'_n) = \min\left(1, \sum_{w' > 0} w'\right) - \min\left(1, \sum_{w' < 0} |w'|\right).$$

The *neural inference mechanism* is based on an analogy with active dynamics of neural networks. To obtain results that correspond to the output of a neuron, the contribution of a rule is computed as a weighted input of the neuron and the global effect of all rules with the same conclusion is computed as piecewise linear transformation of the sum of weighted inputs.

The remaining functions are defined in the same way for all three mechanisms: negation of weight w is evaluated as $-w$, conjunction of weights is evaluated as minimum, and disjunction of weights is evaluated as maximum.

Consultation with the System

NEST offers several modes of consultation. The *dialogue* mode is the classical question/ answer mode when the system selects the current question using backward chaining. The *questionnaire* mode allows answers to be filled in in advance; the system then directly infers the goals using forward chaining. In *dialogue/questionnaire* mode the user can input some volunteer information (using questionnaire mode), and during further consultation the system asks questions if needed.

In each mode, the user answers the questions concerning the input attributes. According to the type of attribute, the user gives the weight (for binary attributes), the value and its weight (for single nominal attributes), a list of values and their weights (for multiple nominal attributes), or the value (for numeric attributes). Questions not answered during consultation get the default answer 'unknown' $[-1, 1]$ or 'irrelevant' $[0, 0]$. Answers can be postponed – the user can return to them after finishing the consultation. The result of the consultation is shown as a list of goals (or all propositions) together with their weights.

Using NEST to Determine Region Similarity

NEST is used for estimation of region dissimilarity. NEST processes each region from the segmented image in the following way:

1. Determine the dominant label(s). This is done by evaluating the initial labels and their confidences of the region A and all its neighbors.
2. Evaluate the similarity between a region A and each of its neighbors.

The overall scheme of this process is shown in Figure 11.5.

NEST is activated repeatedly for different regions in the image. The input into NEST is given as a set of labels (and confidences) of a processed region (let us call this region A) and labels (and confidences) of its neighbors.

Initially, the numeric values of confidences are turned into fuzzy intervals low (red), medium (yellow), high (green). The fuzzy membership functions used are shown in Figure 11.6.

The rule base of NEST evaluates a single tuple of two adjacent regions A and AN (the neighbor of A). The evaluation starts by determining the dominant labels of A and AN. The following rules are used for this purpose:

1. Determine what labels are not dominant using logical rules (if a label has low confidence, then it is not dominant):

```
IF A_X(low) THEN NOT(A_DOMINANT(X))
IF AN_X(low) THEN NOT(AN_DOMINANT(X))
```

where X stands for each label defined for regions of the image.
2. Determine what labels are dominant using compositional rules:
   - if a label has high confidence, then this label is dominant

```
IF A_X(high) THEN A_DOMINANT(X) [1,000]
IF AN_X(high) THEN N_DOMINANT(X) [1,000]
```

Labels and confidences of region *A*
and all of its neighbors

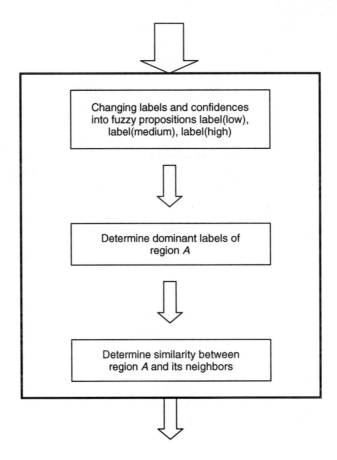

Changing labels and confidences
into fuzzy propositions label(low),
label(medium), label(high)

Determine dominant labels of
region *A*

Determine similarity between
region *A* and its neighbors

Degree of similarity for region *A*
and all of its neighbors

Figure 11.5   Scheme of NEST for image segmentation.

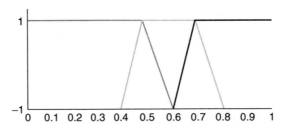

Figure 11.6   Fuzzy confidence.

- if no label has high significance, then dominant labels are given by medium confidence

```
A_NotHigh  : IF A_X(medium) THEN  A_DOMINANT(X) [1,000]
AN_NotHigh : IF AN_X(medium) THEN  AN_DOMINANT(X) [1,000]
```

In the next step, similarity between A and AN is evaluated using two sets of compositional rules:

1. Increase similarity if the regions A and AN share the same dominant label

```
IF A_DOMINANT(X) AND AN_DOMINANT(X) THEN SIMILAR(A,AN)[0,400]
```

2. Decrease similarity if the regions A and AN do not share the same dominant label

```
IF (A_DOMINANT(X) AND NOT(AN_DOMINANT(X)))
   OR (NOT(A_DOMINANT(X)) AND AN_DOMINANT(X))
THEN SIMILAR(A,AN)[-0,200]
```

The result of image processing using NEST is to find the most similar tuple of adjacent regions, which should be merged.

11.4 Reasoning over Resources Complementary to Audiovisual Streams

Multimedia resources such as video or images convey a large amount of information to the consumer. Unfortunately, most of this information represents a big challenge to automated analysis tools. Consequently, additional resources in textual or tabular form, although less profuse in content, can deliver additional information that can be combined with that extracted from the 'core' resource. In addition, even some binary resources may be indirectly associated with the original resources and contribute to their interpretation. All such resources can be considered as complementary.

- *Primary* complementary resources are those 'directly' attached to the multimedia data. Examples are texts around pictures, subtitles of movies, audio tracks, and overlay texts.
- *Secondary* resources are related but not directly attached to the multimedia data under consideration. Examples are a program guide for a TV broadcaster, a structured protocol on events described in a video, or a website displaying similar pictures. These data are of our main interest for this work.

From the above it follows that there are multiple scenarios for mining complementary resources. One is the processing of *textual sources directly attached to multimedia* documents, sounds and images. Such resources are sometimes already available in textual form but often need to be extracted from the images or videos using OCR technology. The region of such textual resources needs first to be detected by means of image segmentation and then processed by OCR. General text detection in video is complex, since text may appear for example on signs (shop name, city name, street names) or on non-rigid

Figure 11.7 Detection of moving objects in soccer broadcasts. In the right-hand image all the moving objects have been removed.

objects (a person's T-shirt). OCR applied to text detected in keyframes of soccer match videos (Declerck and Cobet 2007) is a promising source of information since it provides a way to synchronize video file time with the actual time of the events in the match, using the overlay time counter analysis (as this may significantly vary due to the different start times, or commercials and interviews during the break); see Figure 11.7.

The other scenario is the analysis of the *textual or tabular resources indirectly associated with multimedia*. The whole scope of state-of-the-art data mining and text mining techniques can be exploited in this scenario. These are data available in various forms not directly connected to the media but related by the topic they describe. Examples are:

- (semi-)structured, database-like textual resources containing the summary of statistical, numerical and categorical data connected with events covered by individual video broadcasts (such as soccer matches), and, thanks to their clean structure, providing valuable and easily extractable information;
- unstructured event reports containing detailed descriptions about particular events in the game including time point information.

The quality of video retrieval would thus strongly benefit from the exploitation of complementary textual resources, especially if these are endowed with temporal references. Good examples can be found within the sports domain.

Event Extraction and Semantic Indexing

Unstructured reports usually need to be manually identified and then semi-automatically extracted from websites using wrappers or other extraction methods. Later, using NLP-based information extraction tools such as that described in Drożdżyński et al. (2004), one can extract domain-related ontology concepts from the texts. Combining results from several of these reports (Nemrava et al. 2007) can increase event coverage and also eliminate false positive events with small evidence from the texts. In our case, as a dataset for ontology-based (Oberle et al. 2007) information extraction and ontology learning from text, we used the SmartWeb corpus consisting of a soccer ontology, a corpus of semi-structured and textual match reports and a knowledge base of automatically extracted events and entities. These are extracted from the minute-by-minute reports which are a good example of free text information structured by the time points and containing events that are not covered by structured 'protocols'.

What follows is an alignment of events extracted from multiple minute-by-minute reports with structured reports, synchronizing the video time with the match time (using OCR results) to allow, on the one hand, presentation of the results (a SMIL-based demo) and, on the other hand, further processing of the detected events through the reasoning over complementary resources and cross-media feature selection (cf. Chapter 12).

Fuzzy Reasoning

By reasoning over complementary resources, we refer to the automatic derivation of high-level semantic annotations from primary and secondary complementary resources associated with low-level multimedia data through the utilization of the provided domain knowledge that is soccer in this case.

Our proposal for fuzzy reasoning over complementary resources is based on DL f-$\mathcal{SHIN}$ and FiRE.

The alphabet of the concepts used in the soccer domain is the following set and consist of the data that are extracted by the video and text analysis for a soccer game. The concepts indicated in bold are the ones for which a degree of participation is provided or in other words are the concepts for which we get fuzzy assertions.

Concepts = {*Scoringopportunity Outofplay Handball Kick Scoregoal Cross Foul Clear Cornerkick Dribble Freekick Header Trap Shot Throw Pass Ballpossession Offside Substitution Tackle Double Challenge Charge Lob Nutmeg GoalKeeperDive Block Save Booked* **EndOfField MiddleField Crowd Motion CloseUp Audio**}

Since the features extracted from the text on a soccer game are characterized by coarse-grained minute information, while on the other hand video analysis features characterize every second of the game (bold in the concepts list above), the individuals set consists of the minutes and seconds of the game.

Individuals = {min0 sec20 sec40 sec60 min1 sec80 sec100 sec120 ... },
where min0 corresponds to 1st minute in the game. Each minute is connected by the *consistOf* role with four periods of 20 seconds (the time window consists of 80 seconds, of which 20 seconds are from the previous minute and 60 are from the minute described by the textual data) for which the main video features have been extracted using role assertions like these:

$$(\langle \text{min1}, \text{sec60} \rangle : consistOf) \geq 1$$
$$(\langle \text{min1}, \text{sec80} \rangle : consistOf) \geq 1$$
$$(\langle \text{min1}, \text{sec100} \rangle : consistOf) \geq 1$$
$$(\langle \text{min1}, \text{sec120} \rangle : consistOf) \geq 1$$

The effective extraction of implicit knowledge from explicit knowledge requires an expressive terminology, which is able to define higher concepts. The definitions of some representative event axioms for the soccer domain are presented in Table 11.3.

This modeling associates the domain events with the time at which they occurred. This can prove very useful for various reasons. Firstly, in sports domains such as soccer the exact time at which an event take place is very important. A user, for example, would be

Table 11.3 Knowledge base (TBox): features from text combined with detectors from video

$\mathcal{T} = \{$Goal	$\equiv$	Scoregoal $\sqcap$ ($\exists$consistOfAudio),
LongPass	$\equiv$	(Pass $\sqcup$ Kick $\sqcup$ Shot) $\sqcap$ ($\exists$consistOfMotion),
CornerKick	$\equiv$	Cornerkick$\sqcap$
		($\exists$consistOfMotion) $\sqcap$ ($\exists$consistOfEndOfField)),
SubstitutionD	$\equiv$	Substitution$\sqcap$
		($\exists$consistOfMotion) $\sqcap$ ($\exists$consistOfMiddleField),
HardFoul	$\equiv$	Booked $\sqcap$ (Foul $\sqcup$ Tackle)
		($\exists$consistOf $\sqcap$ (CloseUp $\sqcup$ Audio)),
OffSide	$\equiv$	Offside $\sqcap$ ($\exists$consistOfEndOfField),
ScoringOpportunity	$\equiv$	Scoringopportunity $\sqcap$ ($\exists$consistOfEndOfField)
		$\sqcap$(Clear $\sqcup$ Shot $\sqcup$ Kick$\sqcup$
		GoalKeeperDive $\sqcup$ Block $\sqcup$ Save),
ScoringOpportunityFoul	$\equiv$	ScoringOpportunity $\sqcap$ Foul,
ScoringOpportunity CornerKick	$\equiv$	ScoringOpportunity $\sqcap$ CornerKick$\}$.

able to semantically browse the video, retrieving all the minutes of the game in which goals were scored or hard fouls were made. Additionally, a relation of small periods of times (e.g. 5 minutes) similarly to the way that seconds are related to minutes could produce higher implicit knowledge. Such a period, for example, consisting of minutes with hard fouls and booking events would imply a tough game. Furthermore, this representation permits the modeling of a sequence of soccer events, since they are described together with the possible subsequent events.

12

Multi-Modal Analysis for Content Structuring and Event Detection

Noel E. O'Connor,[1] David A. Sadlier,[1] Bart Lehane,[1] Andrew Salway,[2] Jan Nemrava[3] and Paul Buitelaar[4]

[1]*CLARITY: Centre for Sensor Web Technologies, Dublin City University, Ireland*
[2]*Burton Bradstock Research Labs, UK*
[3]*University of Economics, Prague, Czech Republic*
[4]*DFKI GmbH, Germany*

When dealing with the written word, we are used to two key mechanisms that often help us make sense of and get true value from the content: a *table of contents* and an *index*. The former is typically a hierarchal description of the contents of a book in terms of a well-defined organizational structure to which the reader can easily relate: chapters, sections, subsections. The latter is an organized list of important or useful concepts referenced or described within the content: characters, place names, topics. Both can be considered to provide high-level metadata about the text that facilitates non-linear user-friendly access. Given the obvious benefits, it is little wonder that we would like to be able to map these mechanisms to video content to make it similarly accessible. Salembier et al. (2000) presented this idea and its motivation during the early days of the MPEG-7 standardization effort. In the context of video content, we can draw an analogy between the process of creating a table of contents and what is usually termed content structuring or temporal segmentation. Indeed, this has received much attention from the research community over the last decade. The process of creating an index can be likened to the process of concept detection in video, sometimes termed high-level feature extraction. In this chapter we show that combined audio and visual (and sometimes textual) analysis can assist high-level metadata extraction from video content in terms of content structuring, analogous

Multimedia Semantics: Metadata, Analysis and Interaction, First Edition.
Edited by Raphaël Troncy, Benoit Huet and Simon Schenk.
© 2011 John Wiley & Sons, Ltd. Published 2011 by John Wiley & Sons, Ltd.

to creating a table of contents, and in detection of key events depicted by the content, analogous to creating an index (albeit in each case in a much shallower way than we are used to with a book).

12.1   Moving Beyond Shots for Extracting Semantics

The shot is often considered to be the atomic unit of access to video content, and clearly a shot-level structuring is sometimes the appropriate level of granularity. Consider, for example, journalists who need to retrieve very specific shots from a news archive in order to illustrate the context to a particular news story – for example, shots of the ice floes in the Arctic to illustrate a broader piece on the threat of global warming. Satisfying this information need requires structuring the video content in the library in terms of shots and associating relevant semantic concepts with each shot. There has been significant research into shot boundary detection, resulting in a large array of techniques capable of dealing with the vagaries of various different editing techniques (cut, fade, wipe, etc.) – a comprehensive discussion of the various issues involved is provided in Hanjalic (2002). The set of relevant concepts to be associated with the shots depends on the information needs of the end users for a particular application scenario. In practice, usually a set of useful concepts is defined by domain experts, for example the journalists mentioned above. Significant effort has been devoted to developing generic approaches to concept detection and annotation given a target set of concepts based on machine learning approaches that use shot-level audio and visual features. A good example of a set of concepts defined by domain experts in the case of news video is the *Large-Scale Ontology for Multimedia* (LSCOM) initiative (Naphade et al. 2005). Approaches to automatically labeling shots with such concepts are exemplified by the community's efforts within the high-level feature extraction task in the TrecVid benchmarking activity (Smeaton et al. 2008). Perhaps some of the most comprehensive work in this vein has been done within the MediaMill[1] project (Snoek et al. 2007a), which has successfully developed generic approaches to detect a large number of concepts, often with very impressive success rates. In fact, there has been significant progress made in all stages from feature detection through to representation and learning. Other results worthy of mention represent innovations in the specific areas of object detection and recognition (Fergus et al. 2007), region segmentation and labeling (Shotton et al. 2009), large-scale object and scene recognition (Torralba et al. 2008), scene classification (Bosch et al. 2008), moving object recognition (Ommer et al. 2009), generic video event detection (Xu and Chang 2008), and learning human actions (Niebles et al. 2008).

However, sometimes a shot is not the most appropriate level of granularity at which to provide high-level metadata. Sports fans watching soccer, for example, think about the content in terms of the goals scored, the controversial tackles, the free kicks. Movie buffs and even casual cinema goers think about films in terms of storylines, plot, characters, important dialogues or monologues and exciting events such as car chases. An individual shot cannot adequately convey any of these. Rather, they are associated with sequences of shots and associated audio that collectively reflect the semantics that the director or editor is trying to convey. Thus, we believe that the first step is to structure the content in

---

[1] http://www.science.uva.nl/research/mediamill/

terms of appropriate sequences of shots. Given a temporal segmentation that reflects the underlying semantics, we propose that it is then easier to assign relevant annotations, using complementary data sources if necessary. In this chapter we show two case studies for different content genres, field sports and fictional content, show how both can be more favorably represented by a temporal structure above the shot level using audiovisual processing, and how this structure can then be exploited to extract high-level annotations by considering complementary textual sources.

12.2 A Multi-Modal Approach

Clearly, it behooves us to consider multiple sources of information when attempting to extract information from video content. For example, both audio and visual cues are used by a director when conveying the excitement of a score in a televised sports event – the roar of the crowd, the excitement of the commentator's voice, a reaction shot of the crowd and views of the athlete(s) celebrating. In the case studies that follow, we will show that combined audio and visual analysis on their own can be effectively used in order to construct a useful structuring of the content. However, we will also show that these two data sources on their own are not sufficient for the extraction of annotations that might be used to index the content. For example, a close miss and a score in sports video may be represented in sufficiently similar ways in terms of audio and visual features such that they cannot be easily differentiated. Simply put, very often audiovisual analysis can tell us that *something of interest has happened*, but not *what has happened* To address the latter it is necessary to turn to other information sources beyond the video. Luckily, as explained in Chapter 11, there are a variety of complementary textual sources rich in semantics that can be used to this end (e.g. subtitles, film scripts, news stories, audio description, online blogs, etc) and in Sections 12.4.2 and 12.5.2 we show examples of how a subset of these can be leveraged in the case studies we present. In addition, we will provide initial evidence in Section 12.4.2 that we can then 'close the loop' on this process and use the textual sources to inform the best choice of audiovisual or event-level features for further analysis.

12.3 Case Studies

We have chosen two different case studies, featuring two quite different forms of content, in order to illustrate how high-level metadata can be extracted by first providing access to semantically related sequences of shots. In each case we show how joint audiovisual processing can be used to extract relevant features that can be combined using domain knowledge (implemented via machine learning techniques) to detect events. In both cases this produces a temporal segmentation of the video where each segment reflects a level of semantic meaning, corresponding to exciting events in the case of field sports, and dialogue, action and montage events in the case of movies. We then go on to show how complementary text sources, corresponding to minute-by-minute match reports for soccer and audio descriptions for movies, can be used to extract higher-level concepts associated with this structure. The first process, termed content structuring, is analogous to creating a (shallow) table of contents, whilst the second, termed event detection, is analogous to generating a (non-extensive) index.

A key ancillary aspect of the case studies, above and beyond illustrating the central tenet of this chapter, is to show that whilst the approaches described herein are by necessity specific to the chosen genre, they nonetheless generalize well within this genre. For each study we describe a generic approach to content structuring that can be broadly applied to many sub-genres – soccer, field hockey, Gaelic football and rugby in the case of field sports, Hollywood action 'blockbusters', independent films, and TV dramas in the case of fictional content. This is possible because in each case we build our analysis around key conventions used by editors or directors that remain more or less constant despite the creative process involved in producing the content. In should also be noted that these conventions are quite different for each case study. For field sports, the conventions used reflect the constraints of a live broadcast. Movie directors, on the other hand, have greater scope for creativity. As a result, in terms of algorithmic design, the two case studies described are approached individually based on independent content models, a paradigm which, for clarity, is reflected in their description, and hence the structure of this chapter. Of prime significance, however, is that in each case the approach established can be considered quasi-generic within what are actually quite broad content domains.

12.4 Case Study 1: Field Sports

In this section we describe a case study that illustrates an approach to content structuring and summarization in field-sports video. Specifically, it is shown how audiovisual processing is exploited to construct a temporal segmentation of the video, resulting in a two-tier description, with the upper tier exclusively containing segments classified as *exciting*, and the lower tier occupied with the remaining content. Emphasis is focused on field-sports in particular on the basis that represented under this classification are some of the most conventionally popular sports, including soccer, American football, rugby, Australian rules football, field hockey, Gaelic football, and hurling. Testament to this is a considerable amount of literature already targeting each of these sports genres from a video analysis point of view – albeit with the larger portion focusing on soccer video. Assfalg et al. (2003) and Ekin et al. (2003) exploit visual characteristics such as camera motion, and perform object-tracking towards automatically detecting soccer video highlights. In Andrade et al. (2005) the topic of interaction is discussed whereby the authors propose an object-tracking approach for soccer video based on spatial region detection and matching. Visual analysis methods for soccer video indexing, such as player tracking, are proposed in Utsumi et al. (2002), whilst a multi-modal approach to this problem is presented in Leonardi et al. (2004). Xie et al. (2004) propose visual analysis methods for the structural segmentation of soccer video.

12.4.1 Content Structuring

Exciting Events

One of the most salient indicators of an exciting period of play (event) is the occurrence of action replays. However, whilst generic methods for action replay detection have been researched (Pan et al. 2002), most are rooted in exploitation of slow-motion-based frame repetition. Thus, with the increased use of high-speed camera technology, such methods are undermined. However, another related characteristic is the existence of a lag time that immediately follows the occurrence of an exciting event, before the cut to action replay. The director tends to utilize this *reaction segment* (RS) to capture the responses of players and/or crowds to the significance of the event. So, for example, after a score has taken place, the RS tends to be characterized by: (i) close-up shots of the player(s) and/or relevant parties involved; (ii) camera shots showing the crowd cheering; (iii) an increase in audio activity; (iv) activity in the on-screen graphics (e.g. scoreboard updating); and (v) a surge in motion activity (as the camera zooms in to capture the reaction of the players). Now, whilst it is not uncommon to observe such phenomena on an isolated basis throughout the duration of the broadcasts, underpinning our approach is the idea that when many of them are found to exhibit a high prevalence and/or intensity within a short time frame (akin to the duration of an RS), this likely indicates the occurrence of an exciting event. Based on an inspection of 90 hours of content for four different sports from at least four different broadcasters, it was observed that the typical RS length is approximately 18 seconds in duration (time between event occurrence and cut to replay), whilst the longest duration observed was approximately 25 s (Sadlier and O'Connor 2005a). Considering this 25 s limit as a post-event *reaction segment seek window* (RSSW), the combined detection/quantification of the above-mentioned characteristics within this window will provide a reliable basis for the event detection. Furthermore, in addition to the RS features, it was noted that it is typical for the prior action of a large portion of exciting events to be situated at the *end-zone* region of the playing field. Hence, this attribute was also considered exploitable. To this end, adding the characteristic of pre-event end-zone activity to the set of post-event RS orientated cues represents the complete set of *event features* (EFs) that we use. In the next section we specify the EFs in more detail and explain how audio and visual processing can be used to extract them.

Event Feature Extraction

The first EF corresponds to the detection of close-up/mid-shot images (EF1). The salient characteristics of close-up images are (i) the presence of a face in the top-middle-center region (i.e. the focus) of the frame, together with (ii) a jersey in the bottom-middle region of the frame, occluding the background. Based on an inspection of many more close-up images sampled from the data corpus mentioned above, estimations of the typical regions of expectation for these characteristics were delineated – see Figure 12.1. On this basis, close-up detection is performed according to the degree to which the image has (i) a skin-toned (Terrillon and Akamatsu 1999) entity (i.e. a face) in R1 and (ii) a monochromatic entity (i.e. a jersey) in R2, with a relatively lower occurrence in R3. Specifically, a formula for close-up confidence (CuC) exploiting these perceptions is given by

$$CuC = SPR_{R1} \times (DH^{R2}PR_{R2} - DH^{R2}PR_{R3}), \tag{12.1}$$

where SPR_{R1} represents the skin pixel ratio of R1, $DH^{R2}PR_{R2}$ represents the dominant hue pixel ratio of R2, and $DH^{R2}PR_{R3}$ represents the pixel ratio of the R2 dominant hue in R3.

Crowd view detection (EF2) may be performed by exploiting the inherent characteristic that such images are highly textured compared to the vast majority of the other images constituting field-sports video (which conversely tend to capture large, monochromatic,

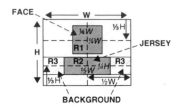

Figure 12.1 Detecting close-up/mid-shot images: best-fit regions for face, jersey, and background. Reproduced by permission of © 2005 IEEE.

homogeneous regions – grassy pitch, players' shirts, etc). Furthermore, the texture density tends to be spatially uniform, and therefore a further heuristic relates to enforcing spatial consistency toward bolstering the discrimination. On this basis, the video frames are divided into five *regions of interest* (ROIs), representing the center and the four extreme corner regions of the image. To quantify the above-mentioned attributes, *edge pixel ratio* (EPR) values are calculated for each ROI. Given the EPR values for each region, the mean value (μEPR) is computed. In addition, a value representing the maximum absolute difference between the EPR values for any two ROIs is computed (ΔEPR). Then *crowd view confidence* (CVC) values are computed using the formula

$$CVC = \mu EPR - \frac{\Delta EPR}{\sum EPR}. \tag{12.2}$$

When an exciting event occurs, the commentator's voice typically increases in volume. To extract the audio volume envelope (EF3), MPEG audio subband scalefactors are extracted and processed to yield audio energy levels at 0.5 s intervals (Sadlier and O'Connor 2005b).

EF4 relates to detecting scoreboard activity. One of the obvious characteristics of a scoreboard graphic is the presence of text, and for text to be visible there must be a strong luminance contrast between the foreground and background. Thus, during the encoding process (MPEG), this text requires a relatively large amount of AC-DCT luminance coefficients to represent it (Sadlier and O'Connor 2005a). Furthermore, for a given broadcast, the location of the graphic is usually static, and it is present on-screen for the main duration of the game. Hence, the pixel blocks that constitute the graphic will exhibit a large number of AC-DCT luminance coefficients consistently over the duration of the game. In contrast, pixel blocks not associated with the graphic will naturally, over the course of the broadcast, constitute many different aspects of the images captured, and will not exhibit a consistently high number of these coefficients. On this basis the pixel block location of the graphic may pinpointed. The task is then to detect when it is updated. It turns out that many broadcasters perform this whilst temporarily withdrawing it from view. A simple template-based image-differencing approach is undertaken to determine when this occurs.

EF5 relates to tracking motion activity. This is based on the analysis of MPEG *motion vectors* (MVs) extracted from the video bitstream. Following this process, for each (motion estimated) image, the *non-zero vector value* (NZVV) is calculated, by counting the number of macroblocks in the frame whose motion vector length is greater than a pre-selected 'zero' threshold. The 'zero' threshold value is chosen to be large enough

Figure 12.2 Goalmouth views capturing events in soccer, rugby, and hockey. Reproduced by permission of © 2005 IEEE.

to ignore slow, smooth motion (which occurs constantly throughout sports-video broadcasts), while detecting jerky motion (i.e. the type of turbulent motion expected during/after exciting moments).

Finally, EF6 detects end-zone views. Due to the fixed position of the camera, the resulting perspective of such views is such that the field lines tend to only assume certain angles, which lie within a particular narrow interval, relative to the point of observation (see Figure 12.2). Again, by examining the data, it was observed that nearly all the angles of such field lines (the most prominent field lines in view) mapped inside the interval [$5°$, $25°$]. Hence, by detecting the field lines and continuously tracking their angle of orientation, it is possible to detect the images that correspond to field end-zone shots. To realize this, pixels representing the playing field (grass) are segmented in the hue space. Segmentation of the field lines from the grass is then achieved via a binarization of the luminance space of the isolated playing field. Edge detection (Roberts 1965) is then utilized to isolate the edges of the binarized images, and then, for each edge-detected binary image, the field lines are extracted by means of the Hough line transform (Terrillon and Akamatsu 1989). For each image processed, the angle of the most prominent detected line becomes the EF6 value.

Event Detection

EF data (extracted at the video frame level) is aggregated into *shot feature vectors* (SFVs), designed to represent the extent to which each shot may be considered 'exciting'. The SFVs then form the input for a higher-level pattern analysis phase in realizing collections of exciting shots, in turn realizing the segmentation of exciting moments at the *super*-shot level. Recalling that EF6 represents the angles of the most prominent field-lines, to quantify the confidence that a particular shot culminates with the camera focused on activity located in goalmouth area, the angle values are examined for the images towards the end of each shot. Then, when the orientations are found to lie in the key range [$5°$, $25°$], the corresponding shot-level confidence value increases accordingly. This confidence value constitutes the first *vector component coefficient* (VCC) of the SFV. For the remaining EFs, the concepts of the RS and RSSW form the basis of the aggregation process. Specifically, VCC2 represents the maximum EF1 close-up confidence value found within the 25 s seek window (RSSW) beginning at the *shot end boundary* (SEB); VCC3 represents the maximum EF2 crowd-view confidence value found within the post-SEB RSSW; VCC4 corresponds to the maximum EF3 audio activity observed; VCC5 corresponds to

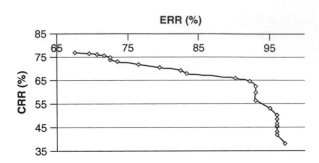

Figure 12.3   ERR vs CRR for rugby video. Reproduced by permission of © 2005 IEEE.

the maximum EF4 scoreboard activity observed. To capture the *extent* of post-SEB visual activity (as opposed to *peak* levels), VCC6 corresponds to a value representing the proportion of post-SEB RSSW images that exhibit EF5 NZVV levels in excess of the mean level. The six-dimensional SFVs constitute the input data for the classifier, which, for the purposes of this analysis, was chosen to be a support vector machine (Burgess 1998).

Illustrative Results

As alluded to earlier, a 100-hour field-sport video corpus was compiled, comprising rugby (18%), hockey (20%), soccer (31%), and Gaelic football (31%). This was then split in half, yielding (i) a 50-hour sub-corpus for support vector machine training, and (ii) a 50-hour sub-corpus for testing. A feature of support vector machines is the *regularization parameter*, C, a user-defined value that can be used during training to tune the classification (Burgess 1998). In effect, varying C during the training phase allows the user to tune the classification, such that an increase should improve precision at the expense of recall (and conversely a decrease should yield higher recall at the expense of precision). Following the shot classification phase, consecutive shots with the same classification (i.e. exciting or non-exciting) were grouped together, yielding content structuring at the level of shot sequences.

   Given the subjectivity associated with ground-truthing exciting moments, qualitative assessment of the approach was carried out by considering its performance in relation to its ability to detect score-related events (goals, tries, conversions, etc.). Considering the performance with respect to rugby video alone, estimations for event retrieval ratio (ERR, a true-positive statistic), and content-rejection ratio (CRR, a true-negative statistic), were computed. Figure 12.3 presents a plot of these statistics for rugby for varying values of C.

   From this it is clear that the maximum ERR value achieved is 97%, and the corresponding CRR value for this particular value of C is 38%. Then, as C varies, the classification performance varies toward a saturation limit of approximately 67% ERR and 77% CRR. Similar statistical analyses were performed for the other field-sport genres comprising the test corpus, and details may be found in Sadlier and O'Connor (2005a). Clearly for every broadcast there is an optimum point where both ERR and CRR are as high as possible. However, we believe that in a practical retrieval system, high recall (ERR) is paramount since a user would be more likely to tolerate the inclusion of false positives in the upper tier, as opposed to relegating true positives to the lower tier. On this basis, Table 12.1

Table 12.1 Performance of event detection across various sports: maximum CRR for 90% ERR

Genre	CRR
Soccer	74%
Hockey	72%
Gaelic football	69%
Rugby	65%

shows the maximum CRR values for a sensibly chosen ERR evaluation point (90%). It is clear that the best overall performing genre is soccer. However, the small variances in these statistics across the different genres suggest that the approach undertaken toward content structuring generalizes well within the domain of field-sports video.

12.4.2 Concept Detection Leveraging Complementary Text Sources

Assuming a temporal segmentation that reflects the semantics of video sequences depicting exciting events, in this section we describe our initial work on how we can associate finer-grained semantics with the video by leveraging complementary sources. Chapter 11 describes the role and potential usefulness of complementary sources in detail. Here we demonstrate the applicability of that discussion to the case study of field-sports video, in the particular case of soccer.

As mentioned in Chapter 11, we can use either structured or unstructured text sources. Unstructured reports, however, require more sophisticated techniques; for example, they need to be extracted from websites using wrappers and using NLP-based information extraction tools, like the one described in Drożdżyński et al. (2004) that can be applied in order to extract domain-related ontology concepts. A dataset for ontology-based information extraction and ontology learning from text (SmartWeb corpus; Oberle et al. 2007) consists of a soccer ontology, a corpus of semi-structured and textual match reports and a knowledge base of automatically extracted events and entities. These are extracted from the minute-by-minute reports, which are a good example of free-text information structured by the time points and containing events that are not covered by structured 'protocols'. Combining several of these reports can increase the probability of covering unidentified concepts from the video analysis and also eliminate false positive events with small evidence from the texts (Nemrava et al. 2007). What follows is an alignment of events extracted from multiple minute-by-minute reports with structured reports, synchronization of the video time with the match time (using OCR results) to allow presentation of the results and also further processing of the detected events through reasoning over complementary resources as described in Chapter 11.

Toward Cross-Media Feature Selection and Extraction

We now consider how textual resources can be used for the *selection* of audiovisual features specific to particular event types that cannot be distinguished using audiovisual processing alone. In other words, we seek to understand if the complementary sources can

Table 12.2 Results of the cross-media feature selection (P, C, N, Previous, Current, Next; M, E, O, Middle, End, Other)

	Crowd			Audio			Motion			Close-up			Field-line		
	P	C	N	P	C	N	P	C	N	P	C	N	M	E	O
Foul	●				●				●			●	●		
Free kick		●			●								●	●	
Header			●					●		●			●	●	
Shot on goal					●				●			●	●	●	●
Foul + shot on goal					●				●			●	●	●	

help indicate what features we should be using to detect specific kinds of events, as oppose to simply building concept detectors blindly as we have done thus far. Identification of characteristic features based on textual evidence is termed *cross-media feature selection and extraction*. Using machine learning techniques, we try to determine discriminative features of selected soccer event types, that is, to identify which video detectors are more prominent for which event class, by analyzing the distribution of video detectors over event classes and identifying significant detectors for each class. Based on this we can design classifiers assigning the appropriate event type to the events detected in Section 12.4.1. We therefore test whether general detectors such as the ones described in Section 12.4.1 are able to classify events of a certain kind. We used decision trees as the machine learning algorithm and built a binary classifier for each of the observed events. The task of the classifier was to decide whether the particular segment is or is not an event of a certain type.

Initial results with cross-media feature selection are presented in Table 12.2. The table shows that different detectors are important for different event types. This potentially allows instances of event types to be detected based on observing only those detectors that are discriminating for them (this assumption is also used by the decision tree algorithm). The letters P, C, N represent the previous, current, next value of the detector for a particular event type. For example, for fouls the discriminating detector values are values of the crowd detector before the foul, audio is discriminating during the foul event and increased motion and close-up were observed after the event occurred in the video stream. (M)idle represents discretized values of the field-line feature, and in the foul example defines that fouls mostly occur mostly in the mid-field region. Referring again to the discussion of Section 12.1, the different event types listed in Table 12.2 effectively constitute an initial coarse index of concepts built around the event-based temporal structure.

12.5   Case Study 2: Fictional Content

In this section we describe the second case study, which illustrates a very similar approach to content structuring and knowledge extraction but adapted to fictional content. Fictional content refers to the filming of a 'story' as opposed to live broadcasts (such as sports or news) or documentary footage. Thus, in this context we are concerned mainly with movies and television dramas. We show how such content can be structured to three kinds of events that viewers understand. We then go on to show initial work in extracting deeper semantics of the content by linking this structure with a text source using relatively shallow textual processing. Again, there is a considerable amount of existing literature relating to the analysis of fictional content that warrants mention. Fitzgibbon and Zisserman (2002) propose an approach for cast listing based on a face clustering paradigm. Meanwhile, Everingham et al. (2006) present encouraging results in the automatic naming of characters in fictional content based on aligning subtitles and transcripts. More recent works relate to proposed solutions for the automatic identification of people, objects and actions in movie content (Laptev et al. 2008), ultimately moving towards other approaches claiming complete movie scene parsing (Cour et al. 2008).

12.5.1 Content Structuring

Dialog, Action and Montage Events

When considering fictional content we might at first consider the notion of a 'scene' as used in the film-making world to be the appropriate level at which to perform content structuring. Indeed, there has been much research focused on scene-level analysis of content in recent years (Cao et al. 2003; Cotsaces et al. 2006; Lin and Kender 2000; Yeung and Yeo 1997). However, scene-level analysis is problematic, not least because very often humans themselves cannot recognize scene boundaries in a diverse and creative medium, where dramatic devices such as flashbacks/flashforwards, dream sequences, and cross-cutting are regularly used. Our studies indicate, on the other hand, that users can strongly relate to structuring of the content at a level below the scene, but above individual shots that we term an event (Lehane et al. 2006), where the specifics of what we consider an event is described in the next paragraph.

We consider three different kinds of events that together typically account for more than 90% of a film/drama, whilst being intuitive for a user to understand. The choice of these is motivated by a study of the principles of film-making (Bordwell and Thompson 1997; Dancyger 2002):

- Dialogue constitutes a significant part of most fictional content, and is the means by which the viewer obtains much information about the plot, story, background, etc. A relaxed filming style is used with a low amount of camera movement, few long shots, plenty of shot repetition as the camera repeatedly switches between the protagonists involved, and clearly audible speech. This is necessary in order to allow the audience to concentrate on what is being said without being distracted.
- In exciting events the film-maker will typically attempt to stimulate the audience by bombarding them with new information via an increased edited pace (shorter shot length), combined with high levels of movement within shots that feature a wide varying camera angles.
- Montage events[2] are typically emotional scenes (such as somebody weeping) and diagetic music (i.e. music being played within the film), and have a more relaxed editing pace.

---

[2] Note that our use of the term *montage* here is somewhat different than the classical interpretation as it refers to the superset of both emotional events and musical events.

Much of the previous work on film indexing has typically concentrated only on the detection of dialogue events (Leinhart et al. 1999; Li and Kou 2003, 2001) or dialogue and a restricted set of specific kinds of action events (Chen et al. 2003), whereas we consider all three. From this discussion, it should be clear that a scene can consist of multiple events – for example, it starts with a discussion between two characters (dialogue), then a fight breaks out (exciting), ending with an emotional montage as one character lies injured.

Audiovisual Feature Extraction

Prior to any audiovisual analysis, the fictional video content is first segmented into shots and keyframes extracted using a standard color histogram technique (Browne et al. 2002). When tested on 378 shots, including fades and dissolves, this method detected achieved recall of 97% and precision of 95%, indicating that it is sufficiently robust for our purposes. We then perform shot clustering based on the technique proposed in Yeung and Yeo (1996) in order to detect changes in the focus on the video and thereby ultimately delineate event boundaries. Clustering also allows us to calculate a measure of shot repetition within an event by calculating the ratio of clusters to shots, termed the *CS ratio*.

Two motion features are extracted. The first is the MPEG-7 motion intensity (Manjunath et al. 2002) within each frame averaged over the entire shot. The second feature coarsely measures the amount of camera movement in each shot by classifying each shot as moving or non-moving based on consideration of runs of zero motion vectors in the key-frame (Lehane and O'Connor 2004) – this is similar in nature to the motion feature used for field sports described in Section 12.4.1.

For audio analysis we chose to detect specific audio classes corresponding to *speech*, *music*, *quiet music*, *silence*, and *other*. To detect these classes, we first extract three different audio features. The first is the *high zero crossing rate ratio* (HZCRR) calculated as the ratio of audio samples above and below 1.5 times the average zero crossing rate. This feature is useful for detecting speech that is characterized by short silences between spoken words (Chen et al. 2003; Li and Kou 2003). The second audio feature is the silence ratio, calculated as the ratio of silent to non-silent sub-segments in a window of audio samples, where silence is indicated by low *root-mean-square* (RMS) values. This feature is useful for distinguishing between music and speech, as speech tends to have a higher silence ratio (Chen et al. 2003). The third audio feature is the short-term energy variation, calculated as the number of samples that have an average energy less than half the overall energy of an audio segment. Using these features, we train a number of support vector machines to recognize each audio class mentioned above. The audio class of each shot is labeled as the dominant audio class of the samples within the shot. Using a sample of 675 shots, we were able to correctly label 90% of the manually labeled shots with the correct audio class.

The result of audiovisual analysis is a feature vector for shots containing the following components: [% speech, % music, % silence, % quiet music, % other audio, % static camera frames per shot, % non-static camera frames per shot, motion intensity, shot length]. For further details on this entire process, see Lehane and O'Connor (2004, 2006) and Lehane et al. (2005).

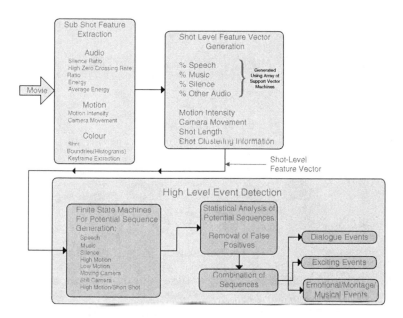

Figure 12.4 Detecting events based on audiovisual features.

Event Detection

The overall process of event detection following audiovisual feature extraction is illustrated in Figure 12.4, which shows that prior to event detection, it is first necessary to
detect sequences of shots where the audiovisual features remain more or less constant.
For this we use a suite of six finite-state machines (FSMs), as shown in Figure 12.5,
to detect sequences predominantly featuring speech, music, non-speech, static camera,
non-static camera and combined high motion and short shots. This choice reflects the
set of tools available to a director to convey the characteristics of the three event types.
The main difference between the different FSMs used was the number of intermediary
states (depicted as the gray states in Figure 12.4) between declaring the start and end
of a sequence when a feature is declared as being dominant. This number of intermediary states was determined empirically. Taking the static camera FSM as an example,
the FSM locates video sequences of arbitrary length where a static camera is used by
the director. This does not mean that every shot need have a static camera, just that this
feature dominates over the sequence. When in the start state (state **S** for each FSM in
Figure 12.4) static camera shots are prevalent, whilst in the stop state (state **E** for each
FSM in Figure 12.4) they are not. The FSM will remain in either state as long as it
encounters shots that have (or do not have) static camera motion. The intermediate state
acts as a buffer as the FSM drifts away from one state to the other (note that movement
toward the end state is depicted by dashed arrows in Figure 12.4). If a change of focus
is detected as a result of the visual similarity clustering algorithm described in Section
12.5.1, then each FSM returns to the start state and is reset, outputting a sequence where
the audiovisual features remain constant.

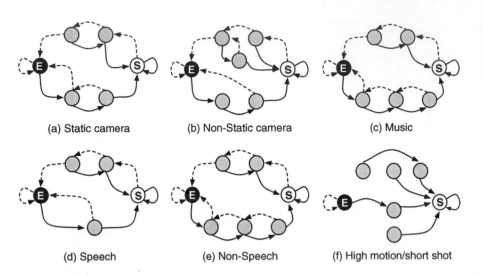

(a) Static camera   (b) Non-Static camera   (c) Music

(d) Speech   (e) Non-Speech   (f) High motion/short shot

Figure 12.5   FSMs used in detecting sequences where individual features are dominant. Reproduced by permission of © 2007 B. Lehane, NE O'Connor, H. Lee and AF Smeaton.

The second step in event detection is to filter and combine the detected sequences using rules that reflect film-making conventions. To detect dialogue events, the speech and static FSMs are used, along with the CS ratio measure of shot repetition described in Section 12.5.1. A potential dialogue sequence from the speech FSM is constrained to have a low CS ratio (i.e. significant repetition) and a high amount of static shots, whilst for the static camera FSM sequences the constraint is a low CS ratio and a high amount of speech shots. Sequences from either FSM retained after this filtering step are combined using a Boolean *OR* operation to produce the output dialogue events.

The high motion and short shot FSM sequences are used to detect exciting events, reflecting the fact that the director typically creates excitement by startling or disorientating the viewer via fast paced editing and lots of motion. Only sequences with a high CS ratio (i.e. little repetition) are retained, as are high-tempo musical sequences.

The audio FSMs are mainly used to detect montage events as these usually have a strong musical accompaniment (Bordwell and Thompson 1997). Although emotional events may contain music, they may also contain silence, so we use the non-speech FSM sequences. We filter sequences with low CS ratios (significant repetition) to take into account dialogue events that have a strong musical background. We then only retain sequences where short shots and high amounts of motion are not present, to reflect the relaxed editing pace typically used by directors.

Illustrative Results

To validate this approach to content structuring, we applied it to a large variety of fictional content. The test corpus consisted of a variety of Hollywood movies (*American Beauty*, *Dumb and Dumber*, *Goodfellas*, *High Fidelity*, *Reservoir Dogs*, *Snatch*, and *Shrek*), less mainstream movies (*Amores Perros*, *Battle Royale*, and *Chopper*) and TV content (three episodes of each of *The Sopranos*, *The Simpsons*, and *Lost*). Precision and recall figures were generated by comparing the output of the event detection process to a manual ground truth created for each event type for each of the mentioned documents, marked up by human viewers. For dialogues we obtained a precision of 81% and a recall of 94%. For exciting events we obtained a precision of 59% and a recall of 95%, and for montages we obtained a precision of 73% and a recall of 91%. These results clearly demonstrate that this approach to content structuring can generalize within the domain of fiction content. It should be noted that system performance is deliberately biased towards recall to reflect the fact that different viewers have different interpretations of the different events (e.g. is a very heated discussion a dialogue or an action event?). Therefore we ensure that all such borderline events are detected in *both* classes, giving higher recall but lower precision for any single interpretation. These results are discussed in greater depth in Lehane et al. (2007).

12.5.2   Concept Detection Leveraging Audio Description

Audio description (AD) is the complementary text data source that we choose to use for semantic concept detection. AD provides a spoken description of what is occurring on screen, for the benefit of blind and visually impaired film and TV audiences. The on-screen action is described by a human in between the existing dialogue via an additional soundtrack. It is scripted in advance with precise time-codes for when a particular piece of description is to be spoken, and descriptions contain information on who is present, what they look like, what they are wearing, their facial expressions, and, of course, what they are doing. In our approach we use AD scripts to determine the on-screen presence of characters and then use this to infer relationships between characters as high-level metadata. See Salway (2007), Salway et al. (2005), Vassiliou (2006), and Tomadaki (2006) for more detailed discussions on AD and its relationship to events, plot summaries and film scripts.

We use relatively straightforward text processing to infer the on-screen presence of characters. We first search for capitalized words that occur frequently in a given description segment that are candidates for characters' names. For common names we then disambiguate the gender by looking up lists of male and female names generated by the US Census.[3] This is followed by simple pronoun resolution that searches backwards from each occurrence of / he | she | he's | she's / to the first character name of matching gender. This simple process is actually quite effective as AD tries to avoid ambiguous pronouns so as not to distract the audience. This results in a list of characters and a time-code each time the character is on-screen.

Character Occurrence

Given the temporal structuring of the data as described in Section 12.5.1, it is now possible to associate characters with events by associating a character with the event containing

---

[3] A list can be found at http://www.census.gov/genealogy/www/freqnames.html. Note that manual intervention was required in our initial experiments for names that would not appear in the census data of any country (e.g. 'Shrek' and 'Stitch'). This was required for half the content in our test set but we plan to automate this in the future using further complementary resources, such as the Internet movie database, http://www.imdb.com.

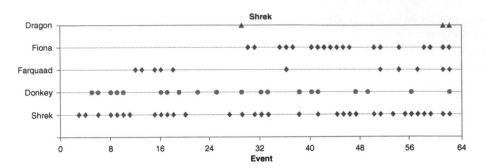

Figure 12.6 An index of character appearances based on dialogues in the movie *Shrek*. Reproduced by permission of © 2007 Association for Computing Machinery, Inc. (ACM).

his/her on-screen time-code with a small offset. This allows us to create an index of when each character is on-screen during a semantically meaningful event. The structuring of the film *Shrek* in terms of dialogue events and the occurrence of the main characters within each dialogue is illustrated graphically in Figure 12.6. This figure attempts to visualize both the table of contents and index at once. From this we can see that this visualization reflects the character development in the movie as well as their relationships. We can see that Shrek and Donkey are the main characters and are present, usually together, throughout the movie. The love interest character, Princess Fiona, appears halfway through the movie and then appears consistently. Lord Farquaad and Dragon are relatively minor characters appearing and disappearing, but all characters come together for the film's denouement. For further examples and discussion, see Salway et al. (2007).

Character Relationships

In fact, using the framework described thus far, we can go further with our analysis of character relationships and take some initial steps toward even higher-level metadata corresponding to relating characters with narrative structure. For example, knowing that two characters are on-screen at the same time during dialogues (and understanding that dialogues are the key technique used by directors to progress the narrative) allows us to objectively measure the co-occurrence between these characters over the course of the entire movie, $C(A, B)$, defined as the total number of events in which the characters appear together, divided by the number of events in which character B appears. This calculation for characters in the film *American Beauty* is shown in Table 12.3. By comparing $C(A, B)$ and $C(B, A)$, we can draw some very interesting conclusions about high-level character relationships. For example, it can be noted that the Carolyn character is very 'important' to the Buddy character since $C(Buddy, Carolyn) = 1$ (meaning that Carolyn is always on-screen when Buddy is on-screen). However, the opposite is not true, $C(Carolyn, Buddy) = 0.23$, indicating that the Buddy character is not particularly important to the Carolyn character in the context of the overall story. In fact, this mirrors exactly their relationship in the movie, whereby their affair forms a minor sub-plot.

The shaded co-occurrence values in Table 12.3 highlight situations when both $C(A, B)$ and $C(B, A)$ are high (calculated as being both above an empirically defined threshold),

Table 12.3 Dual co-occurrence highlighted for different character relationships

	Lester	Carolyn	Jane	Buddy	Angela	Ricky	Frank
Lester	1.00	**0.40**	**0.43**	0.09	0.36	**0.38**	0.21
Carolyn	**0.61**	1.00	0.45	0.23	0.23	0.35	0.13
Jane	**0.49**	0.34	1.00	0.02	0.34	**0.56**	0.12
Buddy	0.57	**1.00**	0.14	1.00	0.00	0.14	0.14
Angela	**0.71**	0.29	0.58	0.00	1.00	0.54	0.29
Ricky	**0.46**	0.28	**0.59**	0.03	0.33	1.00	**0.41**
Frank	0.56	0.22	0.28	0.06	0.39	**0.89**	1.00

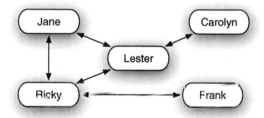

Figure 12.7 Main character interactions in the movie *American Beauty*. Reproduced by permission of © 2007 Association for Computing Machinery, Inc. (ACM).

indicating important dual relationships. These relationships are plotted in Figure 12.7. From this it is clearly seen that Lester is the main character and all other characters are 'defined' in terms of their relationship with him, which reflects the narrative structure of a man reflecting on his life and relationships.

12.6 Conclusions and Future Work

In this chapter we posited that content structuring above the shot level is often necessary for extracting semantics from video content. We validated this by means of two different case studies targeting different kinds of content that include many sub-genres – field sports and fictional content. For each we showed how quasi-generic content structuring at the *event level* can be achieved by using combined audiovisual analysis and a suitable machine learning paradigm. We then showed how complementary textual sources, corresponding to minute-by-minute match reports for soccer and audio description for movies, can be leveraged to extract even higher-level metadata.

Clearly more experimentation is required toward further revealing the benefits of this approach, and as is evident from the preliminary results presented in Sections 12.4.2 and 12.5.2, significant research remains to be carried out in creating the index based on the results of event-level structuring. Also, whilst the results we show do hint at the possibilities of very fine-grained semantic extraction, a comparison against the results of existing approaches is desirable, and much work remains in order to fully automate these processes so that they are robust and more generically applicable. In particular, in the work on fictional content we plan to introduce other textual sources corresponding to plot summaries and character lists. An intriguing possibility in this context is linking across movies based on knowing which actors play which characters. This would ultimately enable us to support complex user information needs, such as 'Show me all the romantic movie scenes featuring Katharine Hepburn and Spencer Tracy'.

13

Multimedia Annotation Tools

Carsten Saathoff,[1] Krishna Chandramouli,[2] Werner Bailer,[3] Peter Schallauer[3] and Raphaël Troncy[4]

[1] *ISWeb – Information Systems and Semantic Web, University of Koblenz-Landau, Koblenz, Germany*
[2] *QMUL Queen Mary University of London, London, UK*
[3] *JOANNEUM RESEARCH – DIGITAL, Graz, Austria*
[4] *Centrum Wiskunde & Informatica, Amsterdam, The Netherlands*

As we have seen in previous chapters, many techniques exist to analyze, annotate and further interpret multimedia content. However, these techniques remain limited with respect to the information they can automatically generate, which makes it important to provide tools that assist users in correcting, refining, and creating annotations manually. These tools are generally referred to as *multimedia annotation tools*. These tools also often offer retrieval functionalities, since the typical purpose of annotations is to provide the means for semantically browsing or querying a repository.

In this chapter we introduce two tools for multimedia annotation, namely the *Semantic Video Annotation Tool* (SVAT) and the *K-Space Annotation Tool* (KAT). Though technically quite similar, these tools target different problems. On the one hand, SVAT is aimed at professional users in the audiovisual media production and archiving domain, and is therefore driven by a quite specific use case. The underlying model is the *MPEG-7 Detailed Audiovisual Profile* (Bailer and Schallauer 2006), and the whole tool is designed to handle videos in terms of manual and automatic annotation techniques.

KAT, on the other hand, is supposed to be a framework for efficient and rich semantic annotation of multimedia content. In contrast to SVAT, its underlying model is the *Core Ontology for Multimedia* (see Chapter 9), covering not only videos, but all kinds of multimedia data. Similarly, KAT is not geared toward either a specific media type or a specific use case, but provides a framework for implementing annotation tools for various kinds of scenarios. KAT therefore abstracts every kind of functionality as a plugin, which includes not only analysis, but also visualization of content items or collections, retrieval, and annotation.

Multimedia Semantics: Metadata, Analysis and Interaction, First Edition.
Edited by Raphaël Troncy, Benoit Huet and Simon Schenk.
© 2011 John Wiley & Sons, Ltd. Published 2011 by John Wiley & Sons, Ltd.

However, both tools try to facilitate efficient annotation by providing the means to integrate automatic and manual annotation in a framework that reduces the effort required of the user, while supporting the annotation of multimedia content with semantic metadata.

This chapter is structured as follows. In Section 13.1 we briefly discuss the current state of the art in multimedia annotation tools. We then introduce SVAT in Section 13.2 and discuss its core functionality. We continue with KAT in Section 13.3, where we focus on semi-automatic annotation and, in particular, discuss the implications of using a generic multimedia ontology like the COMM. We conclude the chapter in Section 13.4.

13.1   State of the Art

There are already several annotation tools available, both for images and videos. In this section we discuss the most prominent tools available, specifically with a focus on semantic media annotation tools as collected by the W3C Multimedia Semantics Incubator Group,[1] and on audiovisual media production and authoring tools. Obviously, these two areas are also targeted by both tools we present in this chapter.

One of the most prominent annotation tools on the Semantic Web is *Ontomat Annotizer* (Handschuh and Staab 2003), which was mainly focused on annotating web pages, but never emphasized the multimedia aspect involved in this task. Based on Ontomat the tool, *M-Ontomat Annotizer* (Bloehdorn et al. 2005) was developed, a tool that focuses on generating prototypical knowledge for the automatic annotation of multimedia content. However, the tool was never extended to a real annotation tool that also stored the annotations and made them available to other applications. KAT was heavily inspired by M-Ontomat Annotizer, and both built originally on Ontomat as a framework. We discuss this aspect in Section 13.3.1.

AKTive Media (Chakravarthy and Lanfranchi 2006) is a tool integrating automatic and manual annotation of multimedia content. It involves annotation with different domain ontologies, and includes methods to support a user in handling complex ontologies. It also supports region-level annotation and relational annotation, to be able to relate entities depicted within the content. During annotation it tries to gather additional contextual information by querying web services and showing related information, for example by using information extraction techniques.

A tool targeted at semantic image annotation and the publishing of the metadata on the web is *PhotoStuff* (Halaschek-Wiener et al. 2005). This tool does not support any other media types, nor does it integrate any advanced analysis. However, it does extract EXIF and Dublin Core metadata from image files if present and transforms it into an ontology-based representation.

Another interesting tool is *Caliph*, which is usually mentioned in conjunction with its retrieval engine *Emir* (Lux et al. 2003). The tool is based on MPEG-7 and is capable of extracting low-level features and indexing them for typical content-based image retrieval. Furthermore, it allows for semantic annotation using the appropriate MPEG-7 descriptor and provides a graphical user interface for creating relational annotations. However, since the tool is based on MPEG-7, the metadata created hardly integrates with other available metadata from the Semantic Web, and even linking metadata across different media items becomes hard.

---

[1] http://www.w3.org/2005/Incubator/mmsem/wiki/Tools_and_Resources

We are not able to discuss all available tools in this chapter, and obviously other tools exist. A more complete overview is provided by the W3C Multimedia Semantics Incubator Group. Web 2.0 platforms might also be considered annotation tools, as they typically provide the means for tagging content with keywords or geolocations. However, Web 2.0 tools usually employ proprietary data models and do not publish their data in a common format. But also the Semantic Web tools discussed typically use custom ontologies for representing the content and annotations internally, which are often rather tool-specific. The exception is Caliph, which is based on MPEG-7, but then suffers from the problems discussed in Chapter 9.

13.2   SVAT: Professional Video Annotation

The Semantic Video Annotation Tool targets professional users in audiovisual media production and archiving and is designed to cover the use case described in Section 2.3. The tool can be integrated with automatic audiovisual and semantic content analysis tools (cf. Chapters 4 and 12). It allows the user to view and validate results of automatic audiovisual and semantic content analysis as well as to manually structure content and to add textual and semantic annotation. If content analysis has not been done, annotation can start 'from scratch'.

The SVAT architecture uses a modular plugin system. All the components described below are plugins which use a framework for data access and synchronization with other components. The components to be loaded are listed in the main application's configuration file, so it is easy to extend the functionality by simply developing new plugins. This also makes it possible to enable and disable functionality in a flexible way depending on the application context. In addition, the user interfaces of components can be hidden/shown and moved freely in accordance with the user's needs.

SVAT uses the MPEG-7 Detailed Audiovisual Profile (Bailer and Schallauer 2006) as metadata representation and both default input and output format. If content analysis results are available in the input MPEG-7 document, they can be displayed and modified. Import from and export to other formats can be supported via plugins.

13.2.1   User Interface

The SVAT user interface (see Figure 13.1) contains:

- a full-featured video player;
- the main timeline for navigation and specification of a temporal section of interest (local time period), and a number of timeline views that display information of the local time period such as a shot editor timeline for navigation and shot editing, visualization of extracted keyframes and stripe images, visual activity, camera motion and speech to text (ASR) transcript;
- the video structure tree view, which displays one or more high-level segmentations of the content and is also synchronized with the other views; and
- the semantic annotation view, which allows annotation of textual information and named entities at segment level.

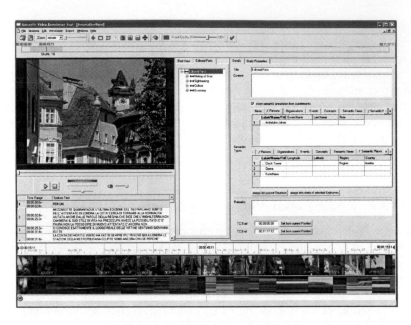

Figure 13.1   SVAT user interface.

All the components are synchronized with respect to the temporal position in the video; thus navigation is possible by clicking or clicking and dragging in any of the timeline views, by selecting segments in the tree view, by changing the temporal resolution of the timeline, or by selecting a phrase in the ASR transcript.

Main Timeline

The main timeline shows the time period from the start position to the end position of the video. By clicking the left mouse button at a specific position in the timeline, the actual position of the video can be changed. The position cursor (blue line) follows the mouse pointer as long as the left mouse button is held down. Each other window in the SVAT application will immediately be updated accordingly.

From the whole time period a temporal section of interest (local time period) can be selected in the toolbar. The zoom slider determines the length of the selected time period. This time period is indicated in the main timeline by a color marked range around the actual cursor. All windows below the main timeline display information only from this selected time period.

Shot Editor Timeline

Shots are the basic building blocks of the visual modality of a video. The visualization of the shots supports the user in efficiently navigating through the video and accessing certain time segments within the video. Shots are also an important basis for manual annotation, as many annotations are at the shot level. Knowledge of the shot structure facilitates building higher-level content structures such as scenes or chapters.

The shot editor timeline provides viewing and simple editing functionality for shot boundaries. If automatic content analysis has been done, the detected shot boundaries are shown in this view. Depending on the results of the automatic shot detection, missing shot boundaries (both hard cuts and gradual transitions) can be inserted and wrongly detected shot boundaries can be removed.

Keyframe, Stripe Image and Visual Activity Views

Keyframe extraction tools identify a number of frames in each shot, which are used as representative proxies of the shot's content and support effective navigation within the video. If automatically extracted keyframes are available, they are displayed in the keyframe view with the corresponding time code overlaid.

Stripe images are spatiotemporal representations of the visual content and hence are an effective way to navigate within video footage. A stripe image is created by extracting one column from each of the frames of the video and putting them together in temporal order. This visualization clearly shows shot boundaries as discontinuities, and allows the user to easily track camera motion and the movements of larger objects.

The visual activity view shows spatiotemporal activity of the video content over time as line graph, if it has been extracted during automatic content analysis. An average value for a short time period (typically a second) is displayed. The measure represents the frame to frame visual activity, caused by either global (camera) motion or local object motion.

Camera Motion View

This view displays the results of automatic camera motion detection and allows the user to add new camera motion annotations on a shot basis. The camera motion detection algorithm identifies segments containing salient camera movements and describes this motion both qualitatively and quantitatively. The original camera motion, however, cannot be reconstructed unambiguously from the analysis of the video, since for example it is not possible to discriminate a translation of the camera (e.g. a track left) from a rotation (e.g. a pan left), if the scenery is far away from the camera. The result of camera motion detection is thus described in terms of four types of camera motion (pan, tilt, zoom, roll) and a rough quantization of the amount of motion. Camera motion annotation can guide the user and support manual annotation. In addition, camera motion also conveys some semantics of the content; for example, zooming in on an object or person can be an indicator of its importance. The view displays the most dominant camera motion for each time segment if there is significant motion. The color indicates the type of motion and if sufficient space is available at the current zoom level, the type of motion is also displayed as text.

Speech to Text Transcript

The ASR component displays the speech to text transcripts if available. The view is time-synchronized with the video player and selects the segment of the current time. Double-clicking on a row makes the video player jump to the start of the segment. The columns and contents of the view that are displayed can be configured.

Video Structure Tree

If the video has been processed with an automatic analysis tool, the video is automatically segmented into shots. This shot list is displayed by the video structure tree view, including the automatically extracted keyframes within the shots. If there are no shots present on loading, they can be manually created by other components within the application (like the shot editor timeline), or the video structure view can be used without shot information.

In order to structure the video, different views can be created. A view is the representation of a logical structure of the content (e.g. the scenes of a movie, the stories of a news broadcast, the takes in rushes). It is possible to create more than one view. This feature allows more than one segmentation to be described. For example, it can be used to display automatically generated segmentations, such as one based on audiovisual features and one on the lexical analysis of the speech transcript. Another use is to describe content structure in terms of different criteria, scenes of a movie according to a script, a segmentation based on appearance of a certain character, etc.

Within the created view, the video can be structured into arbitrary segments. The types of segments in the structure, as well as the accepted types of sub-segments, are defined by an MPEG-7 classification scheme loaded by SVAT. The content structure helps to group sections of the video that belong together and also allows semantic types to be annotated on the created segments.

13.2.2   Semantic Annotation

Detailed annotation can be done on shots or any other segment nodes in the tree structure. The user interface (Figure 13.2) allows the following detailed annotations to be edited: title, content synopsis, named entities and comments.

Named entities can be annotated in the Details tab. With the option to duplicate the semantic annotation control (with the context menu), two different types can be annotated and viewed separately. Annotation can be done on both semantic annotation controls and both are kept up to date.

There is also an option to show the semantic annotation from sub-elements (e.g. the shots within a structure). These entries are displayed as read-only within the Label/Name/ Title column. It is also possible to merge the semantic annotation of one segment into the parent segment and also into the shots of selected keyframes that are displayed in a separate component within the application. The supported types of semantic entities correspond to those defined by the MPEG-7 SemanticDS (MPEG-7 2001): items (objects), persons, organizations, events, concepts, time and place. The properties of these types are also supported.

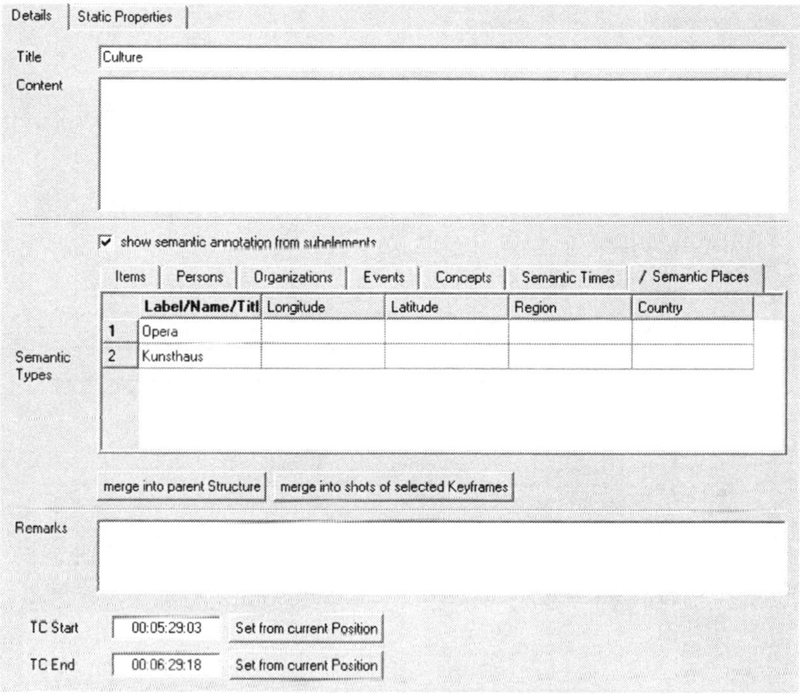

Figure 13.2 Detailed annotation interface for video segments.

Global annotation and production information are related to the entire content. The user interface (see Figure 13.3) allows one to set the title, subtitle, place, production date, creators, abstract, rights and information related to a content management system (CMS) (date imported, CMS reference and tape reference).

The user interface for the annotation of creators is flexible: one can set as many creators as needed, with each having a role that can be selected from a list of roles. Different creators can be specified with the same role.

Controlled vocabulary can be used for the annotation of named entities in both detailed and global annotation. In this case only the URI of the named entity is annotated in the MPEG-7 document, thus establishing a link to the controlled vocabulary. By default MPEG-7 classification schemes are used. SVAT also allows the linking of named entities stored in an external knowledge base. An interface for accessing KIM (Kiryakov et al. 2004) is available, but any other knowledge base can be integrated using the plugin concept.

13.3 KAT: Semi-automatic, Semantic Annotation of Multimedia Content

The K-Space Annotation Tool aims to provide a framework for semi-automatic, semantic annotation of multimedia content, with an emphasis on rich semantic annotations. Compared to pure keyword or concept annotation, rich semantic annotation specifically denotes the

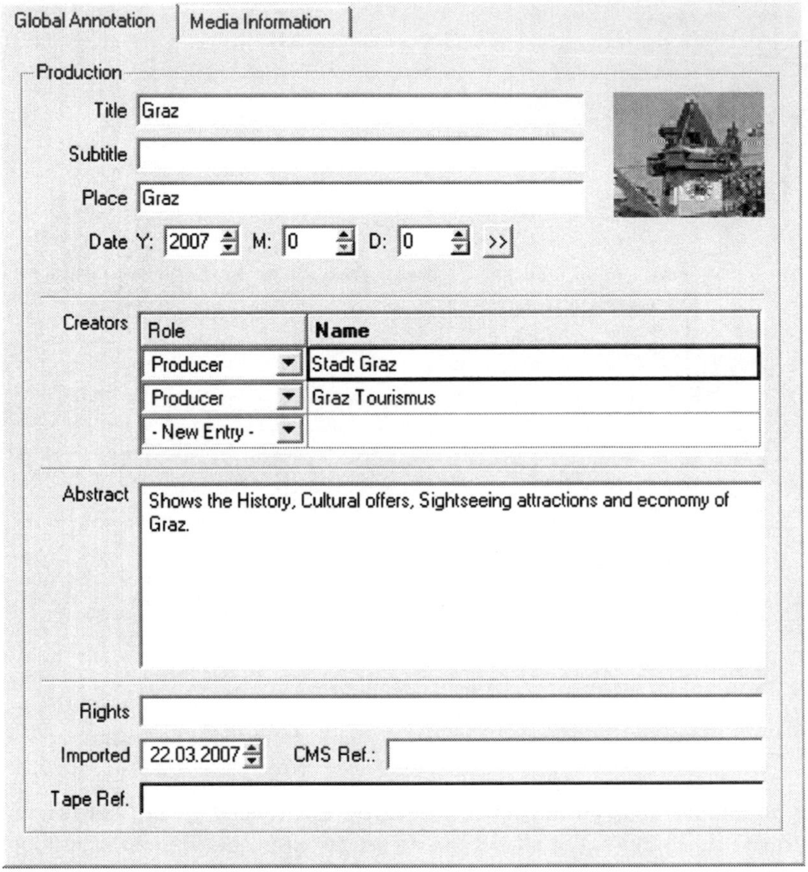

Figure 13.3   Global annotation dialogue.

employment of relations between concepts and individuals, for example, the explicit representation of events as first class objects, linking individuals to such events and describing the roles they play in the context of specific events. This provides the means not only to query for depicted objects, but also to query for content depicting very specific events, to allow for querying for related content and to exploit the relations between real-life entities for browsing multimedia repositories. A KAT screenshot showing an annotated image is shown in Figure 13.4.

KAT is designed with the well-known model–view–controller paradigm in mind. The model is provided by the Core Ontology for Multimedia (cf. Chapter 9) and an underlying RDF repository, Sesame 2.[2] The COMM provides an MPEG-7 based, generic model for multimedia content and can easily be specialized or extended for more specific scenarios. However, due to the formal, ontology-based modeling modules, the generic handling of special content is also possible, since the relationship between different content types is made explicit. For instance, for a medical application, X-ray images stored in a specialized

---

[2] http://www.openrdf.org

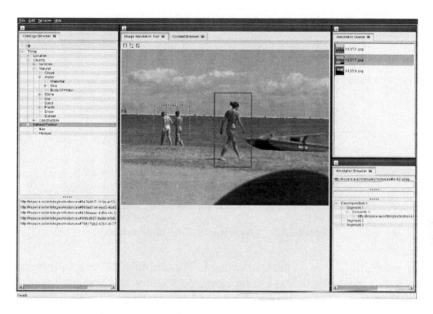

Figure 13.4 KAT screenshot during image annotation.

file format might be important. A generic content browser might not know how to display this file format, but it still knows that the files contain images and might display a generic image icon instead of a thumbnail. Further, a content-independent structure view, such as the annotation browser (cf. Section 13.3.3), can still display the decompositions and annotations, since the representation of a decomposition and an annotation is defined independently of the specific content type, but for images in general. Making these kinds of relations between concepts explicit, and abstracting from specific content and scenarios, is a great advantage of using the COMM and Semantic Web standards for the underlying model in KAT.

KAT is published as open source software under the terms of the LGPLv3 license and hosted on *launchpad* at the URL `http://launchpad.net/kat`. We further provide up-to-date documentation in the Semantic Multimedia Wiki at `http://semantic-multimedia.org/`.

13.3.1 History

The KAT code is based on the Ontomat-Annotatizer (Handschuh and Staab 2003) tool, but has been greatly extended and refactored. In the current version, basically all of the original Ontomat code has been modified, so that at the time of writing the relationship is historical rather than technical.

The development of KAT was inspired by M-Ontomat-Annotizer (Petridis et al. 2006), which provided the means to link ontology concepts with low-level features based on the aceMedia ontology infrastructure (Bloehdorn et al. 2005). These features were called visual prototypes and could be employed for automatic annotation of images. However, M-Ontomat-Annotizer never provided the means to actually annotate the content, either

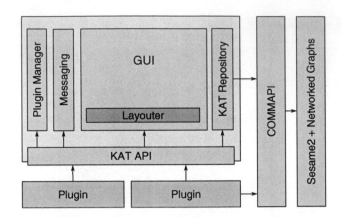

Figure 13.5 Overview of KAT architecture.

manual or automatically, or to retrieve content. The aim of KAT was to integrate feature extraction, automatic methods exploiting extracted features and semantic annotation into a single framework – to provide a tool for semi-automatic annotation, where user input is used by automatic methods to provide or propose annotations. The overall target is to work toward efficient annotation of multimedia content, that is, to reduce the time required by a user to annotate his content with rich semantics.

13.3.2 Architecture

An overview of KAT architecture is presented in Figure 13.5. KAT consists of three main building blocks. The core consists of a set of services and an API that is used by the plugins. The plugins implement all functionality relevant to the end-user, such as the processing of content, retrieval, and visualization. Finally, the COMMAPI and the underlying storage layer provide the common model to all plugins. Basically, all actions that might be performed within KAT are either visualization or modification of COMM objects.

The most important element of the core is the graphical user interface. Within it, plugins might register views that are displayed at certain locations. The available locations are determined by the layouter used. The default layouter provides four locations where a view might be displayed. These locations are indicated in Figure 13.6.

The remaining functionality provided by the core is as follows:

- the `MessageDispatcher`, which delivers messages and handles subscriptions to certain message types;
- the `PluginManager`, which controls import, loading and removal of plugins;
- the `KATRepository`, which provides convenient methods for creating, retrieving and storing COMM objects and other metadata that might be added to the RDF repository;
- the KAT API, which provides the interface for plugins to the KAT core functionality.

We discuss these components below in greater detail.

A plugin implements the functionality that is relevant to an end user: the analysis, visualization, modification or retrieval of content and its metadata. The core component

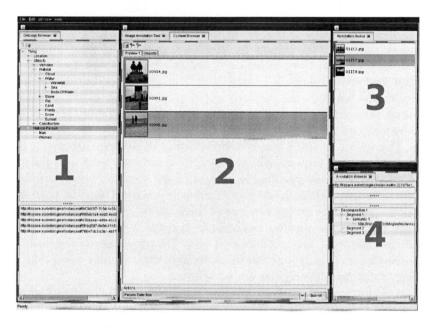

Figure 13.6   Available view positions in the default layout.

of a plugin is the controller, the only class that is directly instantiated by the KAT core. The basic functionality provided by the KAT core to a plugin is as follows:

- Installation, loading, execution and removal of plugins based on stored preferences or user request.
- Registration of views that are displayed in one of the locations of the current layout.
- Execution of workers. Workers correspond to resource-consuming tasks (like most analyses), where usually only one should run at a time in order to not slow down a typical desktop machine too much. Workers are executed by the KAT core sequentially.

In other words, the only required component for a plugin is the controller. The internal structure of a plugin is not further defined. However, controllers, views and workers have to implement the corresponding interfaces.

The model and storage layer of KAT is based on the Core Ontology for Multimedia (Arndt et al. 2008) and the Sesame 2 RDF repository in conjunction with the Networked Graphs Sail (Schenk and Staab 2008). KAT is basically a tool to visualize and modify COMM graphs and the corresponding content. Therefore, the underlying model for all components, and specifically the plugins, is the COMM it self. Communication among components is based on passing around URIs of COMM objects. The semantics of the URI is clearly defined by the COMM based on the available metadata in the repository. The type of message then clearly defines what should be done with the content item or COMM graph, respectively. The Networked Graphs Sail is used to structure the RDF repository into parts that are easier to handle, to allow for the addition of provenance and administrative metadata, and to provide transparent read and write access to more than one underlying repository at once.

13.3.3   Default Plugins

KAT contains a set of default plugins for image annotation. The most important plugins are the *Ontology Browser, Content Browser*, and *Image Annotation Tool*. The Ontology Browser can load and display OWL ontologies. It might be used to create instances of ontology concepts, but does not currently support modification of the ontology. Using drag and drop, the ontology browser is capable of annotating arbitrary content items. The content browser is used to import content, and to display results of queries against the repository. Queries are currently triggered by selecting an ontology concept or individual. In that case, the content browser will display all content annotated with the concept or individual selected. Importing content is done via a file browser. The selected content is added to the repository and displayed with thumbnails in a list. Using the content browser, content can be selected for analysis by one of the installed analysis plugins, chosen from a drop-down box. Double-clicking a content item opens it in the respective annotation tool. For images this is the Image Annotation Tool. The plugin allows for creating regions using bounding boxes or polygons and for annotating these regions using drag and drop. Furthermore, it can visualize regions already present.

Within K-Space, several analysis plugins implementing functionality described in Chapters 10–12 have been implemented, enabling detection of highlights and events in videos, identification of semantic concepts in images, and the like. Furthermore, the *Audio-Visual Annotation Tool* allows the playback of video and audio content, visualization of shots and keyframes, and semantic annotation of videos and audio files.

13.3.4   Using COMM as an Underlying Model: Issues and Solutions

As already explained, the COMM is the common model for all KAT components. It provides the formal model that is used to classify and annotate the content and provides the basic metadata required by plugins to decide whether content should be processed or how it should be processed. From a programming point of view, the COMMAPI abstracts from the storage in the underlying Sesame repository. However, depending on the plugin, more complex queries or custom COMM descriptors might require direct interaction with the repository. In order to acquire an instance of the internal repository correctly configured with respect to remote repositories, one can use the appropriate methods of the KATRepository object. This is the internal repository manager, which takes care of configuring the repositories correctly and provides some primitives to be used for constructing correct queries or storing to the correct remote repository.

The Networked Graphs Sail provides transparent access to multiple repositories, and is used to structure the content using graphs for provenance tracking. Since we also require write access to remote repositories, KAT assumes that every remote repository is also a Sesame 2 installation, since generic write access to remote repositories is not available for RDF at the time of writing.

In order to handle transparent read and write access to remote repositories, the DistributedSail, which is part of the Networked Graphs package, is employed. Each repository is added to the configuration of the distributed sail and mapped to a namespace. All graphs within this namespace are retrieved from the given repository, and write operations are also performed on that repository. This, however, requires the creation

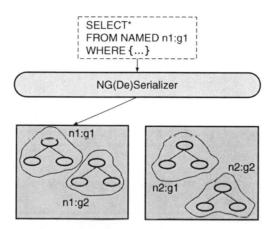

Figure 13.7   Using named graphs to map COMM objects to repositories.

of adequate graphs. In order to achieve this transparently, we provide the NGSerial
izer and NGDeserializer which handle (de)serialization of COMM graphs. These
classes implement the Serializer and Deserializer interfaces provided as part of
the COMM API.

The implementation creates a namespace for the repository and when a new COMM
object is serialized, which is not yet present in the repository, a dedicated named graph is
created within the namespace. This allows the mapping of COMM objects to the repository
storing them. This approach is illustrated in Figure 13.7. For querying and writing we
just have to specify the graph name, and the DistributedSail will map the graphs to
the correct repository.

In a similar fashion we can also store ontologies. We create a named graph for each
imported ontology, and store appropriate metadata for display. This allows easy selection
of all available ontologies, including the source repository. We can also easily extract all
statements belonging to a single ontology.

As specified in Chapter 9, COMM is designed around the numerous MPEG-7 patterns
used to describe spatial, temporal, spatio-temporal and media source decompositions of
multimedia contents into segments. The main issue in implementing COMM for annotating media documents (e.g. for audiovisual documents) for a reasonable video duration
is the memory requirements. In Table 13.1 the number of triples generated for the same
video exported with different metadata is presented. Here COMM is constructed: (1) with
media information (MediaURI, MediaTimeDescriptor, etc.); (2) with media information
and semantic annotation; (3) with media information, semantic annotation (for whole
video, events, spatial objects) and MPEG-7 low-level visual features (Colour Layout
Descriptor, CLD).

In Table 13.2 the number of triples generated for different videos with different numbers
of shots and keyframes is presented. The COMM object is constructed to contain metadata
about the media (e.g. its title, its mime type, its duration) and about low-level visual
features (represented in MPEG-7) which are temporally aligned to the video (media time).

The number of triples generated by COMM is highly influenced by both number of shots
(subsequently keyframes) and the amount of metadata associated with each shot/keyframe.

Table 13.1   Number of RDF triples for MPEG-7 export of the same video with different metadata

Parameter	Inst. 1	Inst. 2	Inst. 3
Total duration (s)	218	218	218
No. of shots	164	164	164
No. of keyframes	164	164	164
Total triples generated	41 454	50 337	70 681

Table 13.2   Number of RDF triples for MPEG-7 export of different videos with the same metadata

Parameter	Video 1	Video 2	Video 3	Video 4
Total duration (s)	305	236	198	290
No. of shots	17	129	131	170
No. of keyframes	17	129	131	170
Total triples generated	7 618	55 666	56 524	73 256

MPEG-7 facilitates the browsing and retrieval of audiovisual content by defining partitions and decompositions and variations of the audiovisual materials. Thus, COMM also enables the description of detailed decomposition of a multimedia document. The granularity of the decomposition can range from a region in a keyframe to an abstract description of a shot (without any keyframes). Thus, increasing the number of triples associated with a COMM object requires an increase in the memory required to handle the object.

However, this can be resolved by exploiting the decomposition patterns defined in COMM (inspired by MPEG-7). The following is the pseudocode for constructing a COMM object in a hierarchical approach.

```
construct VideoData for complete video
export VideoData using COMM Serializer
for shot = 1 to N do
  construct VideoData for shot segments
  for keyframes = 1 to M do
    construct VideoSegmentTemporalDecomposition for all keyframes extracted
    from the shot
  end for
  construct MediaTimeDescriptor for shot segments
  export VideoData and MediaTimeDescriptor to RDF repository using COMM
  Serializer.
end for
apply a patch in order to link complete video uri to shot decompositions.
```

As may be noted, the COMM object is always constructed representing a single 'Video-Data' or 'VideoSegmentTemporalDecomposition', thus significantly reducing the memory constraints. At the time of writing, the approach was used to export TRECVid MPEG-7

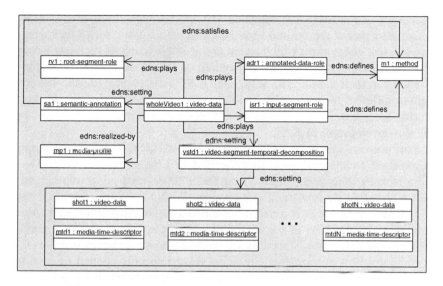

Figure 13.8 COMM video decomposition for whole video.

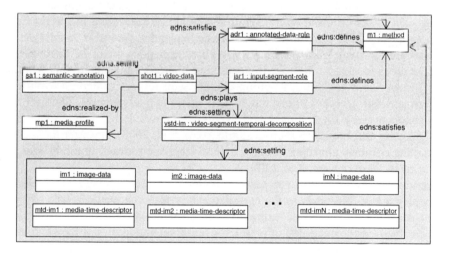

Figure 13.9 COMM video decomposition for video segment.

data to the RDF repository using COMM. The two-stage decomposition represented in the pseudocode is presented in Figures 13.8 and 13.9 for complete video and shot segments, respectively.

13.3.5   Semi-automatic Annotation: An Example

As mentioned earlier, the aim of KAT is to facilitate the annotation of multimedia content with rich semantic metadata using a combination of automatic and manual techniques.

It is well known that creating rich semantic annotations manually is time-consuming and tedious, and thus means are desired to reduce this effort. We refer to techniques of this kind of as *efficient annotation*, which refers to the employment of methods to support the user in the task of annotating content. Depending on the scenario, the concrete techniques or tools used for such a task will vary, and KAT aims to provide the framework for linking these different tools in order to implement applications that can be used within more specific scenarios. In this section we will discuss two scenarios that have been used to inspire the development of KAT.

To support personal content management we can see clear benefits in using automatic image annotation techniques, such as presented in Chapters 11 and 12. As we have shown in a recent publication (Saathoff and Staab 2008), well-performing classifiers can be trained with relatively few training examples. Within KAT this might be used as an automatic analysis plugin that is iteratively trained using manual annotations provided by the user. Obviously, this kind of annotation plugin will usually ship with a set of already available classifiers. However, since it is known that training generic classifiers is hard, an iterative training will 'optimize' the classifiers for the specific content of each user. The manual annotations either might result from a set of manually labeled images, or could be drawn from automatic annotations that have been corrected by the user. On top of these basic annotations, further inferencing can be performed to infer additional annotations. Nevertheless, only simple and basic concepts might be recognized using such automatic methods, such as *sky* or *beach scene*. While this might be interesting in certain situations, annotations of this kind provide only very basic semantics (see also Jaimes and Chang 2000), and are far away from the user's perceived semantics when he browses his own content. In order to support the user in generating higher-level semantic annotations, one can, for instance, use clustering to group content with respect to certain events. A user might then annotate an event in one batch. This offers two advantages. The user does not have to annotate every single image and events are known to be an important primitive when organizing and browsing multimedia (Scherp et al. 2008). In order to free the user from handling ontology concepts directly, natural language processing might also be used, as discussed in Dasiopoulou et al. (2007).

A second scenario is the semi-automatic annotation of sports footage, in which different methods play a key role, such as the complementary resource analysis (see Chapter 12), highlight detection or key-frame extraction (cf. Chapters 4 and 5). The video analysis plugin described earlier is capable of producing temporal boundaries of the video depicting salient events. The shots are further analyzed to extract significant information from such shots, known as keyframes. In terms of high-level semantics, the shots depict an event and the keyframe represents the agents participating in these events. Thus, in order to annotate both events and agents participating in an event, the COMMAPI enables semantic labeling of both VideoSegment and Image classes. Using the video analysis tool, the shots depicting an event and keyframe representing the agents participating in an event can be semantically labeled with instances of any domain ontology. In order to describe this annotation, a two-tier VideoSegmentTemporalDecomposition is necessary. The first decomposition has to be applied to the whole video content in order to decompose it into video shots. Subsequently, the keyframes have to be identified by attaching one VideoSegmentTemporalDecomposition to each video shot. This decomposition objects only contain one segment (a still image), the keyframe.

Additionally, the complementary resource analysis provides a structuring of the content in terms of important events (e.g. a shot on goal) and semantic annotations extracted from text. So, in addition to knowing that some highlight is present at a certain point, the algorithm also proposes the kind of event. This might then be refined or corrected by a user. In the case of a foul the algorithm might, for instance, not know who committed the foul and who was fouled, but it does detect that there is a foul and that two people were involved. The user now only has to specify the role a certain entity plated within this event, which can be done using drag and drop, for instance.

13.4 Conclusions

In this chapter we have presented an overview of available multimedia annotation tools and discussed two of them in detail. The Semantic Video Annotation Tool is targeted at professional audiovisual media production and archiving and provides a framework, based on MPEG-7, for annotating audiovisual media. It integrates different methods for automatic structuring of content and provides the means to semantically annotate the content. The K-Space Annotation Tool is a framework for semi-automatic, semantic annotation of multimedia content based on the COMM. The aim here is to provide a media-independent framework that might be specialized for certain scenarios. Building on the COMM facilitates interoperability between plugins even when specialized descriptors are introduced. The formal model of the COMM explicitly allows for handling specialized descriptions in a generic fashion using the inheritance relations.

Both tools try to tackle similar problems, namely semantically annotating multimedia content in a efficient manner, using different approaches. While one tool employs an existing standard and focuses on a specific, rather broad domain, the other tries to employ Semantic Web standards and formal ontologies across different domains and media.

We have discussed issues that arise when using formal ontologies for data-intensive applications such as multimedia annotation and provided solutions to this problem. However, compared to the known standards and data formats such as MPEG-7 (see also Chapter 8), the greater flexibility provided by Semantic Web standards comes at a cost of increased complexity and currently still reduced scalability. Both certainly have to be tackled in future research, in order to provide tools and frameworks that allow for the employment of Semantic Web standards in real applications.

14

Information Organization Issues in Multimedia Retrieval Using Low-Level Features

Frank Hopfgartner,[1] Reede Ren,[2] Thierry Urruty[3] and Joemon M. Jose[4]

[1]*International Computer Science Institute, Berkeley, CA, USA*
[2]*University of Surrey, Guildford, UK*
[3]*University of Lille, France*
[4]*University of Glasgow, Glasgow, UK*

In this chapter, we address index structures in a large multimedia collection. Multimedia indexing is the backbone of all multimedia information retrieval (MIR) engines, including audio, image, video and even cross-media retrieval. Note that a media document consists of several modalities (e.g. a video document consists of audio tracks, visual streams, caption texts and user annotations). Hence, multimedia indexing has to take numerous modality features into consideration. Moreover, these features are of various natures. For example, the audio feature, baseband energy, is a continuous time sequence; image texture is a direction-based histogram and user annotations are described by a boolean vector which marks the appearance of high-level concepts. As a result, there are two major challenges in the development of a multimedia index structure: (1) multimedia indexing works in an extremely high-dimensional feature space; and (2) feature dimensions are of variable characters (i.e. boolean and multi-value).

We present two multimedia indexing approaches in the following sections. The first divides a feature space into disjoint subspaces by using a pyramid tree. An index function is proposed for efficient document access. The second exploits the discrimination ability of a media collection to partition the document set. A new feature space is proposed to facilitate the identification of effective features as well as the development of retrieval models.

Multimedia Semantics: Metadata, Analysis and Interaction, First Edition.
Edited by Raphaël Troncy, Benoit Huet and Simon Schenk.
© 2011 John Wiley & Sons, Ltd. Published 2011 by John Wiley & Sons, Ltd.

14.1 Efficient Multimedia Indexing Structures

The state of the art in the indexing structure field involves many different methods. The B-Tree family structures (Bayer and McCreight 1972) are one-dimensional structures still widespread in numerous applications. These reported methods require a total order on the key index of data. The study by Ooi and Tan (2002) shows the various adaptations and applications of the B-Tree. The main goal of multidimensional indexing structures is to efficiently access data in a large corpus with respect to their position in the data space. This research field started with the Kd-Tree (Bentley 1979) and the R-Tree (Guttman 1984), which represent two different families of indexing structures: the space partitioning and data partitioning structures, respectively. But most of these methods are not efficient enough for multimedia data whose description contains a high dimensionality, as is the case for low-level visual features. This phenomenon is well known as the 'curse of dimensionality' and has been well studied by Weber et al. (1998).

To overcome this problem, some researches have focused on approximate nearest-neighbor searches. Indyk et al. proposed locality sensitive hashing (LSH) and some extensions (Andoni and Indyk 2008; Indyk and Motwani 1998). The main idea of LSH is to choose a family of hash functions that best represent the data space. Then the query will retrieve a bucket of approximate nearest neighbors for each hash function. This retrieved data will be used to compute the final results. Another family of structures approximate the data space. The VA-File indexing structure (Weber et al. 1998) creates a simple code for each region of the space; this region coding will accelerate and approximate the sequential scan. Other methods are also based on this idea (Cha and Chung 2002; Cha et al. 2002). Another structure is based on a geometric cut of the data space to create a one-dimensional index. The first method of this family is the *pyramid* technique introduced by Berchtold et al. (1998). The main idea of the pyramid technique is to cut the data space into pyramid regions and give each data point a one-dimensional index with respect to the pyramid number and the height from the point hyperplane to the top of its pyramid. Each key index is stored in a B+Tree which is accessed with range queries.

The performance of these algorithms depends on the data distribution and the position of the query in the space. A simple search can generate uncoordinated accesses to a very large volume of data, which considerably impacts the performance and makes the pyramid technique less efficient. Two other methods, *IDistance* (Yu et al. 2001) and *IMinMax* (Ooi et al. 2000), use geometric decomposition of the space but are not suitable for real databases whose distribution is heterogeneous. More recent methods, *P+Tree* (Zhang et al. 2004) and *KpyrRec* (Urruty 2008), attempt to improve the pyramid technique by reducing the space accessed for a search query. The dividing space method of P+Tree and the K-means algorithm used in KpyrRec are not effective enough in a multidimensional space.

Shen et al. (2005) have proposed a new structure called the *ViTri index*. This uses a recursive clustering algorithm based on K-means with $K = 2$ combined with a indexing structure that uses principal component analysis. The first principal component of each cluster is used as reference to create a one-dimensional index which is stored in a B+Tree. While some of the indexing techniques introduced perform well for some databases, others perform poorly.

14.1.1   An Efficient Access Structure for Multimedia Data

We now present an efficient approach for indexing multimedia data. This combines a random projection clustering technique and a recursive space splitting method to index a numerical database. Our main goal is to develop an efficient indexing structure for multimedia data usage. Most existing indexing structures are not well adapted to multimedia data. This is due to two main reasons: the high dimensionality and heterogeneous distribution of data. Thus, an effective clustering technique is needed to ensure good quality of indexing.

Previous results (Urruty 2008) have shown two important points. Firstly, they show that using a clustering algorithm combined with an indexing structure improves the performance of the method; however, the quality of the clustering strongly affects the performance of the retrieval. Thus, we use a clustering algorithm for numerical data that combines ideas from random projection techniques and density-based clustering. The methodology of this clustering algorithm is described by Urruty et al. (2007), who demonstrate the potential of this algorithm when applied to multidimensional databases. The use of this clustering algorithm provides a set of clusters with a better distribution of data than the original dataset. Using our indexing structure on improved clusters increases the effectiveness of the indexing, that is, the precision of the index.

Secondly, we notice that the performance of such structures is affected by the amount of data accessed for a simple query, which impacts the response time. This last point suggests the need to focus on reducing the volume of irrelevant data that is not useful for a query which can be solved by using a clustering algorithm. Our prior experimental results show that the performance can be improved by combining a clustering algorithm with the original pyramid technique. However, redundant data is still present and becomes worse with the increase in dimensionality.

This phenomenon appears because the volume of the space near the base of a pyramid is a lot more important than that near the top of the pyramid with an increase in the number of dimensions. So, when a query is close to the pyramid base, it accesses too much data. In order to diminish the number of useless accesses, we propose a recursive splitting indexing method named *PyrRec* that manages the data space differently than the pyramid technique.

Data Space Division of PyrRec

The general idea of PyrRec is to cut the data space into equal volumes. After partitioning the space into pyramids, we partition each pyramid into N_t trunks of equal volume by considering N_{t-1} dividing planes parallel to the basis of the pyramid. The distance from the top of the pyramid to the ith plane, h_i, is given by

$$h_i = \frac{h \times (\sqrt[D]{i} - \sqrt[D]{i-1})}{\sqrt[D]{N_t}}, \tag{14.1}$$

where $1 \leq i < N_t$, D is the dimension, and h the pyramid height.

Then we apply PyrRec recursively on data inside each trunk. Each recursive call gives a new data space for each trunk with dimension $D - 1$. The dimension removed from the dataset corresponds to the dimension that is orthogonal to the base. A recursive call is triggered only if the number of points exceeds a threshold S which is determined by the user. Generally, the number of recursive calls is not higher than 3, even if we take $S = 1$. For example, with 1 million data points in 50 dimensions, using $N_t = 10$, we have an average of 1000 points per trunk, so a recursive call will yield approximately one point per trunk at the end of the algorithm. Few trunks may need a second and last call. With homogeneous data, the approximate number of recurrences N_r is given by

$$N_r \geq \frac{\ln(N_{\text{data}})}{\ln(2DN_t)}. \tag{14.2}$$

RPyrRec Indexing Algorithm

To construct our RPyrRec indexing structure, we proceed as follows. First, the random projection clustering algorithm is used to create K clusters. Note that K is dynamically chosen by the clustering method. For every cluster, the minimum bounding rectangle is kept in memory and used to normalize the data space. Then, for each point P of every cluster, we find the pyramid P_y that contains the point with a simple computation of distance between the point and every pyramid base center. The next step determines the trunk T with respect to the height of the point from the top of the pyramid containing this point. T and P_y are concatenated to create an (integer) index I_y for each point of data space. A recursive call of PyrRec on all trunks containing more than S points is made. S is a parameter fixed by the user or with respect to the dataset size. The data dimensionality in the recursive algorithm is reduced to $D - 1$. Thus, we have obtained a principal index for each point and multiple indexes for points that passed through a recursive call. Finally, those multiple indices are put in a variant of a *B+tree* structure, that is, a *B+ tree* that can contain more than one index for a point.

Figure 14.1 shows the limited access region for a window query q with the use of a recursive index. We remark that the difference in computing cost between our algorithm and the original Pyramid search algorithm is the cost of the comparison function when

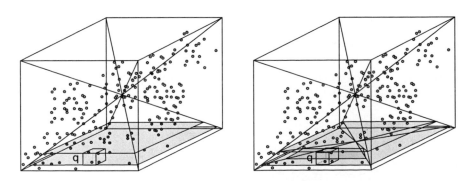

Figure 14.1 Geometrical representation of PyrRec.

a point has multiple indexes. This cost is not important relative to the cost of analysis needed if the point is not included in the set of possible results.

14.1.2   Experimental Results

We performed our experiments with RPyrRec on the TRECVid 2006 corpus (Smeaton et al. 2006). The aim of the TRECVID initiative is to promote progress in content-based video retrieval. The dataset consists of approximately 160 hours of television news in English, Arabic and Chinese, recorded in late 2005. These videos are segmented into shots and keyframes are extracted to represent each video shot. This gives us a database of about 150 000 keyframes. We used the AceToolBox provided by Acemedia (O'Connor et al. 2005) to extract the low-level visual features of each keyframe. We choose to extract the following visual features: color layout (12 dimensions), color structure (256 dimensions), dominant color (6 dimensions), edge histogram (80 dimensions) and homogeneous texture (62 dimensions). Each of these five sets of low-level visual features has different workloads: dimensionality and distribution of data.

The main aim of our experiments is to show the efficiency and effectiveness of indexing structures compared with a whole scan of the database. Thus, we use the first 100 results given by the sequential scan as ground-truth data and compute the precision with respect to this set. We call the size of the window query used to access the structures the *selectivity*. The selectivity could be used as a parameter fixed by the user after a set of experiments.

In our experiments, we compare the performance of our new RPyrRec method, existing indexing techniques, sequential scan and some variations on our method. In the following, we give more details about the evaluation. For the P+Tree (Zhang et al. 2004) and ViTri indexing (Shen et al. (2005)) techniques, the number of clusters is a parameter (as a density/radius threshold), for which we use several different values where we present the best performing. We present some variations on the yramid technique: *DivPyr* is a combination of a simple recursive division of the data space into two new spaces with respect to the number of data and the pyramid. *RPyr* is the combination of our random projections clustering algorithm and the pyramid. PyrRec is our only recursive splitting indexing technique without any prior clustering method. We note also that ViTri results are shown only for time with respect to precision results as the selectivity values do not use the same scale as other methods.

Figures 14.2 and 14.3 present the precision results of the evaluated indexing structures with respect to the selectivity of the window query. We present results on the color layout (12 dimensions) and edge histogram (62 dimensions) features. We can see the influence of the number of dimensions on the precision. First, the selectivity value has to be higher to access data. This is due to the 'curse of dimensionality', that is, the higher the dimension of the data spaces, the more sparse are the data in space and the less data we access with the same selectivity (size) of window query. P+Tree precision results are affected by the nature of the dataset. Other indexing techniques show similar result shapes: a large increase in precision for a small range of selectivity values.

These results are useful to fix the parameter of selectivity with respect to the precision of the results. But a high selectivity value means a high number of data accesses, as Figures 14.4 and 14.5 illustrate, and slower running time, as Figures 14.6 and 14.7 show.

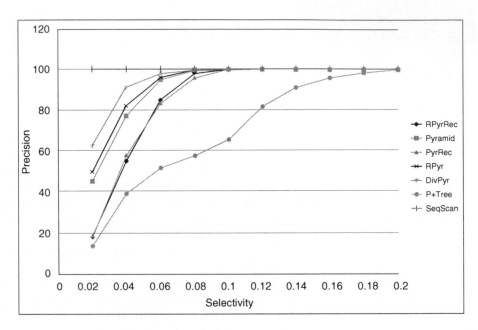

Figure 14.2   Precision with respect to selectivity for color layout feature.

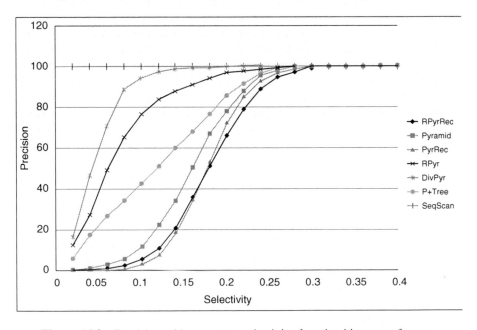

Figure 14.3   Precision with respect to selectivity for edge histogram feature.

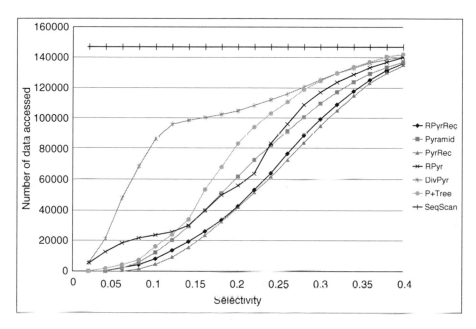

Figure 14.4   Number of data accessed with respect to selectivity for colour structure feature.

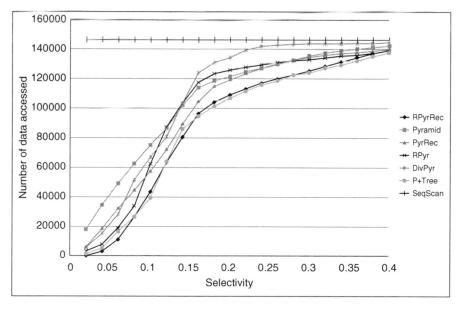

Figure 14.5   Number of data accessed with respect to selectivity for dominant colour feature.

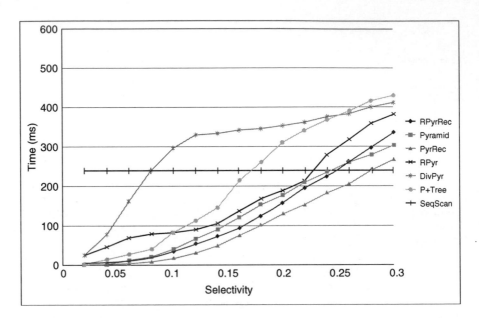

Figure 14.6   Time with respect to selectivity for colour structure feature.

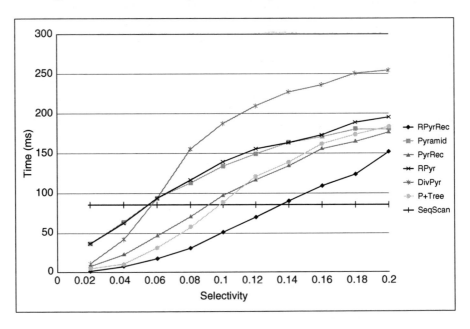

Figure 14.7   Time with respect to selectivity for homogeneous texture feature.

From the experiments shown above, we conclude that the proposed RPyrRec indexing technique performs better over various datasets than other existing methods and is not affected by the distribution or the dimensionality of data. The query processing time is halved for most of the query compared with other state-of-the-art methods, without any precision loss. This shows the potential of our indexing structure to improve multimedia search engine efficiency.

14.1.3   Conclusion

In this section we have introduced *RPyrRec*, a new multidimensional indexing structure that combines a random projection clustering algorithm and a recursive splitting indexing method. This combination allows our indexing structure to be more robust to the heterogeneous distribution of data with high dimensionality. Our experiment on a synthetic database and a real video database showed the effectiveness and efficiency of our proposal. We reduced the search access time without losing precision compared to other state-of-the-art methods with different databases presenting different workloads. We will investigate in the future an incremental indexing structure that performs well with a very large dataset size. We also plan to adapt a feature selection method to reduce the number of dimensions of the low-level visual feature.

14.2   Feature Term Based Index

In the previous section we presented a state-of-the-art index structure, RPyrRec, which partitions an original feature space to speed up neighborhood search. Note that a multimedia collection usually holds an extremely high-dimensional feature space, since a large amount of modality features can be extracted from a multimedia document. The efficiency of an index structure needs further improvement to meet MIR requirements. Moreover, modality features are of various formats. For example, high-level concepts are described by a boolean vector, and many visual features such as color and texture are histogram-based. This points to two challenges in multimedia indexing. On the one hand, there will be redundancy if the same index structure is implied for all features. For instance, a linear index is more efficient for a boolean feature than a tree index. On the other hand, a set of feature related indices hinders the estimation of document relevance. This is because each index could provide a ranked list of relevance and we have to combine the similarity from all indices to estimate the overall relevance. Given the lack of prior knowledge on queries, it is difficult to justify a combination of feature/indexed based similarity. To alleviate these problems, we propose a middle-layer data presentation which transforms all features into a unified feature space. This has many advantages. First, the projection to feature terms allows further reduction of the overall feature space dimension, as it allows the analysis on dimension based feature effectiveness and removes some ineffective dimensions from media indexing. Second, the middle-layer presentation is able to keep and even enlarge the ability to discriminate between media documents, if a proper projection is selected (e.g. principle component analysis). Third, a unified feature space helps relevance estimation as all features are considered at the same time. This will facilitate the formulation of retrieval models.

The remainder of this chapter describes a unified feature space called the 'feature term' and is organized as follows. Section 14.2.1 justifies the proposal of feature terms by addressing index structure and feature characters. Section 14.2.2 surveys the discussion on textual term distribution to guide the computation of feature terms from a media collection. Section 14.2.3 deals with the computation of feature terms and Section 14.2.4 proposes an efficient approach for the removal of ineffective feature dimensions, which uses the number of feature terms to judge the effectiveness for retrieval. Section 14.2.5 briefly describes our retrieval system and related relevance estimation functions. Experimental results on the TRECVid 2006 corpus and conclusions are set out in Sections 14.2.6 and 14.2.7, respectively.

14.2.1   Feature Terms

The idea of feature terms is inspired by data partitioning based indexing (Xu et al. 2008) and the wide employment of high-level concepts in MIR (Snoek et al. 2007b). A high-dimensional index can be categorized into two groups, space and data partitioning. Space partitioning indices recursively divide a feature space into disjoint subspaces. The RPyrRec structure is a typical space-partitioning index. The performance of a space-partitioning index is badly affected by data distributions, as many blank nodes exist in the structure if the data distribution is not uniform. In addition, the retrieval is a statistical decision problem based on the variance of term distributions in both document collection and a query (Zhai and Lafferty 2006). The exploitation of the document distribution will significantly improve the effectiveness as well as the efficiency of a MIR index structure. Data partitioning based indices, such as R-tree and its variants, such as R*-tree, X-tree and SS-tree (Xu et al. 2008), divide a dataset to construct a hierarchy consisting of bounding regions or hyperspheres. The advantages of data partitioning based indices are: (1) the size of hyperspheres can be adjusted by the density of data distribution; and (2) tree nodes allow the overlap of nearby regions. These characters are helpful in information retrieval. This is because information retrieval is a specified neighborhood search. Document relevance is closely associated with the number of relevant documents (Jones et al. 2000) and with the density of document distribution. To ensure enough relevant documents are collected, we seek a changeable hypersphere whose diameters are adaptive to the density of document distribution on a feature dimension. For instance, inverse document frequency (idf) in textual retrieval could be regarded as a parameter which estimates the document distribution on given textual terms. Moreover, the overlapped directory nodes speed up the search for relevant documents in multiple scales. This facilitates not only performance measurement but also the combination of multiple ranked lists (Snoek et al. 2006).

The usage of high-level features is a significant character of modern MIR engines (Snoek et al. 2006, 2007b). The large-scale experiments on the TRECVid corpus prove the effectiveness of high-level features in content-based MIR. As a result, an index for MIR should pay enough attention to the storage as well as the management of high-level features. Moreover, high-level features usually appear as boolean vectors (Snoek et al. 2006). On the one hand, it is possible to transform low-level features into a similar boolean representation. For example, modality features can be grouped into a few clusters. The appearance of a cluster label becomes a boolean description for a modality feature, which indicates whether a document belongs to a cluster. On the other hand, it is difficult to project a boolean like high-level feature into a continuous distributed variable. This is because we can hardly measure the distance among a group of concepts.[1]

We define a *feature term* as an interval within a feature range. A feature term denotes a grid in a large feature space, into which a document falls. There are a few approaches for extracting a feature term from a media collection (e.g. clustering). The extraction process can be described as a projection from a multiple-valued N-dimensional variable to an integer or a boolean vector of integer appearance. For example, the classification of an RGB color into four classes can be descried as $[0, 255]^3 \rightarrow \{0, 1, 2, 3\}$ or $[0, 255]^3 \rightarrow \{0, 1\}^4$ for the appearance of class labels $0, 1, 2, 3$. The extraction function can be symbolized as $\hat{f} : [0, K]^N \rightarrow \{0, 1, \ldots, M - 1\} \sim \{0, 1\}^M$, where K denotes the variable range and M the number of integers. These integers are the labels of feature terms. In addition, if we accept the hypothesis that dimensions in a low-level feature are independent,[2] we can take the one-dimensional case, where $N = 1$. Feature term extraction from a low-level feature is therefore of low computational complexity.

14.2.2   Feature Term Distribution

The feature term approach simulates the bag-of-words approach in textual retrieval. Our method exploits the same principle used in statistical textual retrieval. This means that we take the hypothesis on textual term distribution to decide on the selection of feature terms.

Textual term distribution is an important aspect in term weighting, as a statistical justification of retrieval models (Amati and Rijsbergen 2002; Harter 1975; Jones et al 2000). For example, the effectiveness of a textual term is estimated by the difference from a normal term distribution (Amati and Rijsbergen 2002). Harter (1975) proposed that a term should follow a 2-Poisson distribution, because (1) the appearance of a textual term can be regarded a random arrival with a certain average and (2) the appearance of a textual term is a boolean variable. Margulis (1992) extended this model to the N-Poisson distribution. The authors argue that an N-Poisson distribution provides a more precise estimation than a 2-Poisson one does if a Poisson-like distribution is followed. However, several class numbers, from 2 to 7, are evaluated on real document collections. No specific class number achieves a significant improvement in comparison with other class numbers. Amati and Rijsbergen (2002) conclude these early experiments and simulate a retrieval process by a Bernoulli distribution. The authors suggest a uniform distribution, since the joint probability of text terms is so small that a simple uniform distribution is good enough. In addition, this suggestion is accepted in many successful textual retrieval engines, such as Terrier (He et al. 2008).

In this work, we follow a uniform distribution for the extraction of term-like variants from modality features. This is because the uniform hypothesis of term distribution leads to a superior retrieval performance (Amati and Rijsbergen 2002). Moreover, the computational cost of a uniform distribution is significantly lower than N-Poisson. This is essential in computationally expensive MIR tasks.

---

[1] Some linguistic statistics, for example from WordNet (Miller and Hristea 2006), can give an estimate of the distance between two concepts. However, it remains a difficult task to compare the similarity and dissimilarity between a group of concepts.

[2] This hypothesis is mostly agreed, as a feature dimension would be ignored if it could be computed from remaining dimensions.

14.2.3   Feature Term Extraction

For a collection D, the frequency of a feature term f_t is the number of times that document features fall into the tth range interval:

$$f_t = |D_t|, \quad D_t = \{d \mid \hat{f}(d) = t, d \in D\}, \tag{14.3}$$

where d is a document in D. The probability of a feature term t is

$$p(t) = \frac{f_t}{\sum_{i=0}^{M-1} f_i}. \tag{14.4}$$

Obviously, $\sum_{i=0}^{M-1} p(t_i) = 1$.

Selection Criterion

As many feature quantization methods exist, some statistical criteria are necessary to decide on an optimal solution. Given the uniform assumption on the feature term distribution (Amati and Rijsbergen 2002), we propose three criteria: minimized χ^2 test, maximized entropy and minimized AC-DC rate.

The χ^2 *test* computes the similarity of a sample sequence from a given distribution. Since the optimal term probability is $\hat{p}(t) = \frac{1}{M}$ in the uniform distribution, the χ^2 test is as follows:

$$\chi^2(M) = \sum_{i=0}^{M-1} \frac{(p(t_i) - \hat{p}(t_i))^2}{\hat{p}(t_i)} = \sum_{i=0}^{M-1} \frac{(Mp(t_i) - 1)^2}{M}. \tag{14.5}$$

The minimized χ^2 test criterion is defined as

$$I_{\chi^2} = \arg\min_M \chi^2(M). \tag{14.6}$$

Entropy measures the information gain resulting from a term selection:

$$Entropy_s(M) = -\frac{1}{\sqrt{M-1}} \sum_{i=0}^{M-1} p(t_i) \log(p(t_i)). \tag{14.7}$$

A high entropy indicates a good selection:

$$I_{entropy} = \arg\max_M Entropy_s(M). \tag{14.8}$$

A uniform distribution will maximize the entropy.

The *AC-DC rate* computes signal variance from the average. For a frequency sequence $f_0, f_1, \ldots, f_{M-1}$, the DC parameter,

$$DC = \frac{1}{M} \sum_{n=0}^{M-1} f_n, \tag{14.9}$$

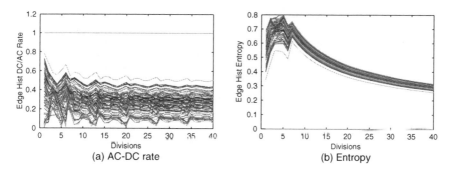

Figure 14.8   Selection criterion distribution for 80-dimensional edge histogram.

denotes the mean while the first AC parameter,

$$AC = \frac{1}{M} \left\| \sum_{n=0}^{M-1} f_n e^{-\frac{2\pi i n}{M}} \right\|, \qquad (14.10)$$

refers to the strongest deviation. The AC-DC rate,

$$R_{AC/DC} = \frac{AC}{DC} \sim \sum_{n=0}^{M-1} f_n e^{-\frac{2\pi i n}{M}}, \qquad (14.11)$$

reflects the bias of the frequency sequence away from the average. A low $R_{AC/DC}$ is preferred.

Figure 14.8 displays the criterion value distribution (y-axis) with different on feature term (x-axis) selections for an 80-dimensional edge histogram in the TRECVid 2006 corpus. Favored maxima/minima appear on all dimensions, which indicates the effectiveness of the respective criterion. In the following section, we will demonstrate how to employ feature terms for retrieval.

14.2.4   Feature Dimension Selection

In this section, we use feature terms for feature dimension selection in order to reduce the dimension number. The number of feature terms reflects the distribution of documents on a feature dimension. A low feature term number shows all media documents are compactly distributed on a feature dimension. This indicates that the ineffectiveness of this feature dimension for retrieval, as media documents can hardly be discriminated. A too high feature term number hints that documents are randomly distributed across the whole feature space. This is also not welcomed in retrieval, as such a feature dimension may introduce too much noise. In addition, it can be regarded as a random event that a document falls into a feature term and is thus labeled by an integer. Note that the uniform hypothesis ensures documents are distributed as evenly as possible. The number of possible slots, that is, the number of feature terms on a dimension, reflects the average of document labels. This indicates that the number of feature terms follows a Gaussian

Table 14.1 Term numbers of homogenous texture in the TRECVid 2006 collection

Criterion	Term number on each feature dimension
Entropy	7,9,10,4,4,10,4,4,3,3,3,3,3,3,3,3,3,3,3,3,3,3,3,3,3,4,3,5 6,3,6,5,3,3,3,3,3,3,3,3,3,3,3,3,3,3,3,3,3,3,3,3,3
χ^2 test	4,4,4,4,4,4,4,4,4,4,4,4,4,4,4,4,4,4,4,4,4,4,4,4,4,4,4,4,4, 4,4,4,4,4,4,4,4,4,4,4,4,4,4,4,4,4,4,4,4,4,4,4,4,4
AC-DC rate	4,4,5,4,4,6,4,4,5,5,5,5,5,5,5,5,5,5,5,5,9,9,5,16,38,6,243, 254,6,253,253,4,4,3,3,5,3,5,5,5,5,5,5,5,6,5,5,6,6,5,12,15

distribution, according to the central limit theorem (Table 14.1). The confidence threshold 0.7 is therefore set to remove ineffective feature dimensions.

14.2.5 Collection Representation and Retrieval System

The framework of our MIR system is shown in Figure 14.9. Four MPEG-7 low-level features are extracted by the ACEToolBox (O'Connor et al. 2005), including color layout (12 dimensions), dominant color (7 dimensions), edge histogram (80 dimensions) and

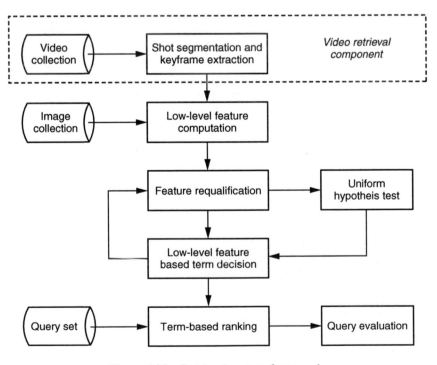

Figure 14.9 Retrieval system framework.

homogeneous texture (53 dimensions). The number of feature terms is computed by the criteria outlined in Section 14.2.3. A boolean vector of feature terms is computed to represent a media document while a frequency vector represents a collection and also generates a query from examples. We then employ the Kullback–Leibler (KL) divergence and the BM25 model to estimate relevance. One advantage of this approach is that no free parameters are necessary for media feature combination.

Document/Query Example Representation

Let $V = \{t_1, t_2, \ldots, t_n\}$ be the vocabulary of feature terms. A media document d is therefore represented by a boolean vector

$$I_d = \{I_{t_1.d}, \ldots, I_{t_n.d}\} \tag{14.12}$$

based on the vocabulary. Here $I_{t.d} = 1$ if and only if $t \in d$, otherwise $I_{t.d} = 0$. A query Q is described by a frequency vector of feature terms which accumulates the appearances of feature terms in all examples q:

$$C_Q = \sum_{q \in Q}' I_q. \tag{14.13}$$

A vector representation of feature terms is defined here for a document/query. We will employ text retrieval ranking functions for relevance estimation.

KL Divergence Ranking

Amati and Rijsbergen (2002) compare term distribution bias between a query Q and a media document d. We can write the negative KL divergence as

$$-D_{KL}(\theta_Q|\theta_d) = H(\theta_Q) - H(\theta_Q, \theta_d) = H(\theta_Q) + \sum_{t \in V} \theta_{t.Q} \log \theta_{t.d},$$

where H is the entropy, t denotes a term in the vocabulary V, and θ_Q and θ_d stand for the representation for the query and a document, respectively. $\theta_{t.Q}$ and $\theta_{t.d}$ are shorthand for $P(t|\theta_Q)$ and $P(t|\theta_d)$. Note that $H(\theta_Q)$ is constant for a given query, so that

$$-D_{KL}(\theta_Q|\theta_d) \sim \sum_{t \in V} \theta_{t.Q} \log \theta_{t.d}. \tag{14.14}$$

Since the appearance of a feature term $I_{t.d}$ is boolean, the relevance status value is defined as

$$RSV(d; Q) = \sum_{t \in V} \theta_{t.Q} \log \theta_{t.d} I_{t.d}. \tag{14.15}$$

BM25 Ranking

We propose an approach based on the BM25 model for text retrieval (Amati and Rijsbergen 2002). For a media document, the frequency of a feature term t in document d, $f_{t,d}$, is the binary $I_{t,d}$. The decision of document length is a research problem in MIR. This is because we can hardly estimate how many high-level concepts or visual words are enough to describe a media collection. It is therefore difficult to decide on the length of a media document (e.g. the number of concepts in an image). In this work, we limit our scope to low-level features and the TRECVid corpus, simply evaluating the effectiveness of the new feature term space for low-level feature based retrieval. As a result, all documents in the experimental collection are of the same size. The relevance status value is

$$RSV(d; Q) = \sum_{t \in V} IDF(t) I_{t,d} C(t, Q), \tag{14.16}$$

where $C(t, Q)$ is the appearance frequency of a feature term t in a query Q (equation (14.13)) and $IDF(t)$ is similar to inverse document frequency,

$$IDF(t) = \log \frac{N - n(t) + 0.5}{n(t) + 0.5}, \tag{14.17}$$

where N is the number of documents in a collection and $n(t)$ is the number of documents with a given feature term t.

14.2.6 Experiment

The TRECVid 2006 video collection is employed to evaluate the effectiveness of the feature term space for content-based image and video retrieval. Twenty-four content-based queries (topics 173–196) are involved, such as *finding shots of Condoleezza Rice* (Topic 194). Each query is represented by 7-11 image examples and other annotations (e.g. text tags and audio clips). Note that low-level features are ineffective for most queries (Hoi et al. 2006). For example, Natsev et al. (2007) treat low-level features as an additional knowledge source and argue that little improvement can be achieved by using low-level features in comparison with text and high-level concepts.

Feature Term Selection

Table 14.1 lists the optimized number of feature terms for each feature dimension in homogeneous texture (53 dimensions), using selection criteria outlined in Section 14.2.3. The most favorable extremes appear in the interval from 3 to 20. This suggests a small number of feature terms for the representation of an individual feature dimension. However, the overall number of feature terms increases when more feature dimensions and more features are included. In addition, these feature terms can be indexed by an inverted file for efficiency.

Content-Based Retrieval

We take two low-level feature-based retrieval methods from an earlier TRECVid workshop (TRECVid 2003) as our baseline, including direct feature comparison and K-nearest-neighbor clustering. Direct feature comparison computes a mixed Euclidean distance to identify the closest or most similar keyframes to a query. The K-NN clustering groups keyframes into K clusters, each of which contains 600 keyframes. The top two closest clusters are returned as results and returned documents are re-ranked by visual similarity to query examples.

The number of relevant documents in the top 1000 returned documents is used for performance evaluation (TRECVid 2003). Table 14.2 lists the performance of dominant color. The set of feature terms collected by the minimized AC-DC criterion achieves the best average performance. In summary, the approaches based on feature terms outperform K-NN and are comparable with the method of direct comparison in performance but with a low computational cost. Table 14.3 displays the average (mean) and standard deviation (Δ^2) of the number of relevant documents in the top 1000 returned documents for other features. A high average number of relevant documents shows the effectiveness of respective features in retrieval, while low standard deviation reflects robustness. The employment of feature terms reduces the standard deviation and improves retrieval robustness of features.

Table 14.2 Number of relevant documents of dominant color, in the top 1000 returned documents, in the TRECVid 2006 collection

Topic	Direct	K-NN	Entropy	χ^2	AC-DC
173	5	1	11	12	13
174	43	29	34	45	51
175	18	5	21	16	24
176	7	3	7	7	7
177	23	3	25	25	7
178	1	1	8	8	13
179	6	0	2	4	2
180	0	1	9	11	2
181	8	0	1	7	4
182	25	2	8	15	17
183	33	7	11	21	22
184	30	6	45	46	51
185	8	1	3	10	10
186	71	12	56	79	57
187	24	2	12	28	50
188	25	2	9	38	29
189	24	0	63	54	74
190	3	1	7	11	11
191	49	5	63	76	75
192	2	8	10	9	1
193	2	0	2	8	9
194	5	0	6	8	9
195	87	0	59	95	101
196	58	26	20	49	46
Average	23.21	4.79	20.50	28.42	28.54

Table 14.3   Average number of relevant documents, in the top 1000 returned documents, of all query topics

	Direct		kNN		Entropy		χ^2		AC-DC	
	Mean	Δ^2	Mean	Δ^2	Mean	Δ^2	Mean	Δ^2	Mean	Δ^2
Homogenous texture	27.583	27.080	13.958	16.096	15.208	24.010	23.083	22.981	40.542	16.897
Color layout	59.625	111.428	42.917	103.484	23.25	25.328	21.833	14.403	22.708	15.149
Edge histogram	51.833	63.748	28.708	78.891	37.083	69.766	49.292	50.122	68.958	58.397

Figure 14.10 shows the mean average precision of the color layout feature in all topics. Max-Max denotes the maximum mean average precision when all relevant documents are found by individual examples and direct comparison (Hoi et al. 2006). Max-Mean is the average mean average precision achieved by individual examples. Max-Max is the upper boundary of performances for late fusion approaches, while Max-Mean denotes the baseline. The difference between Max-Max and Max-Mean reflects the number of examples which contribute positive knowledge. In most topics, the mean average precision of feature term approaches are below Max-Max but above Max-Mean. This proves the effectiveness of feature terms. In topics 174, 176, 177, 179, 187 and 191 our performance exceeds Max-Max. Similar conclusions are found for edge histogram and homogeneous texture.

14.2.7   Conclusion

We propose a term-like representation, the feature term, to improve the effectiveness of media document representation as well as to facilitate the exploration of statistical strategies in MIR. The uniform term distribution is hypothesized for feature term computation. This leads to a new feature space for media document indexing and helps

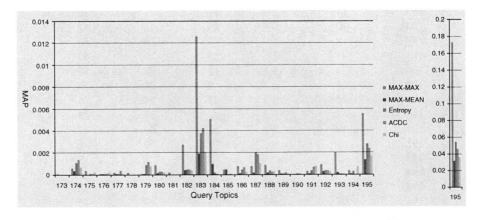

Figure 14.10   Mean average precision (MAP) of color layout query.

the modeling of media collections. Two textual retrieval models, KL divergence and BM25, are adapted to evaluate the effectiveness of feature terms in MIR. Experiments on the TRECVid 2006 video collection show significant improvements on retrieval performance as well as on system robustness. In summary, feature term space brings the following benefits: firstly, we are able to exploit powerful text retrieval models in the multimedia domain; secondly, some efficient access structures are allowed, for example an inverted index, for media data processing; thirdly, we avoid intensive parameter tuning in media combination and feature selection by using the ranking function and accumulated features representation.

This work is limited to low-level features, as high-dimensional low-level features are usually difficult to store, manage and access (Rui et al. (1999)). Note that the representation of feature terms can easily be extended to high-level features. This provides an approach to fuse both low-level and high-level features into a unified feature space for document representation. Another advantage of feature terms is the convenience of aggregation. A long visual frame sequence, such as a shot, can be represented by a frequency vector of feature terms. It is interesting to investigate the effectiveness of shot-based video retrieval, as a shot-based query can define video contents more precisely and robustly than an image does.

14.3 Conclusion and Future Trends

In this chapter, we have discussed two approaches for building an efficient multimedia search engine. The first is based on building an efficient index structure and the second is based on building a set of visual concepts.

The index structure is based on a random projection clustering technique and a recursive space splitting of a numeric dataset. The clustering results in a set of clusters with a better data distribution than the original dataset, hence supporting a more precise index.

The retrieval model is based on a concept called feature terms which is proposed for media documents. It results in an efficient approach for query generation from multiple examples, as well as an effective method of collection modeling. We have adapted two text retrieval models, KL and BM25, for multimedia retrieval and evaluated our approach on the Corel photo collection and the TRECVid 2006 collection. This new approach brings the following benefits: firstly, we are able to exploit powerful text retrieval models in the multimedia domain; secondly, some efficient access structures are allowed (e.g. inverted index for media data processing); and thirdly, we avoid parameter tuning in media combination and feature selection by using ranking functions and aggregated feature representations. Moreover, experimental results show the effectiveness of this approach, in comparison with other popular methods employed in low-level feature-based MIR.

Acknowledgement

The first author is supported by a fellowship within the PostDoc program of the German Academic Exchange Service (DAAD).

15

The Role of Explicit Semantics in Search and Browsing

Michiel Hildebrand,[1]* Jacco van Ossenbruggen[2] and Lynda Hardman[2]
[1]*Vrije Universiteit Amsterdam, Amsterdam, The Netherlands*
[2]*Centrum Wiskunde & Informatica, Amsterdam, The Netherlands*

In recent years several Semantic Web applications have been developed that support some form of search. These applications provide different types of search functionality and make use of the explicit semantics present in the data in various ways. The aim of this chapter is to give an overview of the semantic search functionalities provided by the current state of the art in Semantic Web applications. Excluding search based on structured query languages such as SPARQL (Prud'homme and Seaborne 2007), we focus on queries based on simple textual entry forms and queries constructed by navigation (e.g. faceted browsing). We provide a systematic understanding of the different design dimensions that play a role in supporting search on Semantic Web data. Using the required insights, we describe the design decisions made in two of our own tools with practical use cases.

Section 15.1 establishes the basic search terminology used throughout the chapter. Section 15.2 provides an in-depth analysis of the different types of search functionality based on a survey of 35 Semantic Web applications. We argue that search is far from trivial, and list the different roles semantics currently plays throughout the various phases of the search process. Then we show with two use cases how 'semantic' functionality can support the user while searching for museum objects in Section 15.3 and use faceted navigation to explore a collection of news images in Section 15.4. We summarize our main conclusions in Section 15.5.

15.1 Basic Search Terminology

We introduce the basic definitions used throughout the paper.

---

* At the time of writing Michiel Hildebrand was employed by CWI, Amsterdam, The Netherlands.

Multimedia Semantics: Metadata, Analysis and Interaction, First Edition.
Edited by Raphaël Troncy, Benoit Huet and Simon Schenk.
© 2011 John Wiley & Sons, Ltd. Published 2011 by John Wiley & Sons, Ltd.

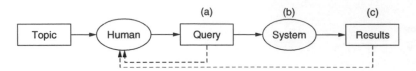

Figure 15.1 High level overview of text-based query search: (a) query construction; (b) search
algorithm of the system; (c) presentation of the results. Dashed lines represent user feedback. Figure
adapted from TRECVID (2008).

Search process. We follow the TRECVID 2007 Evaluation Guidelines (TRECVID
2008) and divide the search process into three different phases: query construction, execution of the core search algorithm and presentation and organization of the results. We
also take into account user feedback on the query and on the results (see also Figure 15.1).

Semantic search. We use the term 'semantic search' when semantics is used during
any of the phases in the search process.

Type of results. Traditional search engines on the web typically assume the result of a
search to be a set of web resources (e.g. text documents, images or video). This also holds
for some Semantic Web applications. Others, however, return sets of matching URIs, sets
of matching triples (an RDF sub-graph) or a combination of these. Often the behavior of
a system depends on the type of result it assumes, so we make this explicit wherever this
is relevant.

Overview pages and **surrogates**. We refer to presentation of the (*k* first) results as the
overview page, which typically represent results using *surrogates*. For example, resulting
HTML pages can be represented by their title and a text snippet, while image or video
results are often represented by thumbnails.

Local view page. Surrogates typically provides links to the full presentation of the
result they represent. The latter is either the full presentation of an information resource
(e.g. the full HTML page that is linked from the snippet presented on the overview page),
some human-readable representation of metadata associated with the result, or a human-
readable representation of a resulting subgraph. Following Rutledge et al. (2005), we refer
to the latter presentation as a *local view page*.

Syntactic matching. Syntactic search matches the query against the textual content of
the resources (if applicable), the literals in the RDF metadata, the URIs in the system, or
a combination of these.

Semantic matching. We use the term 'semantic matching' for those algorithms that,
in addition to syntactic matching, use the graph structure of the RDF metadata and/or its
semantics to find results.

15.2   Analysis of Semantic Search

We systematically scanned all proceedings of the International and European Semantic
Web Conference series as well as the *Web Semantics* journal, to compile a list of end-user
applications described or referred to. For each system we collected basic characteristics
such as the intended purpose, intended users, the scope, the triple store and the technique
or software used for literal indexing. The resulting document was made available online[1]

---
[1] http://www.webscience.org/swuiwiki?title = Semantic_Search_Survey

Table 15.1   Functionality and interface support in the three phases of semantic search

Query construction	
Feature: Functionality	**Interface Components**
Free text input: Keywords, natural language	Single text entry, property-specific fields
Operators: Boolean constructs, syntactic disambiguation, semantic constraints on input/output	Application-specific syntax
Controlled terms: Disambiguate input, restrict output, select predefined queries	Value lists, faceted browser, graph
User feedback: Pre-query disambiguation	Autocompletion
Search algorithm	
Syntactic matching: Exact, prefix or substring match, minimal edit distance, stemming	Not applicable
Semantic matching: graph traversal, query expansion, spread activation, RDFS/OWL reasoning	Not applicable
Result presentation	
Data selection: Selected property values, class-specific template, display vocabularies	Visualized by text, graph, tagcloud, map, timeline, calendar
Ordering: Content and link structure based ranking	Ordered list
Organization: Clustering by property, by result path or dynamic	Tree, nested box structure, clustermap
User feedback: Post-query disambiguation, recommendation of related resources	Facets, tagcloud, value list

and announced on three public mailing-lists.[2] Additionally, we sent personal emails to the authors of papers and developers of all systems included. This resulted in an updated version of the online document. This update was based on 15 email threads, in which additional information was provided for 11 systems and 6 additional systems within the scope of the survey were recommended, giving a total of 35 systems.

Based on the data resulting from the survey, we perform a more thorough analysis of the three individual phases in the search process of Figure 15.1. For each of these we consider the underlying functionality and the features of the corresponding user interface. The results of this analysis are summarized in Table 15.1, and the table also provides the structure of the remainder of this section. We discuss query construction in Section 15.2.1, the search algorithms in Section 15.2.2 and the presentation of the results in Section 15.2.3. Note that the examples and references merely serve as illustrations; the full analysis with references is available in the online survey.

15.2.1   Query Construction

The search process starts with the user constructing a query that reflects his or her information needs. We describe the functionality for this process as provided by the systems in the survey and how this functionality is supported at the interface.

---

[2] `public-xg-mmsem@w3.org, semantic-web@w3.org, public-semweb-ui@w3.org`

Functionality

Constructing a query in free text requires little knowledge of the system and data structure. The price users have to pay is ambiguity: words can have multiple meanings (lexical ambiguity) and a complex expression can have multiple underlying structures (structural ambiguity). Ambiguous input often leads to irrelevant search results. To reduce ambiguity several systems allow additional query constructs beyond free text input: structural and semantic operators and controlled terms. We describe free text input and the additional query constructs and the role of user feedback in the process of matching free text with controlled terms.

Free text input is supported in existing systems in three ways. First, full text search allows the user to find all resources with matching textual content or metadata. In many semantic search engines full text search is the main entry point into the system (Celino et al. 2006; DERI 2005; Ding et al. 2005; ;Guha et al. 2003; Schreiber et al. 2006). Second, free text input can be restricted to match a value of a specific property. In faceted browsers this is the case when searching for a value within a particular facet (Hildebrand et al. 2006; Schraefel et al. 2005; SIMILE 2003). Finally, systems such as AquaLog (Lopez et al. 2005) and Ginseng (Bernstein et al. 2006) support free text input in the form of natural language expressions.

Syntactic operators explicitly define the interpretation of complex search terms. Well-known examples are the boolean operators AND and OR. Several applications employ third-party search libraries such as Apache Lucene,[3] which typically provide an extensive collection of syntactic control structures.

Semantic operators add explicit meaning to a query. In SemSearch, for example, the user specifies to which RDFS or OWL class a result should belong. The authors illustrate this with the example `article:motta` for which the system retrieves all resources that are of type `article` and match the search term `motta` (Lei et al. 2006).

Controlled terms provide the use of predefined concepts. In QuizRDF (Davies and Weeks 2004) the user selects an RDF class to determine the type of the search term. Other systems provide autocompletion to support users with keyboard-based input of controlled terms (Hildebrand et al. 2006; Hyvönen and Mäkelä 2006; Kiryakov et al. 2004; Schraefel et al. 2005; SIMILE 2003). A different approach is seen in DBin (Tummarello et al. 2006) and Haystack (Quan and Karger 2004), which allow the user to select predefined queries.

User feedback on the input is useful when there are multiple controlled terms that match with the free text input. Several systems allow the user to select the intended term before it is processed by the search algorithm (Celino et al. 2006; Hildebrand et al. 2006; Hyvönen and Mäkelä 2006). This form of user feedback allows pre-query disambiguation. In contrast, post-query disambiguation is performed on the results of the search algorithm.

Interface

The basic interface components to enter or construct a query are text entry boxes and value selection lists. These components are used in various designs, of which we mention three. If applicable, we describe the link from the interface components to the underlying

---

[3] http://lucene.apache.org/java/docs/queryparsersyntax.html

data structure. Furthermore, we describe the interface aspects of user feedback on the input and several proposals for more advanced query construction.

A *single text entry field* is sufficient for free text input, for example, Google and several systems that we analyzed (Celino et al. 2006; Davies and Weeks 2004; Ding et al. 2004; Guha et al. 2003; Schreiber et al. 2006). Additional features included by some systems are selectable result types (Ding et al. 2004) and options for the search algorithm or presentation of the results (Davies and Weeks 2004).

Property-specific search fields support query construction guided by a specific set of possible search values (Heflin and Hendler 2000; Kiryakov et al. 2004; Metaweb 2007; Mika 2006). The value sets are typically defined by the range of the corresponding RDF property.

Faceted browsing allows the user to constrain the set of results within a particular facet. Typically, facets are directly mapped to properties in RDF. Alternatively, the mapping is made by projection rules. The advantage of an indirect mapping is that this allows the developer to define facets that match the user's needs while keeping the data structure unchanged (Suominen et al. 2007). Faceted browsing is applied to Semantic Web data by various authors (Fluit et al. 2003; Hildebrand et al. 2006; Hyvönen et al. 2005; Oren et al. 2006; Schraefel et al. 2005; SIMILE 2003; Suominen et al. 2007) as well as by the company Siderean[4] in the Seamark Navigator.

User feedback is typically provided after the query has been entered, or dynamically during the construction of the query as a form of *semantic autocompletion*. The former method is used in Squiggle (Celino et al. 2006) and MuseumFinland (Hyvönen et al. 2005) where the disambiguation of the matching query terms is presented after submitting the query. In semantic autocompletion the system suggests controlled terms with a label prefix that matches the text already typed in. Hyvönen and Mäkelä (2006) describe the idea of semantic autocompletion and several implementations. In faceted interfaces autocompletion is often used within a single facet (Hildebrand et al. 2006; Kiryakov et al. 2004; Schraefel et al. 2005; SIMILE 2003).

In general there is a clear need for simple interfaces, such as the single text entry field. On the other hand, interface designs that support more complex interaction styles potentially give the user more control, which is useful for the formulation of more precise information needs.

15.2.2   Search Algorithm

All text-based search involves some form of syntactic matching of the query against textual content and/or metadata, an aspect well covered in information retrieval (Baeza-Yates and Ribeiro-Neto 1999). Semantic matching can extend syntactic matching by exploiting the typed links in the semantic graph. Before we describe semantic matching we briefly describe syntactic matching, focusing on the indexing functionality and the support that is already provided by the low-level software on which various systems are built.

---

[4] http://www.siderean.com/

Syntactic Matching

All systems in our study index the textual data in their collection for performance reasons. Which textual data is indexed (e.g. the content, the metadata or the URIs) is important for the search functionality of the system. In an ontology search engine such as Swoogle, users might want to search on URIs (Ding et al. 2005). In annotated image collections the metadata forms the primary source for indexing (Celino et al. 2006; Hyvönen et al. 2005; Schreiber et al. 2006). For images that occur in web pages the contextual text provides an alternative source (Celino et al. 2006; Falcon-S 2005). Indices can be based on the complete word or on a stemmed version. Some interface functionalities require additional features. Autocompletion interfaces, for example, require efficient support for prefix matching. Some triple stores provide built-in support for literal indexing, for example, OpenLink Virtuoso[5] and SWI Prolog's Semantic Web library.[6] Alternatively, a search engine, such as Lucene, can be used together with a triple store.

Semantic Matching

After syntactic matching, the structure and formal semantics of the metadata can be used to extend, constrain or modify the result set. Note that in a connected RDF graph, any two nodes are connected by a path. Naive approaches to semantic search are computationally too expensive and dramatically increase the number of results. Systems thus need to find a way to reduce the search space and to determine which semantically related resources are really relevant.

Inspired by the semantic continuum described by (Uschold 2003), we distinguish three levels of semantic matching: graph traversal, explicit use of thesaurus relations and inferencing based on the formal semantics of RDF, RDFS and OWL.

Graph traversal takes only the structure of the graph into account. Several techniques are in use to constrain graph search algorithms. In Tap, constraints define which relations to traverse for the instances of a particular class (Guha et al. 2003). Alternatively, a weighted graph search algorithm may constrain the possible path structures and path length. Such an algorithm requires the assignment of weights to the edges in the graph, where the weights reflect the importance of the corresponding RDF relations. In e-Culture, weights are manually assigned to RDF relations (Schreiber et al. 2006). SemRank automatically computes weights based on statistics derived from the graph structure (Anyanwu et al. n.d.). Spread activation (Rocha et al. 2004) is another computationally attractive technique for graph traversal, which can incorporate weights as well as the number of incoming links.

Thesaurus relations are sometimes used for query expansion. With the acceptance of SKOS (Miles et al. 2005) as a standard representation for thesauri, semantic matching with hierarchical broader term, narrower term, and the associative related term can be implemented in a generic way. The Squiggle framework is an example in which this is done (Celino et al. 2006). Facet browsers typically rely on hierarchical thesaurus relations to restrict their result sets (Hildebrand et al. 2006; Hyvönen et al. 2005). Within the FACET project the integration of thesauri in the search process is studied extensively. This resulted in a demonstrator as well as a proposal for a semantic expansion

---

[5] http://www.openlinksw.com/
[6] http://www.swi-prolog.org/packages/semweb.html

service (Binding and Tudhope 2004), which in turn formed the basis for the experimental SKOS API.[7]

RDFS/OWL reasoning can also influence the search results. Several systems support RDFS subsumption once an RDF class is selected in the interface (Auer et al. 2006; Duke et al. 2007; Lei et al. 2006; Tummarello et al. 2006). In Dose, specialization and generalization over the subclass hierarchy are used dynamically according to the number of search results (Bonino et al. 2004). Some systems support partial OWL reasoning, and process, for example, only the OWL identity relations. In Flink (Mika 2005) and SWSE (DERI 2005) these are extensively used to model the identity between extracted entities.

15.2.3   Presentation of Results

We describe how explicit semantics is used to extend the baseline functionality in the presentation of search results, and the techniques that are used to visualize the results in the interface. As a baseline we consider the presentation of search results by popular search engines such as Google: the information selected for presentation is the URI or label of the result, surrogates of the content (e.g. text snippets or image thumbnails) and, optionally, additional information such as the file size. The results are typically organized in a plain list and ordered by relevance. We describe, for each aspect, how additional semantics are used to extend this baseline.

Functionality

We consider three aspects of the presentation: selecting what data to present, organizing the results and ordering the results. In addition, we discuss the function of user feedback on the results.

Selecting what to present is tightly bound with what the search engine considers to be a 'result'. If the result is a Web page or other information resource, traditional surrogates are typically used. When the search result is a set of URIs referring to nodes in an RDF graph, or a set of RDF triples, systems need to invent new ways to represent the results in their overview page. In most systems we studied, the decision on what (meta)data is used for the surrogates is hardwired into the system. QuizRDF supports template definitions for each RDF class (Davies and Weeks 2004). Dbin (Tummarello et al. 2006) creates templates for specific user tasks and domains. Display vocabularies such as Fresnel (Bizer et al. 2005), as used by Longwell (SIMILE 2003), provide full control over what data to select for presentation and how to present it.

Turning to *organizing the results*, semantics can also play a role in grouping semantically similar results together in the presentation, a feature commonly referred to as *clustering*. Assuming that users are interested in the results of only one cluster, clustering can also be considered as a form of post-query disambiguation. In our study we found several forms of clustering. In many systems, the values of a particular property are used to group the result set on common characteristics within a particular dimension. In Guha et al. (2003) results with similar types are clustered together. In faceted browsers similar behavior is found: systems described in SIMILE (2003), Hildebrand et al. (2006), and

---

[7] http://www.w3.org/2001/sw/Europe/reports/thes/skosapi.html

Hyvönen et al. (2005) all support clustering on the values of a particular facet. Noadster uses concept lattices to determine dynamically which properties to use for a given result set (Rutledge et al. 2005). In e-Culture (Schreiber et al. 2006) the RDF path between the literal that is syntactically matching the query and the result may span more than one property. Clustering the results on these paths illustrate the interpretations of the query.

The *ordering of results* can be done using different techniques. Ranking of results based on relevance is covered well in information retrieval (Baeza-Yates and Ribeiro-Neto 1999). Numerous algorithms have been developed, evaluated and applied in successful applications. Term frequency-inverse document frequency (tf-idf) is a syntactic measure often used to determine the importance of a word based on the number of occurrences in a document relative to the number of occurrences in the entire collection. Many systems in our study use Lucene, which provides ranking based on tf-idf. In addition to textual content, the link structure is another source for ranking. Swoogle uses a variant of PageRank (Page et al. 1998) to measure the relevance of RDF documents. PageRank was adapted to compensate for different types of relations that link RDF documents and terms (Ding et al. 2005). In SWSE (Hogan et al. 2006) a variant of PageRank based on the principle of focused subgraphs (Kleinberg 1999) is used.

In our study, we did not find instances where matching and ranking algorithms are influenced by *user feedback*. We mainly encountered user feedback in disambiguating, specializing, generalizing or expanding the result set. Most systems support expansion of a query by adding a keyword or by selecting a value from a property field or facet. In several systems, post-query disambiguation of free text input is supported through the selection of an RDF type (Berlin 2007; Davies and Weeks 2004; DERI 2005; Duke et al. 2007). Alternatively, queries can be specialized or generalized with concepts from narrower or broader thesaurus relations (Celino et al. 2006; Hildebrand et al. 2006; Hyvönen et al. 2005). An unwanted side effect of query refinement is the risk of ending up with no results. This can be avoided by restricting the user beforehand to the use of only those terms that lead to results. This is one of the principles behind faceted browsing interfaces (Yee et al. 2003).

We observed that domain-specific applications use the semantics to organize the search results into clusters. Domain-independent search engines typically rely on ranking techniques for effective presentation of the search results.

Interface

Most systems provide a straightforward interface that directly reflects the structure of the data selected and how it is organized. Typical examples include numbered lists for a linearly ranked set of results or visual grouping of clustered results in nested box layout structures. Since RDF is represented as a graph, visualizing the data as a graph may seem a straightforward choice. However, from a user interface perspective, 'big fat graphs' quickly become unmanageable (Schraefel and Karger 2006), with only a few exceptions, including the visualization of social networks between small groups of people (Mika 2005). Other visualization techniques (Geroimenko and Chen 2003b) we encountered include the following:

- *Tagclouds* that indicate the importance of textual metadata with variations in the font size. OpenAcademia (Mika 2006) visualizes the concepts related to research publications and search. DBpedia.org (Berlin 2007) presents the available RDF types of the search results.
- *Clustermaps* that visualize the overlap between classes of instances, without needing an explicit concept representing this overlap (Geroimenko and Chen 2003a). For example, in AutoFocus a clustermap visualizes the results of individual constraints as well as result sets that satisfy multiple constraints (Fluit et al. 2003).
- *Data type-specific* visualizations are used in several systems to present space and time on a geographical map, timeline or calendar. The Simile timeline,[8] several map visualization tools and Google Calendar can be used through publicly available APIs. Hence, we do not list the individual systems that make use of these.
- *Local view pages* provide a detailed presentation of the metadata associated with single URI. Systems such as Tabulator (Berners-Lee et al. n.d.) and Disco (Bizer and Gauss 2007) are based on the notion of a *concise bounded description* (CBD)[9] and present the statements where the current focus URI is a subject. Others systems' local view pages may contain all statements in which the URI occurs either as a subject or object. Sesame's URI explorer pages,[10] Noadster (Rutledge et al. 2005) and E culture (Schreiber et al. 2006) also include statements where the URI plays the role of the property.

15.2.4   Survey Summary

The survey shows that explicit semantics can play a role in all three parts of the search process. In the query construction controlled terms are used to create unambiguous queries. Terms are either presented in a selection list or made available on demand by autocompletion. Faceted interfaces give the user even more control, allowing the user to define the particular role of a query term. Natural language input also allows the user to construct more precise queries. In this case the system has to do the interpretation of the query.

Current search applications make only limited use of OWL reasoning. Only equivalence reasoning is used in several applications. Lightweight reasoning, based on thesaurus relations, is a more common feature in search algorithms. The SKOS relations for broader term and related term are used both for query expansion as well as to recommend related results. More generic graph search algorithms have the advantage that all relations in the data can be considered at the cost of returning less relevant results.

In the result presentation standard representations of space and time properties make visualizations on a map, timeline or calendar a common feature. The link structure of the graph provides valuable information for ranking algorithms as well for clustering algorithms. Determining the effectiveness of these techniques remains, however, unclear. Note that, for example in information retrieval, there is a long tradition of evaluating

---

[8] http://simile.mit.edu/timeline/
[9] http://www.w3.org/Submission/CBD/
[10] http://www.openrdf.org/

the quality of retrieval systems. Conference series such as TREC and INEX contribute to an (evolving) community consensus about which dimensions to evaluate, and how to measure a system's performance on that dimension. It is safe to say that within the Semantic Web community, we have not yet developed a similar consensus about the use of explicit semantics to improve search, and how to evaluate and to compare semantic search systems.

We now turn to a detailed description of two types of semantic search supported by our own tool, ClioPatria (Wielemaker et al. 2008). We first describe semantic keyword search with a cultural heritage use case. Then we describe how faceted navigation can be used for the exploration of a collection of news images.

15.3   Use Case A: Keyword Search in ClioPatria

The search facility of the MultimediaN E-Culture demonstrator[11] allows the user to search for museum objects that are semantically related to one or more keywords. A typical scenario for this type of semantic search is the exploitation of vocabulary alignments. Consider a use case in which the user is interested in finding works from the Japanese Tokugawa style period. We describe the search process when using the E-Culture demonstrator to find these artworks.

15.3.1   Query Construction

To support keyword search a simple Google-like interface is sufficient. The user types the keyword 'Tokugawa' into the text entry field and submits this query. Instead of submitting a keyword query, the user can choose to first disambiguate the query. The system uses autocompletion to suggest artworks or thesaurus terms while the user is typing. Figure 15.2 shows the suggestions given when the user typed 'toku'. The suggestions are organized into different clusters, grouping similar suggestions together.

15.3.2   Search Algorithm

Consider the situation in Figure 15.3, which is based on real-life data. The user has submitted the query 'tokugawa'. The Dutch ethnographic museum in Leiden actually has works in this style in its digital collection, such as the work shown on the right of the figure. However, the Dutch ethnographic thesaurus SVCN, which is being used by the museum for indexing purposes, only contains the label 'Edo' for this style. Fortunately, another thesaurus in our collection, Getty's Art and Architecture Thesaurus[12] (AAT), does contain the same concept with the alternative label 'Tokugawa'.

To exploit all possible relations available in the heterogeneous data a generic graph traversal algorithm is used. Either the keyword query or the disambiguated thesaurus term is the starting point of the graph traversal.

Figure 15.4 shows the result of a graph search for the keyword query 'tokugawa'. The three rectangular boxes on the left show the literals that matched the keyword. The

---

[11] http://e-culture.multimedian.nl/demo/search
[12] http://www.getty.edu/research/conducting_research/vocabularies/aat/

Figure 15.2 Autocompletion suggestions are given while the user is typing. The partial query 'toku' is contained in the title of three artworks, there is one matching term from the AAT thesaurus and the artist Ando Hiroshige is found as he is also known as Tokubel.

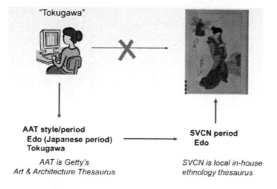

Figure 15.3 A user searches for 'tokugawa'. The Japanese painting on the right matches this query, but is indexed with a thesaurus that does not contain the synonym 'Tokugawa' for this Japanese style. Through a 'same-as' link with another thesaurus that does contain this label, the semantic match can be made.

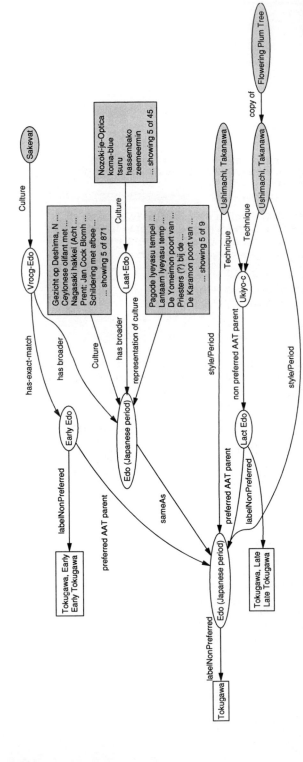

Figure 15.4 Result graph of the E-Culture search algorithm for the query 'Tokugawa'. The rectangular boxes on the left contain the literal matches, the colored boxes on the left contain a set of results, and the ellipses a single result. The ellipses in between are the resources traversed in the graph search.

algorithm found that these literals are labels of three thesaurus terms. All three are from the AAT thesaurus. Note that the three terms are related. The terms 'Early Edo' and 'Late Edo' are specializations of the Japanese period 'Edo'. Let us follow the generic term 'Edo' two steps further. Through an `owl:sameAs` relation the algorithm has found the equivalent term from the Dutch ethnographic thesaurus. This term leads us to many museum objects, the first five of which are shown in each of the rectangular boxes on the right of the figure. The AAT term 'Edo' leads to other interesting results. For two paintings the term was used for the 'Style/Period' property. Both are painted by the famous Japanese artist Ando Hiroshige, titled *Plum Estate* and *Ushimachi*. Note that one of these paintings is also found because it is a woodblock print (*ukiyo-e*), which is as specialization of the 'Late Edo' style. This print also leads to another object, namely *Flowering Plum Tree* by Vincent van Gogh. This painting is found because van Gogh made a copy of the Japanese original.

The entire graph contains some RDFs and OWL properties (or subproperties thereof). Most properties are, however, domain-specific properties used within the thesauri or to index the museum objects. The generic traversal algorithm allows the system to exploit all these properties.

15.3.3 Result Visualization and Organization

Different paths in the search graph indicate different semantic relations. In other words, a path reflects an interpretation of the query. In the E-Culture demonstrator the paths are, therefore, used to disambiguate the search results. Figure 15.5 shows the presentation of

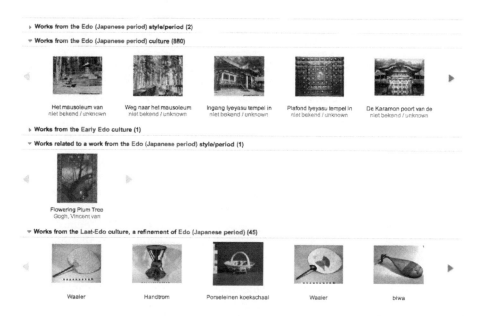

Figure 15.5 Presentation of the search results for the query 'Togukawa' in the E-Culture demonstrator. The results are presented in five groups (the first and third groups have been collapsed). Museum objects that are found through a similar path in the graph are grouped together.

the search results for the query 'Tokugawa'. The museum objects with a similar search path are grouped together. Five clusters are shown. For readability the first and third clusters have been collapsed. An additional abstraction step is performed to merge similar clusters together. In this step the resources in the graph path are abstracted over the subproperty and subclass relations. The seven result groups (colored boxes on the right) in the graph of Figure 15.4 are merged into five, because the property 'representation of culture' is a subproperty of 'culture'. The results 'Ushimachi, Takanawa' and 'Plum Estate, Kameido' share the same shortest path.

15.4   Use Case B: Faceted Browsing in ClioPatria

Within K-Space, ClioPatria is used to support search and browsing of news items. These news items are described with multimedia standards, news codes from the IPTC standard and additional metadata from various thesauri. The additional metadata is acquired through extraction of named entities such as persons, organizations and locations, from the textual stories. The extracted named entities are mapped to existing resources available on the Web, such as locations from Geonames, and individuals from DBpedia. The dataset uses in this demonstration consists of news items from 2006, including the soccer World Cup.

Consider a use case in which the user wants to selects a number of images that reflect different aspects of the French soccer player Zinedine Zidane. Keyword search allows us to find a set of news items related to Zidane, but for a more thorough exploration of these results additional control is required. Faceted presentation of the metadata gives the user additional control in the formulation of her request. The facets allows the user to specify relatively complex queries by simply following navigation links.

15.4.1   Query Construction

As in the previous use case, we start with a simple keyword search to find news items that are related to 'zidane'. The results contain both textual news messages as well as photos. Among the results there are news items with the keyword 'zidane' in the title/headline, photos showing Zidane, photos with his team mates and many photos with happy as well as sad fans. On the result page we can now activate the faceted navigation component and use it to further explore the result set.

A screenshot of the faceted interface from ClioPatria is shown in Figure 15.6. The top part contains four facets: document type, creation site, event and person. The result viewer, visible below the facets, contains news items related to the keyword 'zidane'. The current query is shown in the header of the result viewer. The user can extend the query by selecting values from the facets. In this case the value 'photo' is selected from the document type facet. The other facets only contain values that correspond to the current result set. Note that this prevents the user from constructing queries that lead to an empty result set.

The person facet contains, besides Zidane himself, other persons who are related to the photos. Most of the persons are soccer players, but the list also contains the former French president Jacques Chirac. Selecting the value corresponding to this person constrains the result set to only those images related to both Zidane and Chirac. In this case the photos

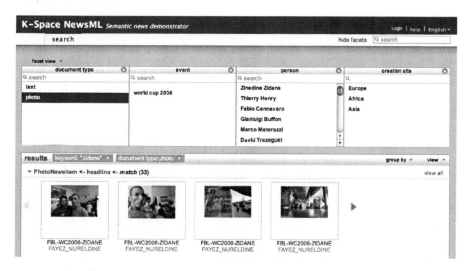

Figure 15.6 Faceted interface of the NewsML demonstrator. Four facets are active: document type, creation site, event and person. The value 'photo' is selected from the document type facet. The full query also contains the keyword 'Zidane', as is visible in the header above the results.

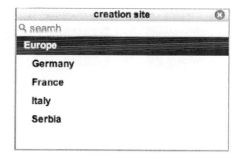

Figure 15.7 Hierarchical organization of the values in the creation site facet. The value 'Europe' is selected and below it the four countries in which photos related to Zidane are created.

show Zidane visiting the French president after the lost final of the 2006 World Cup. Clicking on a facet value that is selected will deselect this value and remove it from the query.

The screenshot in Figure 15.7 shows the creation site facet in which the value 'Europe' is selected and below it the four European countries in which the photos are created. We can navigate further into this hierarchy by selecting one of the countries. Selecting the value 'France' constrains the result set to photos of soccer fans in different regions of France. Selecting the value 'Germany' constrains the result set to all cities where Zidane played during the World Cup. By selecting the city Berlin, the result set now only contains photos of the World Cup final. Finally, we constrain the result set to photos of the infamous head-butt incident by selecting the Italian player 'Marco Materazzi' from the person facet.

15.4.2 Search Algorithm

The values selected from a facet constrain the results to those news items that are indexed with that particular value. Each additional constraint adds a new conjunct to this this query. Values selected from a hierarchical facet require an additional step, as the news items may not be directly indexed with a value selected from the hierarchy. A news item is considered a result when the value it is indexed with is reachable from the selected value through some relation. In a SKOS thesaurus this relation corresponds to `skos:broader` or `skos:narrower`.

To support the combination of semantic keyword search and faceted navigation, the graph search algorithm has to be combined with facet constraints. In ClioPatria this is solved by executing the graph search and using the facet constraints to determine if a resource encountered during the search is a result.

15.4.3 Result Visualization and Organization

After the initial query of photos related to Zidane the `creation site` facet contained three continents: Europe, Africa and Asia. To get a view on the photos for each continent

Figure 15.8 Grouped presentation of search results. The photos related to Zidane are presented in groups with the same creation site.

each facet value can be selected in turn, but to compare different groups of photos it is much more convenient to group the results. The grouping can be selected from the drop-down menu in the result viewer. The screenshot in Figure 15.8 illustrates that the photos made in Europe primarily show soccer matches and French supporters, the photos in Africa show Zidane's home country Algeria, and the photos in Asia depict French soldiers stationed in Afghanistan while watching the World Cup on television.

The organization and visualization of the facet values is important as well. The hierarchical organization of the values in the creation site facet provides different geographical perspectives on the data, by continent, country, region or city. An alternative organization is to group similar facet values together. For example, grouping persons by their nationality would allow the user to select all news items related to German persons. The visualization of facet values is important when there are ambiguous terms. Instead of presenting just the label, further information can be added. For example, for persons we can present the nationality and birth date alongside the person's name.

15.5   Conclusions

We have analyzed the state of the art in search as currently implemented in Semantic Web applications. We have identified the various roles played by semantics in query construction, the core search algorithm and presentation of search results. Our study shows that many systems already support aspects of semantic search. For some, it is even the main entry point into the underlying data. The quality of the search functionality and its user interface thus has a significant impact on the overall usability of these applications. Working toward generic libraries and publicly available services that support the individual processes will help developers to incorporate more semantic search functionality into applications. To improve the uptake of Semantic Web applications by end users, we feel our community should strive to make this research area more mature and start to develop methods for objective and systematic evaluation of the functionality and interface aspects of end-user applications.

16

Conclusion

Raphaël Troncy,[1] Benoit Huet[2] and Simon Schenk[3]
[1]*Centrum Wiskunde & Informatica, Amsterdam, The Netherlands*
[2]*EURECOM, Sophia Antipolis, France*
[3]*University of Koblenz-Landau, Koblenz, Germany*

New approaches need to be developed to appropriately scale up current systems for use with very large multimedia databases. The diversity of data both in terms of content (audio, video, text, pictures, etc.) and quality (coding artifacts, live vs studio recordings, etc.) is also continuously challenging current approaches. New types of data will also need to be better exploited in future systems. For example, there is nowadays a significant amount of rich metadata content accessible from social networks that could be used despite its inherently approximate nature. But, more generally, it seems that the major improvements in multimedia data indexing will now come from the association of machine learning techniques with high-level semantics approaches. Indeed, as exemplified in this book, this will permit us to successfully bridge the so-called *semantic gap* between the low-level description that can be obtained by machines and the high-level information that is commonly used by humans.

This book has covered a broad range of topics with various level of granularity to enable the semantic web enthusiast to be used to basic machine learning and multimedia analysis techniques while educating the multimedia researcher with methods and tools developed by the Semantic Web community. It is far from being exhaustive and key research results related to 3D content analysis and indexing, multimedia content sharing within social networks or multimedia content search on mobile devices to name a few could also have been presented in depth in their own chapters. The analysis of the different metadata standards shows that many of them target certain types of application, stages of the media production process or types of metadata. There is no single standard or format that satisfactorily covers all aspects of audiovisual content descriptions; the ideal choice depends on type of application, process and required complexity.

Multimedia Semantics: Metadata, Analysis and Interaction, First Edition.
Edited by Raphaël Troncy, Benoit Huet and Simon Schenk.
© 2011 John Wiley & Sons, Ltd. Published 2011 by John Wiley & Sons, Ltd.

References

Abiteboul S 1997 Querying semi-structured data, pp. 1–17.

Abiteboul S, Quass D, McHugh J, Widom J and Wiener JL 1997 The Lorel query language for semistructured data. *International Journal on Digital Libraries* **1**(1), 68–88.

Achermann B and Bunke H 1996 Combination of classifiers on the decision level for face recognition. *Technical report, Bern University*.

Adalí S, Sapino M and Subrahmanian V 2000 An algebra for creating and querying multimedia presentations. *Multimedia Systems* **8**(3), 212–230.

Adamek T, O'Connor N and Murphy N 2005 Region-based segmentation of images using syntactic visual features *In Proc. of Workshop on Image Analysis for Multimedia Interactive Services*, Montreux, Switzerland.

Akrivas G, Stamou G and Kollias S 2004 Semantic association of multimedia document descriptions through fuzzy relational algebra and fuzzy reasoning. *IEEE Transactions on Systems, Man, and Cybernetics* **34**(2), 190–196.

Akrivas G, Wallace M, Andreou G, Stamou G and Kollias S 2002 Context-sensitive semantic query expansion *In Proc. of the IEEE International Conference on Artificial Intelligence Systems*, Divnomorskoe, Russia.

Aleman-Meza B, Bojārs U, Boley H, Breslin JG, Mochol M, Nixon LJ, Polleres A and Zhdanova AV 2007 Combining RDF vocabularies for expert finding *In Proc. of the 4th European Semantic Web Conference (ESWC)* pp. 235–250.

Alkhateeb F, Baget JF and Euzenat J 2009 Extending SPARQL with regular expression patterns (for querying RDF). *Web Semantics: Science, Services and Agents on the World Wide Web* **7**(2), 57–73.

Alvestrand H 2001 RFC 3066: Tags for the identification of languages. IETF Network Working Group.

Amati G and Rijsbergen CJV 2002 Probabilistic models of information retrieval based on measuring the divergence from randomness. *ACM Transactions on Information Systems* **20**(4), 357–389.

Andoni A and Indyk P 2008 Near-optimal hashing algorithms for approximate nearest neighbor in high dimensions. *Communications of the ACM* **51**(1), 117–122.

Andrade E, Woods J, Khan E and Ghanbari M 2005 Region-based analysis and retrieval for tracking of semantic objects and provision of augmented information in interactive sport scenes. *IEEE Transactions on Multimedia* **7**(6), 1084–1096.

Angles R and Gutierrez C 2005 Querying RDF data from a graph database perspective *In Proc. of the 2nd European Semantic Web Conference (ESWC)*, pp. 346–360.

Angles R and Gutierrez C 2008 The expressive power of SPARQL *In Proc. of the 7th International Semantic Web Conference (ISWC)*, pp. 114–129.

Anyanwu K, Maduko A and Sheth A 2005 SemRank: Ranking complex relationship search results on the semantic web. *In Proc. of the 14th World Wide Web Conference (WWW)*

Arndt R, Troncy R, Staab S, Hardman L and Vacura M 2008 COMM: designing a well-founded multimedia ontology for the web *In Proc. of the 6th International Semantic Web Conference, 2nd Asian Semantic Web Conference* pp. 30–43.

Assfalg J, Bertini M, Colombo C, del Bimbo A and Nunziati W 2003 Semantic annotation of soccer videos: automatic highlights identification. *Computer Vision and Image Understanding* **92**(2-3), 285–305.

Multimedia Semantics: Metadata, Analysis and Interaction, First Edition.
Edited by Raphaël Troncy, Benoit Huet and Simon Schenk.
© 2011 John Wiley & Sons, Ltd. Published 2011 by John Wiley & Sons, Ltd.

Athanasiadis T, Mylonas P, Avrithis Y and Kollias S 2007 Semantic image segmentation and object labeling. *IEEE Transactions on Circuits and Systems for Video Technology* **17**(3), 298–312.

Athanasiadis T, Tzouvaras V, Petridis K, Precioso F, Avrithis Y and Kompatsiaris Y 2005 Using a multimedia ontology infrastructure for semantic annotation of multimedia content *In Proc. of 5th International Workshop on Knowledge Markup and Semantic Annotation*, Galway, Ireland.

Athineos M, Hermansky H and Ellis D 2004 Lp-trap: Linear predictive temporal patterns *International Conference on Spoken Language Processing*.

Atkinson M, Bancilhon F, DeWitt D, Dittrich K, Maier D and Zdonik S 1989 The object-oriented database system manifesto, pp. 223–240.

Auer S, Dietzold S and Riechert T 2006 Ontowiki – a tool for social, semantic collaboration *In Proc. of the 5th International Semantic Web Conference (ISWC)*, Athens, GA, USA.

Azzag H, Monmarche N, Slimane M, Venturini G and Guinot C 2003 Anttree: A new model for clustering with artificial ants *In Proc. of the Congress on Evolutionary Computation*, pp. 2642–2647. IEEE Press.

Baader F, Calvanese D, McGuinness DL, Nardi D and Patel-Schneider PF 2003 *The Description Logic Handbook: Theory, Implementation and Applications*. Cambridge University Press.

Baeza-Yates R and Ribeiro-Neto B 1999 *Modern Information Retrieval*. Addison-Wesley.

Bahlmann C, Haasdonk B and Burkhardt H 2002 On-line handwriting recognition with support vector machines – a kernel approach *In Proc. of the 8th Workshop on Frontiers in Handwriting Recognition (IWFHR)*, pp. 49–54.

Bailer W and Schallauer P 2008 Metadata in the audiovisual media production process In *Multimedia Semantics – The Role of Metadata* vol. 101 of *Studies in Computational Intelligence*.

Bailer W and Schallauer P 2006 The Detailed Audiovisual Profile: Enabling interoperability between MPEG-7 based systems.

Bailer W, Schallauer P, Messina A, Boch L, Basili R, Cammisa M and Popov B 2007 Integrating audiovisual and semantic metadata for applications in broadcast archiving *Workshop Multimedia Semantics – The Role of Metadata*, pp. 81–100, Aachen, Germany.

Bailey BP, Konstan JA and Carlis. JV 2001 Supporting multimedia designers: Towards more effective design tools *In Proc. of Multimedia Modeling: Modeling Mutlimedia Information and Systems (MMM'01)*, pp. 267–286.

Baker T, Dekkers M, Fischer T and Heery R 2005 Dublin Core application profile guidelines http://dublincore.org/usage/documents/profile-guidelines.

Bandyopadhyay S and Maulik U 2002 Genetic clustering for automatic evolution of clusters and application to image classification. *IEEE Pattern Recognition* **35**, 1197–1208.

Bayer R and McCreight EM 1972 Organization and maintenance of large ordered indices. *Acta Informatica* **1**, 173–189.

Bechhofer S, van Harmelen F, Hendler J, Horrocks I, McGuinness DL, Patel-Schneider PF and Stein LA 2004 OWL Web Ontology Language Reference. W3C. http://www.w3.org/TR/owl-ref/.

Beckett D 2002 The design and implementation of the Redland RDF application framework. *Computer Networks and ISDN Systems* **39**(5), 577–588.

Beckett D and Berners-Lee T 2008 Turtle – Terse RDF Triple Language. Technical report.

Beckett D and Jeen Broekstra 2008 SPARQL Query Results XML Format. W3C Recommendation.

Benmokhtar R and Huet B 2006 Classifier fusion: combination methods for semantic indexing in video content *In Proc. of the International Conference on Artificial Neural Networks*, pp. 65–74.

Bentley JL 1979 Multidimensional binary search trees in database applications. *IEEE Transactions on Software Engineering* **5**, 333–340.

Berchtold S, Bohm C and Kriegel HP 1998 The pyramid technique: Towards breaking the curse of dimensionality *In Proc. of the ACM SIGMOD International Conference on Management of Data*, pp. 142–153, Seattle, Washington, USA.

Berglund A, Boag S, Chamberlin D, Fernández M, Kay M, Robie J and Siméon J 2007 XML Path Language (XPath) 2.0. W3C Recommendation.

Berka P, Ferjenčík J and Ivánek J 1992 *Expert System shell SAK based on complete many-valued logic and its application in territorial planning* Academia, Prague and Kluwer, Dodrecht pp. 67–74.

Berlin F 2007 Search DBpedia.org. http://dbpedia.org/search/.

Berners-Lee T 2001 An RDF language for the Semantic Web.

Berners-Lee T 2006 Linked data – design issues. W3C. http://www.w3.org/DesignIssues/LinkedData.html.

Berners-Lee T and Fischetti M 1999 *Weaving the Web*. Harper, San Francisco.

Berners-Lee T, Chen Y, Chilton L, Connolly D, Dhanaraj R, Hollenbach J, Lerer A and Sheets D 2006 Tabulator: Exploring and analyzing linked data on the semantic web. *In Proc. of the 3rd International Semantic Web User Interaction Workshop (SWUI)*.

Berners-Lee T, Fielding R and Masinter L 1998 RFC 2396: Uniform resource identifiers (URI): Generic syntax. IETF, Network Working Group.

Berners-Lee T, Hendler J and Lassila O 2001 The Semantic Web. *Scientific American* **284**(5), 34–43.

Bernstein A, Kaufmann E, Kaiser C and Kiefer C 2006 Ginseng: A Guided Input Natural Language Search Engine for Querying Ontologies *Jena User Conference*, Bristol, UK.

Berretti S, Bimbo AD and Vicario E 2001 Efficient matching and indexing of graph models in content-based retrieval. *IEEE Transactions on Circuits and Systems for Video Technology* **11**(12), 1089–1105.

Beucher S and Meyer F 1993 *The Morphological Approach to Segmentation: The Watershed Transformation*. Marcel Dekker, NY.

Bicego M and Murino V 2004 Investigating hidden Markov models' capabilities in 2D shape classification. *IEEE Transactions on Pattern Analysis and Machine Intelligence* **26**(2), 281–286.

Bicego M, Castellani U and Murino V 2005 A hidden Markov model approach for appearance-based 3D object recognition. *Pattern Recognition Letters* **26**(16), 2588–2599.

Bilmes J 2002 What HMMs can do. Technical report, University of Washington.

Binding C and Tudhope D 2004 KOS at your service: Programmatic access to knowledge organisation systems. *Journal of Digital Information*.

Biron P, Permanente K and Malhotra A (eds) 2004 XML Schema Part 2. Datatypes Second Edition. W3C Recommendation.

Bizer C 2007 The TriG syntax http://www.wiwiss.fu-berlin.de/suhl/bizer/TriG/Spec/.

Bizer C and Gauss T 2007 Disco – hyperdata browser. http://sites.wiwiss.fu-berlin.de/suhl/bizer/ng4j/disco/.

Bizer C, Lee R and Pietriga E 2005 Fresnel – a browser-independent presentation vocabulary for RDF *In Proc. of the 2nd International Workshop on Interaction Design and the Semantic Web*, Galway, Ireland.

Bloch I 2003 *Fusion d'informations en traitement du signal et des images*. Hermes-Science, Lavoisier.

Bloehdorn S, Petridis K, Saathoff C, Simou N, Tzouvaras V, Avrithis Y, Handschuh S, Kompatsiaris Y, Staab S and Strintzis MG 2005 Semantic annotation of images and videos for multimedia analysis *In Proc. of the 2nd European Semantic Web Conference (ESWC)*, Heraklion, Crete, Greece, pp. 592–607.

Bock H 2007 *Selected contributions in data mining and classification* Springer pp. 161–172.

Bonino D, Corno F, Farinetti L and Bosca A 2004 Ontology Driven Semantic Search. *WSEAS Transactions on Information Science and Application* **1**(6), 1597–1605.

Bordwell D and Thompson K 1997 *Film Art: An Introduction*. McGraw-Hill.

Borgo S and Masolo C 2009 Foundational choices in DOLCE. S. Staab and R. Studer (eds), *Handbook on Ontologies*, pp. 135–152. Springer.

Bosch A, Zisserman A and Munoz X 2008 Scene classification using a hybrid generative/discriminative approach. *IEEE Transactions on Pattern Analysis and Machine Intelligence* **30**(4), 712–727.

Bray T, Paoli J, Sperberg-McQueen CM, Maler E and Yergeau F 2006 Extensible markup language (xml) 1.0 (fourth edition). W3C.

Breiman L 1996 Bagging predictors. *Machine Learning* **24**(2), 123–140.

Brickley D and Guha RV (eds) 2004 *RDF Vocabulary Description Language 1.0: RDF Schema*. W3C Recommendation.

Brickley D and Miller L 2010 FOAF Vocabulary Specification 0.97. http://xmlns.com/foaf/spec/.

Broekstra J and Kampman A 2003 The SeRQL query language. Technical report, Aduna.

Broekstra J, Kampman A and van Harmelen F 2002 Sesame: A generic architecture for storing and querying RDF and RDF Schema *In Proc. 1st International Semantic Web Conference (ISWC)*, pp. 54–68.

Brown JC 1998 Musical instrument identification using autocorrelation coefficients *International Symposium on Musical Acoustics*.

Brown JC, Houix O and McAdams S 2000 Feature dependence in the automatic identification of musical woodwind instruments. *Journal of the Acoustical Society of America* **109**, 1064–1072.

Browne P, Smeaton A, Murphy N, O'Connor NE, Marlow S and Berrut C 2002 Evaluating and combining digital video shot boundary detection algorithms *Irish Machine Vision and Image Processing Conference*.

Buneman P 1997 Semistructured data *In Proc. ACM Symposium on Principles of Database Systems*, pp. 117–121.

Burgess C 1998 A tutorial on support vector machines for pattern recognition. *Data Mining and Knowledge Discovery* **2**(2), 121–167.

Campbell W, Campbell J, Reynolds D, Singer E and Torres-Carrasquillo P 2006 Support vector machines for speaker and language recognition. *Computer Speech and Language* **20**(2-3), 210–229.

Campedel M 2007 Performance evaluators for relevance feedback and classifiers. Technical Report K-Space, Telecom ParisTech.

Campedel M, Kyrgyzov I and Maître H 2007 Sélection non supervisée d'attributs – application à l'indexation d'images satellitaires *Soc. Française de Classification*, Paris.

Cao Y, Tavanapong W, Kim K and Oh J 2003 Audio-assisted scene segmentation for story browsing *In Proc. of the International Conference on Image and Video Retrieval (CIVR)*.

Carroll J, Bizer C, Hayes P and Stickler P 2005 Named graphs, provenance and trust *In Proc. of the14th World Wide Web Conference (WWW)*.

Celino I, Valle E, Cerizza D and Turati A 2006 Squiggle: a semantic search engine for indexing and retrieval of multimedia content *1st International Workshop on Semantic-Enhanced Multimedia Presentation Systems (SEMPS)*, pp. 20–34, Athens, Greece.

Cha GH and Chung CW 2002 The GC-tree: a high-dimensional index structure for similarity search in image databases. *IEEE Transactions on Multimedia* **4**(2), 235–247.

Cha GH, Zhu X, Petkovic P and Chung CW 2002 An efficient indexing method for nearest neighbor searches in high-dimensional image databases. *IEEE Transactions on Multimedia* **4**(1), 76–87.

Chakravarthy FCA and Lanfranchi V 2006 Cross-media document annotation and enrichment *Proceedings of the 1st Semantic Authoring and Annotation Workshop (SAAW2006)*.

Chandramouli K and Izquierdo E 2006 Image classification using self organizing feature maps and particle swarm optimization *7th International Workshop on Image Analysis for Multimedia Interactive Services*.

Chang CC and Lin CJ 2001 *LIBSVM: a library for support vector machines*.

Chapelle O and Vapnik V 2001 Choosing multiple parameters for support vector machines. *Advances in Neural Information Processing Systems*.

Chen CY, Abdallah A and Wolf W 2006 Audiovisual gunshot event recognition, *IEEE International Conf. on Systems, Man and Cybernetics*. **34**(2), 190–196.

Chen L, Rizvi SJ and Ötzu M 2003 Incorporating audio cues into dialog and action scene detection *Proceedings of SPIE Conference on Storage and Retrieval for Media Databases*, pp. 252–264.

Chen PPS 1976 The entity-relationship model – toward a unified view of data. *ACM Transactions on Database Systems* **1**(1), 9–36.

Chibelushi C, Mason J and Deravi N 1997 Integrated person identification using voice and facial features. *IEE Colloquium on Image Processing for Security Applications* pp. 4/1–4/5.

Chou K, Tu L and Shyu I 1994 Performances analysis of a multiple classifiers system for recognition of totally unconstrained handwritten numerals. *4th International Workshop on Frontiers of Handwritten Recognition* pp. 480–487.

Cieplinski MI 2000 Results of core experiment CT4 on dominant color extension ISO/IEC/JTC1/SC29/WG11/M5775.

Clark K, Feigenbaum L and (eds) ET 2008 SPARQL Protocol for RDF. W3C Recommendation.

Clavel C, Ehrotte T and Richard G 2005 Events detection for an audio-based surveillance system, *In Proc. of the International Conference on Multimedia and Expo (ICME)*.

Clavel C, Vasilescu I, Devillers L, Richard G and Ehrette T 2008 Fear-type emotion recognition for future audio-based. *Speech Communication* **50**(6), 487–503.

Codd E 1970 A relational model of data for large shared data banks. *Communications of the ACM* **13**(6), 377–387.

Copeland G and Maier D 1984 Making Smalltalk a database system *Proceedings of the 1984 ACM SIGMOD International Conference on Management of Data*, pp. 316–325.

Cortes C and Vapnik V 1995 Support-vector networks. *Machine Learning* **20**(3), 273–297.

Cotsaces C, Nikolaidis N and Pitas I 2006 Video shot detection and condensed representation: a review. *IEEE Signal Processing Magazine* **23**, 28–37.

Cour T, Jordan C, Miltsakaki E and Taskar B 2008 Movie/script: Alignment and parsing of video and text transcription *ECCV08*, pp. IV: 158–171.

Cover TM and Thomas JA 2006 *Elements of Information Theory*. Wiley.

Cuturi M, Vert JP, Birkenes O and Matsui T 2007 A kernel for time series based on global alignments *IEEE International Conference on Acoustics, Speech and Signal Processing (ICASSP)*, vol. 2, pp. 413–416.

Cyganiak R 2005 A relational algebra for SPARQL. Technical Report HPL-2005-170, HP Labs.

Dancyger K 2002 *The Technique of Film and Video Editing. History, Theory and Practice*. Focal Press.

Darrell T, Fisher J and Viola P 2000 Audio-visual segmentation and 'the cocktail party effect' *Third International Conference on Advances in Multimodal Interfaces (ICMI)*, pp. 32–40.

Dasiopoulou S, Heinecke J, Saathoff C and Strintzis MG 2007 Multimedia reasoning with natural language support *In Proc. of the International Conference on Semantic Computing (ICSC)*, pp. 413–420.

Datta A, Shah M and Da Vitoria Lobo N 2002 Person-on-person violence detection in video data *Proc. 16th International Conference on Pattern Recognition*, vol. 1, pp. 433–438.

Davies J and Weeks R 2004 QuizRDF: Search Technology for the Semantic Web *37th International Conference on System Sciences (HICSS)*, Hawai, USA.

DCMI 2003 Information and documentation–The Dublin Core metadata element set ISO 15836.

de Bruijn J, Franconi E and Tessaris S 2005 Logical reconstruction of normative RDF *OWL: Experiences and Directions Workshop (OWLED-2005)*.

Debuse JCW and Rayward-Smith VJ 1997 Feature subset selection within a simulated annealing datamining algorithm. *Journal of Intelligent Information Systems* **9**(1), 57–81.

Decker S, Melnik S, van Harmelen F, Fensel D, Klein M, Broekstra J, Erdmann M and Horrocks I 2000 The Semantic Web: the roles of XML and RDF. *IEEE Internet Computing* **4**(5), 63–73.

Declerck T and Cobet A 2007 Towards a cross-media analysis of spatially co-located image and text regions in tv-news *In Proc. of the 2nd international conference on Semantics And digital Media Technologies (SAMT)*, pp. 288–291.

Delaney B and Hoomans B 2004 Preservation and digitisation plans. Overview and analysis. Technical Report Deliverable 2.1 User Requirements Final Report, PrestoSpace.

Delany B and Bauer C 2007 Report on usability testing in PrestoSpace. Deliverable D20.4 http://www.presto space.org/project/deliverables/D20.4.pdf.

Demirekler M and Haydar A 1999 Feature selection using genetics-based algorithm and its application to speaker identification *Proc. IEEE International Conference on Acoustics, Speech and Signal Processing (ICASSP)*, vol. 1, pp. 322–329, Phoenix, AZ, USA.

Dempster A 1967 Upper and lower probabilities induced by a multivalued mapping. *Annals of Mathematical Statistics* **38**(2), 325–339.

Dempster AP, Laird NM and Rubin DB 1977 Maximum likelihood from incomplete data via the EM algorithm. *Journal of the Royal Statistical Society, Series B* **39**, 1–38.

DERI 2005 SWSE. http://www.swse.org/.

Di Gesu V and Maccarone M 1985 Feature selection and possibility theory. *Pattern Recognition* **19**(1), 63–72.

Diday E 1970 Une nouvelle méthode de classification automatique et reconnaissance des formes: la méthodes des nuées dynamiques. *Revue de Statistique Appliquée* **19**(2), 19–33.

Ding L, Finin T, Joshi A, Pan R, Cost R, Peng Y, Reddivari P, Doshi V and Sachs J 2004 Swoogle: A search and metadata engine for the semantic web *Proceedings of the Thirteenth ACM Conference on Information and Knowledge Management*.

Ding L, Pan R, Finin T, Joshi A, Peng Y and Kolari P 2005 Finding and ranking knowledge on the Semantic Web *In Proc. of the 4th International Semantic Web Conference (ISWC)*, pp. 156–170, Galway, Ireland.

DMS-1 2004 Material Exchange Format (MXF) – Descriptive Metadata Scheme-1 SMPTE 380M.

Dorigo M and Gambardella LM 1997 Ant colony system: A cooperative learning approach to the traveling salesman problem. *IEEE Transactions on Evolutionary Computing* **1**, 53–66.

Drożdżyński W, Krieger HU, Piskorski J, Schäfer U and Xu F 2004 Shallow processing with unification and typed feature structures – foundations and applications. *KI 1/2004*.

Duda R, Hart PE and Stork DG 2001 *Pattern Classification*. New York: Wiley.

Duke A, Glover T and Davies J 2007 Squirrel: An advanced semantic search and browse facility *In Proc. of the 4th European Semantic Web Conference (ESWC)*, Innsbruck, Austria.

Durkin J 1994 *Expert Systems, Design and Development*. New York: Macmillan.

Eberhart R and Kennedy J 1995 A new optimizer using particle swarm theory *6th International Symposium on Micro Machine and Human Science*, pp. 39–43.

EC-M E 2010 EBU Core Metadata Set, (EBUCore) v1.2 EBU Tech 3293.

Eide E and Gish H 1996 A parametric approach to vocal tract length normalization, *International Conf. on Acoustics, Speech, and Signal Processing*.

Eidenberger H 2003 How good are the visual MPEG-7 features? *SPIE & IEEE Visual Communications and Image Processing Conference*, pp. 476–488.

Eiter T, Ianni G, Schindlauer R and Tompits H 2006 dlvhex: A system for integrating multiple semantics in an answer-set programming framework *Proceedings of the Workshop on Logic Programming and Constraint Systems*, pp. 206–210.

Ekin A, Tekalp A and Mehrotra R 2003 Automatic soccer video analysis and summarization. *Image Processing* **12**(7), 796–807.

Eleftherohorinou H, Zervaki V, Gounaris A, Papastathis V, Kompatsiaris Y and Hobson P 2006 Towards a common multimedia ontology framework http://www.acemedia.org/aceMedia/files/multimedia_ontology/cfr/MM-Ontologies-Reqs-v1.3.pdf.

Eronen A 2001 Automatic musical instrument recognition. Technical report, Master's thesis, Tampere University of Technology.

Eronen A 2003 Musical instrument recognition using ICA-based transform of features and discriminatively trained HMMs *7th International Sumposium on Signal Processing and its Applications*.

Essid S 2005 *Classification automatique des signaux audio-fréquences: reconnaissance des instruments de musique* PhD thesis Université Pierre et Marie Curie.

Essid S, Richard G and David B 2006a Instrument recognition in polyphonic music based on automatic taxonomies. *IEEE Transactions on Audio, Speech, and Language Processing* **14**(1), 68–80.

Essid S, Richard G and David B 2006b Musical instrument recognition by pairwise classification strategies. *IEEE Transactions on Audio, Speech, and Language Processing* **14**(4), 1401–1412.

Euzenat J and Shvaiko P 2007 *Ontology Matching*. Springer Verlag.

Everingham M, Sivic J and Zisserman A 2006 'Hello! My name is . . . Buffy' – automatic naming of characters in TV video *Proceedings of the British Machine Vision Conference*.

Falcon-S 2005. http://www.falcons.com.cn/.

Fallside DC and Walmsley P 2004 XML Schema Part 0: Primer second edition. W3C.

Fergus R, Perona P and Zisserman A 2007 Weakly supervised scale-invariant learning of models for visual recognition. *International Journal of Computer Vision* **71**(3), 273–303.

Fielding R, Gettys J, Mogul J, Frystyk H, Masinter L, Leach P and Berners-Lee T 1999 RFC 2616: Hypertext transfer protocol – HTTP/1.1. IETF, Network Working Group.

Fisher W 1958 On grouping for maximum heterogeneity. *Journal of the American Statistical Association* **53**, 789–798.

Fitzgibbon AW and Zisserman A 2002 On affine invariant clustering and automatic cast listing in movies *ECCV '02: Proceedings of the 7th European Conference on Computer Vision – Part III*, pp. 304–320.

Fluit C, Sabou M and van Harmelen F 2003a Supporting user tasks through visualisation of light-weight ontologies *Handbook on Ontologies in Information Systems*, pp. 415–434.

Fluit C, Sabou M and van Harmelen F 2003b Ontology-based information visualization. In *Visualizing the Semantic Web – XML-based Internet and Information*, pp. 36–48. Springer-Verlag.

Fogel LJ, Owens AJ and Walsh MJ 1966 *Artificial Intelligence through Simulated Evolution*. New York: John Wiley & Sons, Inc.

Freud Y and Schapire R 1996 Experiments with a new boosting algorithms. *In Proc. of the 13th International Conference on Machine Learning*.

Furui S 2001 *Digital Speech Processing, Synthesis and Recognition* 2nd edn. New York: Marcel Dekker.

Gangemi A and Presutti V 2009 Ontology design patterns. S. Staab and R. Studer (eds), *Handbook on Ontologies*, pp. 221–243. Springer.

Gangemi A, Borgo S, Catenacci C and Lehmann J 2004 Task taxonomies for knowledge content. Technical report, Metokis Deliverable 7.

Gangemi A, Guarino N, Masolo C, Oltramari A and Schneider L 2002 Sweetening ontologies with DOLCE *13th International Conference on Knowledge Engineering and Knowledge Management (EKAW'02)*, pp. 166–181, Siguenza, Spain.

Garcia R and Celma O 2005 Semantic integration and retrieval of multimedia metadata *5th International Workshop on Knowledge Markup and Semantic Annotation*.

Geroimenko V and Chen C 2003b *Visualizing the Semantic Web – XML-based Internet and Information Visualization*. Springer-Verlag.

Geurts J, van Ossenbruggen J and Hardman L 2005 Requirements for practical multimedia annotation *Workshop on Multimedia and the Semantic Web*.

Ghahramani Z 2001 An introduction to hidden markov models and bayesian networks. *Journal of Pattern Recognition and Artificial Intelligence* **15**(1), 9–42.

Gil Y and Ratnakar V 2002 A comparison of (semantic) markup languages *Proceedings of the Fifteenth International Florida Artificial Intelligence Research Society Conference (FLAIRS)*, pp. 413–418.

Gillet O and Richard G 2004 Automatic transcription of drum loops *IEEE International Conference on Acoustics, Speech and Signal Processing (ICASSP)*, Montréal, Canada.

Gillet O and Richard G 2005 Automatic transcription of drum sequences using audiovisual features. *IEEE International Conference on Acoustics, Speech, and Signal Processing, 2005* **3**, 205–208.

Gillet O, Essid S and Richard G 2007 On the correlation of automatic audio and visual segmentations of music videos. *IEEE Transactions on Circuits and Systems for Video Technology* **17**(3), 347–355.

Goecke R and Millar J 2003 Statistical analysis of the relationship between audio and video speech parameters for Australian English *ISCA Tutorial and Research Workshop on Auditory-Visual Speech Processing AVSP 2003*, pp. 133–138.

Goldberg D and Deb K 1991 A comparative analysis of selection schemes used in genetic algorithms. In *Foundations of Genetic Algorithms*, pp. 69–93. San Mateo, CA.

Granitzer M, Lux M and Spaniol M (eds) 2008 *Multimedia Semantics – The Role of Metadata* vol. 101 of *Studies in Computational Intelligence*. Springer.

Grosof BN, Horrocks I, Volz R and Decker S 2003 Description logic programs: Combining logic programs with description logic *12th International World Wide Web Conference (WWW)*, Budapest, Hungary.

Guha R, McCool R and Miller E 2003 Semantic search *12th International World Wide Web Conference (WWW)*, pp. 700–709, Budapest, Hungary.

Guttman A 1984 R-trees: a dynamic index structure for spatial searching *Proceedings of ACM SIGMOD International Conference on Management of Data*, pp. 47–57. ACM Press, Boston.

Guyon I and Elisseeff A 2003 An introduction to feature and variable selection. *Journal of Machine Learning Research* **3**, 1157–1182.

Haarslev V and Möller R 2003 Racer: A core inference engine for the semantic web *2nd International Workshop on Evaluation of Ontology-based Tools (EON2003) at ISWC2003*, Sanibel Island, Florida.

Haase P, Broekstra J, Eberhart A and Volz R 2004 A comparison of RDF query languages.

Hájek P 1985 Combining functions for certainty factors in consulting systems. *International Journal of Man-Machine Studies*.

Halaschek-Wiener C, Golbeck J, Schain A, Grove M, Parsia B and Hendler J 2005 Photostuff – an image annotation tool for the semantic web *4th International Semantic Web Conference – Poster Paper*.

Halasz F and Schwartz M 1994 The Dexter Hypertext Reference Model. *Communications of the ACM* **37**(2), 30–39.

Handl J, Knowles J and Dorigo M 2003 Ant-based clustering: a comparative study of its relative performance with respect to k-means, average link and 1D-som. Technical Report TR-IRIDIA-2003-24, Université Libre de Bruxelles, Belgium.

Handschuh S 2005 *Creating Ontology-Based Metadata by Annotation for the Semantic Web* PhD thesis Universität Karlsruhe.

Handschuh S and Staab S 2003 CREAM – creating metadata for the Semantic Web. *Computer Networks* **42**, 579–598. Elsevier.

Hanjalic A 2002 Shot-boundary detection: unraveled and resolved?. *IEEE Transactions on Circuits and Systems for Video Technology* **12**(2), 90–105.

Hardman L 1998 *Modelling and Authoring Hypermedia Documents* PhD thesis University of Amsterdam.

Harris S and (eds) AS 2010 SPARQL 1.1 Query Language. W3C Working Draft 1 June 2010.

Harris S and Shadbolt N 2005 SPARQL query processing with conventional relational database systems *Proceedings of the International Conference on Web Information Systems Engineering (WISE)*, pp. 235–244.

Harris S, Lamb N and Shadbolt N 2009 4store: The design and implementation of a clustered RDF store *In Proc. of the Scalable Semantic Web Knowledge Base Systems Workshop (SSWS)*, Washington DC.

Harter S 1975 A probabilistic approach to automatic keyword indexing, Part I: On the distribution of speciality words in a technical literature. *Journal of the ASIS* **26**, 197–216.

Harth A and Decker S 2005 Optimized index structures for querying RDF from the web *Proceedings of the Third Latin American Web Congress*, pp. 71–80.

Harth A, Umbrich J, Hogan A and Decker S 2007 YARS2: A federated repository for querying graph structured data from the Web *In Proc. of the 6th International Semantic Web Conference (ISWC)*.

Hastie T, Tibshirani R and Friedman J 2001 *The Elements of Statistical Learning*. Springer-Verlag.

Hausenblas M 2008 *Building Scaleable and Smart Multimedia Applications on the Semantic Web* PhD thesis Graz University of Technology.

Hausenblas M and Halb W 2008 Interlinking Multimedia Data *Linking Open Data Triplification Challenge at the International Conference on Semantic Systems (I-Semantics'08)*.

Hausenblas M, Boll S, Bürger T, Celma O, Halaschek-Wiener C, Mannens E and Troncy R 2007 Multimedia vocabularies on the semantic web. W3C Incubator Group report, W3C Multimedia Semantics Incubator Group.

Hausenblas M, Troncy R, Raimond Y and Bürger T 2009 Interlinking Multimedia: How to Apply Linked Data Principles to Multimedia Fragments *2nd Workshop on Linked Data on the Web (LDOW'09)*, Madrid, Spain.

Hayes P (ed.) 2004 *RDF Semantics*. W3C Recommendation.

He B, Macdonald C, He J and Ounis I 2008 An effective statistical approach to blog post opinion retrieval *CIKM '08: Proceedings of the 17th ACM Conference on Information and Knowledge Management*, pp. 1063–1072. ACM, New York.

Heery R and Patel M 2004 Application profiles: mixing and matching metadata schemas http://www.ariadne. ac.uk/issue25/app-profiles.

Heflin J 2004 OWL web ontology language use cases and requirements. W3C.

Heflin J and Hendler J 2000 Searching the Web with SHOE *AAAI Workshop on Artificial Intelligence for Web Search*, pp. 35–40, Menlo Park, CA, USA.

Hermansky H 1990 Perceptual linear predictive (PLP) analysis of speech. *Journal of the American Statistical Asociation* **87**(4), 1738–1752.

Herrera P, Dehamel A and Gouyon F 2003 Automatic labeling of unpitched percussion sounds *Audio Engineering Society, 114th Convention*.

Hildebrand M, van Ossenbruggen J and Hardman L 2006 /facet: A Browser for Heterogeneous Semantic Web Repositories *In Proc. of the 5th International Semantic Web Conference (ISWC)*.

Hitzler P, Krötzsch M, Parsia B, Patel-Schneider P and Rudolph S (eds) 2009 *OWL 2 Web Ontology Language: Primer*. W3C Recommendation. http://www.w3.org/TR/owl2-primer/.

Ho T 1992 *A theory of multiple classifier systems and its application to visual and word recognition* PhD thesis New York University.

Hogan A, Harth A and Decker S 2006 ReConRank: A scalable ranking method for Semantic Web data with context *2nd International Workshop on Scalable Semantic Web Knowledge Base Systems*.

Hoi SC, Wong LL and Lyu A 2006 Chinese University of Hong Kong at TRECVID 2006: Shot boundary detection and video search *TRECVid 2006 Workshop*, pp. 76–86 NIST, Gaithersburg, Maryland, USA.

Holland JH 1975 *Adaptation in Natural and Artificial Systems*. Ann Arbor: University of Michigan Press.

Hollink L, Nguyen G, Schreiber G, Wielemaker J, Wielinga B and Worring M 2004 Adding spatial semantics to image annotations. *Proc. of the 4th International Workshop on Knowledge Markup and Semantic Annotation*.

Holtman K and Mutz A 1998 RFC 2295: Transparent Content Negotiation in HTTP. IETF, Network Working Group.

Hönig F, Stemmer G, Hacker C and Brugnara F 2005 Revising perceptual linear prediction (PLP). *Proc. of Interspeech'05* pp. 2997–3000.

Horridge M and Patel-Schneider PF 2009 OWL 2 Web Ontology Language Manchester Syntax. W3C.

Horrocks I, Kutz O and Sattler U 2006 The even more irresistible SROIQ. *In Proc. of the 10th International Conference on Principles of Knowledge Representation and Reasoning*, pp. 57–67.

Horrocks I, Patel-Schneider PF, Boley H, Tabet S, Grosof B and Dean M 2004 SWRL: A Semantic Web rule language combining OWL and RuleML. W3C.

Horrocks I, Sattler U and Tobies S 2000 Reasoning with Individuals for the Description Logic $\mathcal{SHIQ}$ In *CADE-17: 17th International Conference on Automated Deduction*, pp. 482–496.

Hunter J 2001 Adding multimedia to the Semantic Web – building an MPEG-7 ontology *1st International Semantic Web Working Symposium*, pp. 261–281.

Hunter J 2002 Combining the CIDOC/CRM and MPEG-7 to describe multimedia in museums *6th Museums and the Web Conference*, http://www.archimuse.com/mw2002/papers/hunter/hunter.html.

Hunter J and Armstrong L 1999 A comparison of schemas for video metadata representation *8th International World Wide Web Conference*, pp. 1431–1451.

Hustadt U, Motik B and Sattler U 2004 Reducing SHIQ description logic to disjunctive datalog programs *Proc. of the 9th International Conference on Knowledge Representation and Reasoning (KR2004)*, pp. 152–162.

Hyvärinen A, Karhunen J and Oja E 2001 *Independent Component Analysis*. New York: John Wiley & Sons, Inc.

Hyvönen E and Mäkelä E 2006 Semantic Autocompletion *1st Asian Semantic Web Conference (ASWC)*, pp. 739–751, Beijing.

Hyvönen E, Junnila M, Kettula S, Mäkelä E, Saarela S, Salminen M, Syreeni A, Valo A and Viljanen K 2005 MuseumFinland – Finnish museums on the Semantic Web. *Web Semantics* 3(2-3), 224–241.

Hyvönen E, Salminen M, Junnila M and Kettula S 2004 A content creation process for the Semantic Web *Proceedings of OntoLex 2004*, Lisbon, Portugal.

Idrissi K, Ricard J and Baskurt A 2002 An objective performance evaluation tool for color based image retrieval systems *Proceedings of the International Conference on Image Processing*, vol. 2, pp. 389–392.

Indyk P and Motwani R 1998 Approximate nearest neighbors: Towards removing the curse of dimensionality. *In Proc. of the 30th annual ACM symposium on Theory of computing*, pp. 604–613.

International Standards Organisation 1986 ISO 2788-1986 documentation – Guidelines for the establishment and development of monolingual thesauri.

Iria J 2009 A core ontology for knowledge acquisition *In Proc. of the 6th European Semantic Web Conference*, pp. 233–247.

Isaac A and Summers E 2009 SKOS Primer. W3C Recommendation.

Isaac A and Troncy R 2004 Designing and using an audio-visual description core ontology *Workshop on Core Ontologies in Ontology Engineering*.

Isaac A, Phipps J and Rubin D 2009 SKOS use cases and requirements. W3C.

Iverson P and Krumhansl C 1994 Isolating the dynamic attributes of musical timbre. *Journal of the Acoustic Society of America*.

Iyengar G, Nock H and Neti C 2003 Audio-visual synchrony for detection of monologues in video archives. *Proceedings of the International Conference on Multimedia and Expo* 1, 329–332.

Jaimes A and Chang SF 2000 A conceptual framework for indexing visual information at multiple levels *IS&T/SPIE Internet Imaging*, vol. 3964.

Jain A and Dubes R 1988 *Algorithms for Clustering Data*. Englewood Cliffs, NJ: Prentice Hall.

Jain A, Duin R and Mao J 2000 Statistical pattern recognition: A review. *IEEE Transactions on Pattern Analysis and Machine Intelligence* 22(1), 4–37.

Joder C, Essid S and Richard G 2009 Temporal integration for audio classification with application to musical instrument classification. *IEEE Transactions on Audio, Speech, and Language Processing* 17(1), 174–186.

Jones KS, Walker S and Robertson S 2000 A probability model of information retrieval: development and comparative experiments. *Information Processing and Management: Part I* 36, 779–808.

Karvounarakis G, Alexaki S, Christophides V, Plexousakis D and Scholl M 2002 RQL: A declarative query language for RDF *Proceedings of the International World-Wide Web Conference*, pp. 592–603.

Kasutani E and Yamada A 2001 The MPEG-7 color layout descriptor: a compact image feature description for high-speed image/video segment retrieval *Proceedings of the International Conference on Image Processing*, pp. 674–677.

Kim HG, Moreau N and Sikora T 2005 *MPEG-7 Audio and Beyond: Audio Content Indexing and Retrieval*. Chichester: John Wiley & Sons, Ltd.

Kim W 1990a Object-oriented databases: Definition and research directions. *IEEE Transactions on Knowledge and Data Engineering* **2**(3), 327–341.

Kim W 1990b Research directions in object-oriented database systems *PODS '90 Proceedings of the Ninth ACM SIGACT-SIGMOD-SIGART Symposium on Principles of Database Systems*, pp. 1–15.

Kim Y, Street W and Menczer F 2002 Evolutionary model selection in unsupervised learning. *Intelligent Data Analysis* **6**, 531–556.

King R, Popitsch N and Westermann U 2004 METIS: a flexible database foundation for unified media management *Proc. of the 12th Annual ACM Int. Conf. on Multimedia*, pp. 744–745. ACM.

Kiryakov A, Ognyanov D and Manov D 2005a OWLIM – a pragmatic semantic repository for OWL *Proceedings of the International Conference on Web Information Systems Engineering (WISE)*, pp. 182–192.

Kiryakov A, Ognyanov D and Manov D 2005b OWLIM – a pragmatic semantic repository for OWL *Web Information Systems Engineering (WISE 2005) Workshops*, pp. 182–192. Springer, New York.

Kiryakov A, Popov B, Terziev I, Manov D and Ognyanoff D 2004 Semantic annotation, indexing, and retrieval. *Web Semantics* **2**(1), 49–79.

Kleinberg J 1999 Authoritative sources in a hyperlinked environment. *Journal of the ACM* **46**(5), 604–632.

Klir G and Yuan B 1995 *Fuzzy Sets and Fuzzy Logic, Theory and Applications*. Upper Saddle River, NJ: Prentice Hall.

Klyne G and Carroll JJ (eds) 2004 *Resource Description Framework: Concepts and Abstract Syntax*. W3C Recommendation.

Kochut KJ and Janik M 2007 SPARQLeR: Extended SPARQL for semantic association discovery *Proc. of the 4th European Semantic Web Conference (ESWC)*, pp. 145–159.

Koenderink J 1990 *Solid Shape*. Cambridge, MA: MIT Press.

Koren Y and Carmel L 2004 Robust linear dimensionality reduction. *IEEE Transactions on Visualization and Computer Graphics*.

KRSS Group 1993 Description-logic knowledge representation system specification. Technical report, The Arpa Knowledge Sharing Effort.

Krumhansl CL 1989 *Structure and Perception of Electroacoust Sound and Music* Nielzen and Oisson (Elsevier, Amsterdam) chapter Why is musical timbre so hard to understand?

Kuhn HW 1955 The Hungarian method for the assignment problem. *Naval Research Logistics Quarterly* **2**, 83–97.

Kuncheva L 2003 'Fuzzy' versus 'nonfuzzy' in combining classifiers designed by boosting. *IEEE Transactions on Fuzzy Systems* **11**(6), 729–741.

Kuncheva L, Bezdek J and Duin R 2001 Decision templates for multiple classifier fusion: an experimental comparison. *Pattern Recognition* **34**(2), 299–314.

Kyrgyzov I 2008 *Mining Satellite Image database of landscapes and application to urban zone: clustering, consensus and categorisation* PhD thesis TELECOM ParisTech.

Kyrgyzov I, Kyrgyzov O, Maître H and Campedel M 2007a Kernel MDL to determine the number of clusters *International Conference on Machine Learning and Data Mining*, pp. 203–217.

Kyrgyzov I, Maitre H and Campedel M 2007b A method of clustering combination applied to satellite image analysis. *International Conference on Image Analysis and Processing* pp. 81–86.

Labroche N, Monmarche N and Venturini G 2003 Antclust: Ant clustering and web usage mining *In Genetic and Evolutionary Computation Conference*, pp. 25–36, Chicago.

Lanzi PL 1997 Fast feature selection with genetic algorithms: a filter approach *IEEE International Conference on Evolutionary Computation*, pp. 537–540, Indianapolis, USA.

Laptev I, Marszalek M, Schmid C and Rozenfeld B 2008 Learning realistic human actions from movies *In Proc. of the 21st IEEE Conference on Computer Vision and Pattern Recognition*, pp. 1–8.

Lee H, Smeaton AF, O'Connor NE, Jones G, Blighe M, Byrne D, Doherty A and Gurrin C 2008 Constructing a SenseCam visual diary as a media process. *Multimedia Systems* **14**(6), 341–349.

Lee W, Bürger T, Sasaki F and Malaise V. 2010a Use Cases and Requirements for Ontology and API for Media Resources 1.0 W3C Working Draft. http://www.w3.org/TR/media-annot-reqs/.

Lee W, Bürger T, Sasaki F, Malaisé V, Stegmaier F and (eds.) JS 2010b Ontology for Media Resources 1.0 W3C Working Draft. http://www.w3.org/TR/mediaont-10/.

Lehane B and O'Connor N 2004 Action sequence detection in motion pictures *The International Workshop on Multidisciplinary Image, Video, and Audio Retrieval and Mining*.

Lehane B and O'Connor N 2006 Movie indexing via event detection *7th International Workshop on Image Analysis for Multimedia Interactive Services, Incheon, Korea, 19-21 April*.

Lehane B, O'Connor N and Murphy N 2005 Dialogue scene detection in movies *International Conference on Image and Video Retrieval (CIVR), Singapore, 20-22 July 2005*, pp. 286–296.

Lehane B, O'Connor N, Lee H and Smeaton A 2007 Indexing of fictional video content for event detection and summarisation. *EURASIP Journal on Image and Video Processing*.

Lehane B, O'Connor NE, Smeaton AF and Lee H 2006 A system for event-based film browsing *3rd International Conference on Technologies for Interactive Digital Storytelling and Entertainment*.

Lei Y, Uren V and Motta E 2006 SemSearch: A search engine for the Semantic Web *15th International Conference on Knowledge Engineering and Knowledge Management: Managing Knowledge in a World of Networks (EKAW)*, Podebrady, Czech Republic.

Leinhart R, Pfeiffer S and Effelsberg W 1999 Scene determination based on video and audio features *Proceedings of IEEE Conference on Multimedia Computing and Systems*, pp. 685–690.

Leonardi R, Migliorati P and Prandini M 2004 Semantic indexing of soccer audio-visual sequences: a multi-modal approach based on controlled markov chains. *IEEE Transactions on Circuits and Systems for Video Technology*.

Li Y and Kou CCJ 2003 *Video Content Analysis using Multimodal Information*. Kluwer Academic Publishers.

Li Y and Kou CJ 2001 Movie event detection by using audiovisual information *Proceedings of the Second IEEE Pacific Rim Conferences on Multimedia: Advances in Multimedia Information Processing*.

Liu D, Lu L and Zhang HJ 2003 Automatic mood detection from acoustic music data, *International Conf. on Music Information Retrieval*.

Liu H and Motoda H 2000 *Feature selection for knowledge discovery and data mining* 2nd edn. Kluwer Academic Publishers.

Liu T and Kender JR 2000 Proceedings of the IEEE Workshop on Content-Based Access on Image and Video Libraries *A Hidden Markov Approach to the Structure of Documentaries*.

Lloyd S 1982 Least squares quantization in PCM. *IEEE Transactions on Information Theory* 28(2), 129–137.

Lopez V, Pasin M and Motta E 2005 AquaLog: An ontology-portable question answering system for the Semantic Web *European Semantic Web Conference (ESWC)*, pp. 546–562, Crete, Greece.

Louradour J, Daoudi K and Bach F 2007 Feature space Mahalanobis sequence kernels: Application to SVM speaker verification. *IEEE Transactions on Audio, Speech, and Language Processing* 15(8), 2465–2475.

Lu C, Drew MS and Au J 2002 An automatic video classification system based on a combination of HMM and video summarization. *International Journal of Smart Engineering System Design*.

Luo J, Boutell M and Brown C 2006 Pictures are not taken in a vacuum. *IEEE Signal Processing Magazine* 23(2), 101–114.

Lux M, Becker J and Krottmaier H 2003 Caliph & emir: Semantic annotation and retrieval in personal digital photo libraries *Proceedings of CAiSE'03 Forum at 15th Conference on Advanced Information Systems Engineering*.

Manjunath BS and Ma WY 1996 Texture features for browsing and retrieval of image data. *IEEE Transactions on Pattern Analysis and Machine Intelligence* 18(8), 837–842.

Manjunath BS, Ohm JR, Vasudevan VV and Yamada A 2001 Color and texture descriptors. *IEEE Transactions on Circuits and Systems for Video Technology* 11(6), 703–715.

Manjunath BS, Salembier P and Sikora T 2002 *Introduction to MPEG-7: Multimedia Content Description Interface*. Chichester: John Wiley & Sons, Ltd.

Margulis EL 1992 N-Poisson document modelling *In Proc. of the 15th SIGIR Conference on Research and Development in Information Retrieval*, pp. 177–189.

McAdams S 1993 *Thinking in Sound: The Cognitive Psychology of Human Audition* Oxford University Press chapter Recognition of Auditory Sound Sources and Events.

McGuinness DL and van Harmelen F (eds) 2004 *OWL Web Ontology Language*. W3C Recommendation.

McKinney MF and Breebart J 2003 Features for audio and music classification *International Symphosium on Music Information Retrieval*, pp. 151–158.

Meng A 2006 *Temporal Feature Integration for Music Organisation* PhD thesis Informatics and Mathematical Modelling, Technical University of Denmark, DTU.

Messina A, Boch L, Dimino G, Bailer W, Schallauer P, Allasia W, Groppo M, Vigilante M and Basili R 2006 Creating rich metadata in the TV broadcast archives environment: the PrestoSpace project *2nd International Conference on Automated Production of Cross Media Content for Mulit-channel Distribution*, Leeds, UK.

Messing D, van Beek P and Errico J 2001 The MPEG-7 colour structure descriptor: image description using colour and local spatial information *Proceedings of the International Conference on Image Processing*, vol. 1, pp. 670–673.

MET 2006 METS Profile Components http://www.loc.gov/standards/mets/profile_docs/components.html.

MET 2008 METS Schema 1.7 http://www.loc.gov/standards/mets/mets-schemadocs.html.

freebase Metaweb 2007 Freebase. http://www.freebase.com/.

Mezaris V, Kompatsiaris I and Strintzis M 2004 Still image segmentation tools for object-based multimedia applications. *International Journal of Pattern Recognition and Artificial Intelligence* 18(4), 701–725.

Mezaris V, Kompatsiaris I and Strintzis M 2006 Segmentation of images and video. *Encyclopedia of Multimedia*, B. Furht (Editor), Springer.

Michaud P and Marcotorchino JF 1979 Modèles d'optimisation en analyse des données relationnelles. *Mathématiques et Sciences Humaines* 67, 7–38.

Mika P 2005 Flink: Semantic Web technology for the extraction and analysis of social networks. *Web Semantics* 3(2-3), 211–223.

Mika P 2006 OpenAcademia. http://www.openacademia.org/.

Miles A and Bechhofer S 2009 SKOS Reference. W3C.

Miles A, Matthews B, Beckett D, Brickley D, Wilson M and Rogers N 2005 SKOS: A language to describe simple knowledge structures for the Web *Proceedings of XTech 2005*.

Miller GA and Hristea F 2006 Wordnet nouns: Classes and instances. *Computational Linguistics* 32(1), 1–3.

Millet C, Bloch I, Hede P and Moellic P 2005 Using relative spatial relationships to improve individual region recognition. *Proc. 2nd Eur. Workshop Integration Knowledge, Semantics and Digital Media Technol* pp. 119–126.

Mitchel M 1996 *An Introduction to Genetic Algorithms*. Cambridge, MA: MIT Press.

Mitra P, Murthy C and Pal S 2002 Unsupervised feature selection using feature similarity. *IEEE Transactions on Pattern Analysis and Machine Intelligence* 24(3), 301–312.

Miyamoto S 1990 *Fuzzy Sets in Information Retrieval and Cluster Analysis*. Kluwer Academic Publishers.

Molau S, Pitz M and Ney H 2001 Histogram based normalization in the acoustic feature space, *In Proc. of IEEE Automatic Speech Recognition and Understanding Workshop (ASRU)*.

Moore B, Glasberg BR and Baer T 1997 A model for the prediction of thresholds, loudness and partial loudness. *Journal Audio Engineering Society* 45, 224–240.

Morris O, Lee M and Constantinides A 1986 Graph theory for image analysis: An approach based on the shortest spanning tree. *Institute of Electrical Engineering, Part F* 133(2), 146–152.

Motik B, Grau BC, Horrocks I, Wu Z, Fokoue A and Lutz C 2009a OWL 2 Web Ontology Language Profiles. W3C.

Motik B, Parsia B and Patel-Schneider PF 2009b OWL 2 Web Ontology Language XML Serialization. W3C.

Motik B, Patel-Schneider PF and Grau BC 2009c OWL 2 Web Ontology Language Direct Semantics. W3C.

Motik B, Patel-Schneider PF and Parsia B 2009d OWL 2 Web Ontology Language Structural Specification and Functional-Style Syntax. W3C.

Motik B, Sattler U and Studer R 2005 Query answering for OWL-DL with rules. *Web Semantics* 3(1), 41–60.

Motik B, Shearer R and Horrocks I 2009e Hypertableau Reasoning for Description Logics. *Journal of Artificial Intelligence Research* 36, 165–228.

MPEG Requirements Group 2001 MPEG-7 Interoperability, Conformance Testing and Profiling, v.2. ISO/IEC JTC1/SC29/WG11N4039.

MPEG-7 2001 Multimedia Content Description Interface Standard No. ISO/IEC 15938.

MPEG-7 2005 Multimedia Content Description Interface – Part 9: MPEG-7 Profiles and levels ISO/IEC 15938-9:2005.

MPEG-21 2001 Multimedia Framework – Part 1: Vision, Technologies and Strategy ISO/IEC 21000-1.

MPEG-21 2006 Part 17: Fragment Identification of MPEG Resources Standard No. ISO/IEC 21000-17.

Mühling M, Ewerth R, Stadelmann T, Freisleben B, Weber R and Mathiak K 2007 Semantic video analysis for psychological research on violence in computer games *CIVR '07: Proceedings of the 6th ACM international conference on Image and video retrieval*, pp. 611–618.

Muñoz S, Pérez J and Gutierrez C 2007 Minimal deductive systems for RDF *Proc. of the 4th European Semantic Web Conference (ESWC)*.

MXF 2004 Material Exchange Format (MXF) – File Format Specification (Standard) SMPTE 377M.

Mylonas P and Avrithis Y 2005 Context modeling for multimedia analysis and use *In Proc. of 5th International and Interdisciplinary Conference on Modeling and Using Context*, Paris, France.

Nack F and Lindsay AT 1999 Everything you wanted to know about MPEG-7 (Parts I & II). *IEEE Multimedia*.

Nack F, van Ossenbruggen J and Hardman L 2005 That obscure object of desire: Multimedia metadata on the Web (Part II). *IEEE Multimedia*.

Nam J, Alghoniemy M and Tewfik A 1998 Audio-visual content-based violent scene characterization *Proceedings of the International Conference on Image Processing*, vol. 1, pp. 353–357.

Naphade MR, Kennedy L, Kender JR, Chang SF, Smith JR, Over P and Hauptmann A 2005 A light scale concept ontology for multimedia understanding for TRECVID 2005. Technical report, IBM.

Natsev A, Tešić J, Xie L, Yan R and Smith JR 2007 IBM multimedia search and retrieval system *CIVR 2007*, pp. 645.

Nemrava J, Svatek V, Buitelaar P, Declerck T, Sadlier D, Cobet A, Zeiner H and Petrak J 2007 Architecture for mapping between results of video analysis and complementary resource analysis. K-Space Public Deliverable 5.10. Technical report.

Neumann B and Möller R 2006 *Cognitive Vision Systems* Springer chapter On Scene Interpretation with Description Logics, pp. 247–275.

Neuschmied H, Trichet R and Merialdo B 2007 Fast annotation of video objects for interactive TV *MULTIMEDIA '07: Proceedings of the 15th International Conference on Multimedia*, pp. 158–159, Augsburg, Germany.

Niebles JC, Wang H and Fei-Fei L 2008 Unsupervised learning of human action categories using spatial-temporal words. *International Journal of Computer Vision* **79**(3), 299–318.

Oberle D, Ankolekar A, Hitzler P, Cimiano P, Schmidt C, Weiten M, Loos B, Porzel R, Zorn HP, Micelli V, Sintek M, Kiesel M, Mougouie B, Vembu S, Baumann S, Romanelli M, Buitelaar P, Engel R, Sonntag D, Reithinger N, Burkhardt F and Zhou J 2007 DOLCE ergo SUMO: On foundational and domain models in SWIntO (SmartWeb Integrated Ontology). *Web Semantics* **5**, 156–174.

Oberle D, Lamparter S, Grimm S, Vrandecic D, Staab S and Gangemi A 2006 Towards ontologies for formalizing modularization and communication in large software systems. *Journal of Applied Ontology* **1**(2), 163–202.

O'Connor N, Cooke E, le Borgne H, Blighe M and Adamek T 2005 The ace-toolbox: low-level audiovisual feature extraction for retrieval and classification *2nd IEE European Workshop on the Integration of Knowledge, Semantic and Digital Media Technologies*.

Oh IS, Lee JS and Moon BR 2002 Local search-embedded genetic algorithms for feature selection *16th IEEE International conference on pattern recognition*, pp. 148–151.

Oldakowski R, Bizer C and Westphal D 2005 RAP: RDF API for PHP.

Ommer B, Mader T and Buhmann JM 2009 Seeing the objects behind the dots: Recognition in videos from a moving camera. *International Journal of Computer Vision* **83**(1), 57–71.

Omran M, Salman A and Engelbrecht AP 2002 Image classification using particle swarm optimization. *In Conference on Simulated Evolution and Learning* **2**, 370–374.

Ooi BC and Tan KL 2002 B-trees: Bearing fruits of all kinds In *Thirteenth Australasian Database Conference (ADC)*, vol. 5. Australian Computer Society, Melbourne.

Ooi BC, Tan KL, Yu C and Bressan S 2000 Indexing the edges – a simple and yet efficient approach to high-dimensional indexing *19th ACM SIGMOD SIGACT SIGART*, pp. 166–174, Dallas, USA.

Oren E, Delbru R and Decker S 2006 Extending faceted navigation for RDF data *5th International Semantic Web Conference (ISWC)*, pp. 559–572, Athens, GA, USA.

Oren E, Heitmann B and Decker S 2008 ActiveRDF: embedding Semantic Web data into object-oriented languages. *Web Semantics* **6**(3), 191–202.

(P. Hayes (ed.) 2004 RDF semantics http://www.w3.org/TR/rdf-mt/.

Pachet F 2005 Idea Group Hershey, PA chapter Knowledge management and musical metadata.

Page L, Brin S, Motwani R and Winograd T 1998 The PageRank citation ranking: bringing order to the Web. Technical report, Stanford Digital Library Technologies Project.

Pan H, Li B and Sezan I 2002 Automatic detection of replay segments in broadcast sports programs by detection of logos in scene transitions *Proc. IEEE ICASSP 2002*.

Papadopoulos G, Mezaris V, Kompatsiaris I and Strintzis M 2007 Combining global and local information for knowledge-assisted image analysis and classification. *EURASIP Journal on Advances in Signal Processing*.

Papadopoulos G, Saathoff C, Grzegorzek M, Mezaris V, Kompatsiaris I, Staab S and Strintzis M 2009 Comparative evaluation of spatial context techniques for semantic image analysis *Proc. 10th International Workshop on Image Analysis for Multimedia Interactive Services (WIAMIS09)*, pp. 161–164, London, UK.

Papakonstantinou Y, Garcia-Molina H and Widom J 1995 Object exchange across heterogeneous information sources *Eleventh International Conference on Data Engineering*, pp. 251–260.

Park DK, Jeon YS and Won CS 2000 Efficient use of local edge histogram descriptor *MULTIMEDIA'00: Proceedings of the ACM Workshops on Multimedia*, pp. 51–54. ACM, New York.

Parreiras FS, Saathoff C, Walter T, Franz T and Staab S 2009 Apis a gogo: Automatic generation of ontology apis *International Conference on Semantic Computing*. IEEE.

Patel-Schneider PF, Hayes P and Horrocks I 2004 OWL Web Ontology Language Semantics and Abstract Syntax. W3C.

Peeter G, Burthe AL and Rodet X 2002 Toward automatic music audio summary generation from signal analysis *International Conference on Music Information Retrieval*.

Peeters G 2004 A large set of audio features for sound description (similarity and classification) in the CUIDADO project. Technical report, IRCAM.

Pereira F, Vetro A and Sikora T 2008 Multimedia retrieval and delivery: Essential metadata challenges and standards. *Proceedings of the IEEE* **96**, 721–744.

Pérez J, Arenas M and Gutierrez C 2006a Semantics and Complexity of SPARQL. *In Proc. of the 5th International Semantic Web Conference (ISWC)*, pp. 30–43.

Pérez J, Arenas M and Gutierrez C 2006b Semantics and complexity of SPARQL.

Pérez J, Arenas M and Gutierrez C 2010 nSPARQL: A navigational language for RDF. *Web Semantics* **8**(2), 255–270.

Petridis K, Bloehdorn S, Saathoff C, Simou N, Dasiopoulou S, Tzouvaras V, Handschuh S, Avrithis Y, Kompatsiaris Y and Staab S 2006 Knowledge representation and semantic annotation of multimedia content. *IEE Proceedings on Vision, Image and Signal Processing* **153**(3), 255–262.

Piatrik T and Izquierdo E 2006 Image classification using an ant colony optimization approach *1st International Conference on Semantic and Digital Media Technologies*, pp. 159–168.

Pitz M and Ney H 2005 Vocal tract normalization equals linear transformation, in *cepstral space*. *IEEE Transactions on Speech and Audio Processing* **13**, 930–944.

PME 2007 P_Meta 2.0 Metadata Library EBU-TECH 3295-v2.

Polleres A 2007 From SPARQL to rules (and back) *Proceedings of the International World-Wide Web Conference*.

Polleres A, Scharffe F and Schindlauer R 2007 SPARQL++ for mapping between RDF vocabularies *OTM'07: Proceedings of the 2007 OTM Confederated International Conference on the Move to Meaningful Internet Systems*, pp. 878–896.

Potamianos G, Neti C, Luettin J and Matthews I 2004 *Issues in Visual and Audio-Visual Speech Processing* G. Bailly and E. Vatikiotis-Bateson and P. Perrier, MIT Press chapter 10.

Prud'hommeaux E 2007 Federated SPARQL. W3C.

Prud'hommeaux E and Seaborne A 2010 SPARQL 1.1 Federation Extensions. W3C.

Prud'hommeaux E and Seaborne A (eds) 2007 SPARQL query language for RDF http://www.w3.org/TR/rdf-sparql-query/.

Quackenbush S and Lindsay A 2001 Overview of MPEG-7 audio. *IEEE Transactions on Circuits and Systems for Video Technology* **11**(6), 725–729.

Quan D and Karger D 2004 How to make a Semantic Web browser *13th International World Wide Web Conference (WWW)*, New York.

Quilitz B and Leser U 2008 Querying Distributed RDF Data Sources with SPARQL *In Proc. of the 5th European Semantic Web Conference (ESWC)*.

Rabiner L 1989 A tutorial on hidden Markov models and selected applications in speech recognition. *Proceedings of the IEEE* **77**, 257–286.

Rabiner L 1993 *Fundamentals of Speech Processing*. Englewood Cliffs, NJ: PTR Prentice Hall.

Raimond Y, Abdallah S, Sandler M and Giasson F 2007 The music ontology *International Conference on Music Information Retrieval (ICMIR'07)*, pp. 417–422.

Raudys S and Roli F 2003 The behavior knowledge space fusion method: Analysis of generalization error and strategies for performance improvement *In Proc. Int. Workshop on Multiple Classifier Systems*, pp. 55–64.

Rechenberg I 1965 *Cybernetic solution path of an experimental problem* vol. Library Translation No.1122. Royal Aircraft Establishment.

Reynolds DA and Rose RC 1995 Robust text-independent speaker identification using Gaussian mixture speaker models. *IEEE Transactions on Speech and Audio Processing* **3**, 72–83.

Rissanen J 1978 Modeling by shortest data description. *Automatica* **14**, 465–471.

Ro YM, Kyung H, Kim M, Kang HK, Manjunalh BS and Kim J 2001 MPEG-7 homogeneous texture descriptor. *ETRI Journal* **23**, 41–51.

Roberts L 1965 Machine perception of 3-D solids. *Optical and Electro optical Information Processing* pp. 159–197.

Rocha C, Schwabe D and de Aragao M 2004 A hybrid approach for searching in the Semantic Web *13th International World Wide Web Conference (WWW)*, pp. 374–383, New York.

Rodet X and Jaillet F 2001 Detection and modeling of fast attack transients *International Computer Music Conference*.

Ross K, Westermann GU and Popitsch N 2004 METIS – a flexible database solution for the management of multimedia assets *Proc. of the 10th Int. Workshop on Multimedia Information Systems*.

Rua EA, Bredin H, Mateo CG, Chollet G and Jimenez DG 2008 Audio-visual speech asynchrony detection using co-inertia analysis and coupled hidden Markov models. *Pattern Analysis and Applications*.

Rui Y, Huang TS and Chang SF 1999 Image retrieval: Current techniques, promising directions, and open issues. *Journal of Visual Communication and Image Representation* **10**(1), 39–62.

Rutledge L, van Ossenbruggen J and Hardman L 2005 Making RDF presentable – integrated global and local Semantic Web browsing *14th International World Wide Web Conference (WWW)*, pp. 199–206, Chiba, Japan.

Saathoff C and Staab S 2008 Exploiting spatial context in image region labelling using fuzzy constraint reasoning *WIAMIS: Ninth International Workshop on Image Analysis for Multimedia Interactive Services*.

Sadlier D and O'Connor N 2005a Event detection in field sports video using audio-visual features and a support vector machine. *IEEE Transactions on Circuits and Systems for Video Technology* **15**(10), 1225–1233.

Sadlier D and O'Connor NE 2005b Event detection based on generic characteristics of field sports *ICME 2005 – Proceedings of the IEEE International Conference on Multimedia and Expo*, pp. 759–762.

Salembier P, Qian R, O'Connor N, Correia P, Sezan I and van Beek P 2000 Description schemes for video programs, users and devices. *Signal Processing: Image Communication* **16**(1-2), 211–234.

Salway A 2007 A corpus-based analysis of the language of audio description *Selected Proceedings of Media for All*.

Salway A, Lehane B and O'Connor. NE 2007 Associating characters with events in films *CIVR 2007 – ACM International Conference on Image and Video Retrieval*.

Salway A, Vassiliou A and Ahmad K 2005 What happens in films? *In Proc. of the International Conference on Multimedia and Expo (ICME)*.

Sandhaus P, Thieme S and Boll S 2008 Processes of photo book production. *Multimedia Systems* **14**(6), 351–357.

Sauermann L and Cyganiak R 2008 *Cool URIs for the Semantic Web*. W3C Interest Group Note.

Scaringella N and Zoia G 2005 On the modeling of time information for automatic genre recognition systems in audio signals *International Sumposium on Music Information Retrieval*.

Schallauer P, Fürntratt H and Bailer W 2007 MPEG-7 for video quality description and summarisation *Multimedia Metadata Applications Workshop at I-MEDIA'07*, pp. 183–185.

Scheirer E and Slaney M 1997 Construction and evaluation of a robust multifeature speech/music discriminator *Proc. of International Conf. on Acoustics, Speech, and Signal Processing*.

Schenk S 2007 A SPARQL semantics based on Datalog *Proceedings of KI2007, 30th German Conference on Artificial Intelligence* LNAI. Springer.

Schenk S 2008 On the semantics of trust and caching in the Semantic Web *In Proc. of the 7th International Conference on the Semantic Web (ISWC)*, pp. 533–549.

Schenk S and Staab S 2008 Networked graphs: a declarative mechanism for SPARQL rules, SPARQL views and RDF data integration on the Web In *WWW*, pp. 585–594. ACM.

Schenk S, Saathoff C, Staab S and Scherp A 2009 SemaPlorer – interactive semantic exploration of data and media based on a federated cloud infrastructure. *Web Semantics* **7**(4), 298–304.

Scherp A, Agaram S and Jain R 2008 Event-centric media management *IS&T/SPIE's 20th Annual Symposium Electronic Imaging Science and Technology*.

Schoelkopf B and Smola A 2002 *Learning with Kernels*. MIT Press.

Schraefel M and Karger D 2006 The pathetic fallacy of RDF *3rd International Semantic Web User Interaction Workshop (SWUI)*, Athens, GA, USA.

Schraefel M, Smith D, Owens A, Russell A, Harris C and Wilson M 2005 The evolving mSpace platform: leveraging the Semantic Web on the trail of the Memex *Hypertext*, pp. 174–183, Salzburg, Austria.

Schreiber G, Amin A, van Assem M, de Boer V, Hardman L, Hildebrand M, Hollink L, Huang Z, van Kersen J, de Niet M, Omelayenjko B, van Ossenbruggen J, Siebes R, Taekema J, Wielemaker J and Wielinga B 2006 MultimediaN E-Culture Demonstrator *5th International Semantic Web Conference (ISWC)*, pp. 951–958, Athens, GA, USA.

Schwarz G 1978 Estimating the dimension of a model. *Annals of Statistics* **6**(2), 461–464.

Seaborne A 2004 *RDQL – A Query Language for RDF*. W3C Member Submission.

Shadbolt N, Berners-Lee T and Hall W 2006 The Semantic Web revisited. *IEEE Intelligent Systems* **21**(3), 96–101.

Shafer G 1976 *A Mathematical Theory of Evidence*. Princeton University Press.

Shawe-Taylor J and Cristianini N 2004 *Kernel Methods for Pattern Analysis*. Cambridge University Press.

Shen HT, Ooi BC and Zhou X 2005 Towards effective indexing for very large video sequence database. *SIGMOD*, pp. 730–741. ACM, Baltimore, MD, USA.

Shimodaira H, Noma K, Nakai M and Sagayama S 2002 Dynamic time-alignment kernel in support vector machine In T.G. Dietterich, S. Becker and Z. Ghahramani, eds, *Advances in Neural Information Processing Systems* 14. Cambridge, MA: MIT Press.

Shotton J, Winn J, Rother C and Criminisi A 2009 TextonBoost for image understanding: Multi-class object recognition and segmentation by jointly modeling texture, layout, and context. *International Journal of Computer Vision* **81**(1), 2–23.

SIMILE 2003 Longwell RDF Browser. http://simile.mit.edu/longwell/.

Sintek M and Decker S 2002 TRIPLE – A query, inference, and transformation language for the Semantic Web.

Sirin E and Parsia B 2004 Pellet: An OWL DL reasoner.

Sirin E and Parsia B 2007 SPARQL-DL: SPARQL Query for OWL-DL *Proc. of OWLED*.

Sirin E, Parsia B, Grau BC, Kalyanpur A and Katz Y 2007 Pellet: A practical OWL-DL reasoner. *Web Semantics* **5**(2), 51–53.

Skurichina M and Duin R 1998 Bagging for linear classifiers. *Pattern Recognition* **31**(7), 909–930.

Slaney M 2002 Semantic-audio retrieval *IEEE International Conference on Acoustics, Speech and Signal Processing*, vol. 4, pp. 4108–4111.

Slimane N, Monmarche N and Venturini G 1999 AntClass: discovery of clusters in numeric data by an hybridization of an ant colony with kmeans algorithm. Internal report 213, Laboratoire d'Informatique de l'Université de Tours.

Smaragdis P and Casey M 2003 Audio visual independent components *International Symposium on Independent Component Analysis and Blind Signal Separation*, pp. 709–714.

Smeaton AF, Over P and Kraaij W 2006 Evaluation campaigns and trecvid *MIR '06: Proceedings of the 8th ACM international workshop on Multimedia information retrieval*, pp. 321–330. ACM, New York.

Smeaton AF, Over P and Kraaij W 2008 *Multimedia Content Analysis:Theory and Applications (in press)* chapter High-level Feature Detection from Video in TRECVid: a 5-Year Retrospective of Achievements.

Smets P 1994 The transferable belief model. *Artificial Intelligence* **66**(2), 191–234.

SMIL 2001 Synchronized Multimedia Integration Language (SMIL 2.0) Specification W3C Recommendation.

SMP 2001 Data Encoding Protocol using Key-Length-Value SMPTE 336M.

SMP 2004 Metadata Dictionary Registry of Metadata Element Descriptions SMPTE RP210.8.

Snoek CG, van Gemert JC, Gevers T, Huurnink B, Koelma DC, van Liempt M, de Rooij O, van de Sande KE, Seinstra FJ, Smeulders, Thean AH, Veenman CJ and W.M MW 2006 The MediaMill TRECVID 2006 semantic video search engine *Proceedings of the 4th TRECVID Workshop* NIST, Gaithersburg,USA.

Snoek CGM, Huurnink B, Hollink L, de Rijke M, Schreiber G and Worring M 2007a Adding semantics to detectors for video retrieval. *IEEE Transactions on Multimedia* **9**(5), 975–986.

Snoek CGM, Worring M, Koelma D and Smeulders AWM 2007b A learned lexicon-driven paradigm for interactive video retrieval. *IEEE Transactions on Multimedia* **9**(2), 280–292.

Song Y, Chen Z and Yuan Z 2007 New chaotic PSO-based neural network predictive control for nonlinear process. *IEEE Transactions on Neural Networks* **18**(2), 595–601.

Souvannavong F, Merialdo B and Huet B 2005 Multi modal classifier fusion for video shot content retrieval. *Proceedings of WIAMIS*.

Staab S and Studer R 2004 *Handbook on Ontologies*. Berlin: Springer-Verlag.

Stoilos G, Stamou G, Tzouvaras V, Pan J and Horrocks I 2005 A fuzzy description logic for multimedia knowledge representation *Proc. of the International Workshop on Multimedia and the Semantic Web*.

Stoilos G, Stamou G, Tzouvaras V, Pan J and Horrocks I 2007 Reasoning with very expressive fuzzy description logics. *Journal of Artificial Intelligence Research* **30**, 273–320.

Straccia U 2001 Reasoning within fuzzy description logics. *Journal of Artificial Intelligence Research* **14**, 137–166.

Strehl A and Ghosh J 2003 Cluster ensembles – a knowledge reuse framework for combining multiple partitions. *Journal of Machine Learning Research* **3**, 583–617.

Sun X, Manjunath BS and Divakaran A 2002 Representation of motion activity in hierarchical levels for video indexing and filtering.

Suominen O, Viljanen K and Hyvönen E 2007 User-centric faceted search for semantic portals *In Proc. of the 4th European Semantic Web Conference (ESWC)*, Innsbruck, Austria.

Sure Y, Staab S and Studer R 2009 Ontology engineering methodology. S. Staab and R. Studer (eds), *Handbook on Ontologies*, pp. 135–152. Springer, Berlin.

Tanghe K, Degroeve S and Baets BD 2005 An algorithm for detecting and labeling drum events in polyphonic music *Proc. of First Annual Music Information Retrieval Evaluation eXchange*.

ter Horst H 2005 Combining RDF and part of OWL with rules: Semantics, decidability, complexity *In Proc. of the 4th International Semantic Web Conference (ISWC)*, pp. 668–684.

Terrillon JC and Akamatsu S 1989 Hough transform for line recognition *Proc. Computer Vision and Image Processing*, vol. 46, pp. 327–345.

Terrillon JC and Akamatsu S 1999 Comparative performance of different chrominance spaces for colour segmentation and detection of human faces in complex scene images *Proc. 12th Conf. on Vision Interface*, vol. 2, pp. 180–187.

Theodoridis S and Koutroumbas K 1999 *Pattern Recognition*. Academic Press.

Theodoros G, Dimitrios K, Andreas A and Sergios T 2006 Violence content classification using audio features.

Tomadaki E 2006 Cross-document coreference between different types of collateral texts for films *PhD dissertation, Dept. of Computing, University of Surrey*.

Topchy A, Jain A and Punch W 2004 A mixture model for clustering ensembles *SIAM International Conference on Data Mining*, pp. 379–390.

Torralba A, Fergus R and Freeman WT 2008 80million tiny images: A large data set for nonparametric object and scene recognition. *IEEE Transactions on Pattern Analysis and Machine Intelligence* **30**, 1958–1970.

TRECVID 2003 http://www-nlpir.nist.gov/projects/tv2003/tv2003.html.

TRECVID 2008 Guidelines for the trecvid 2007 evaluation. http://www-nlpir.nist.gov/projects/tv2007/tv2007.html.

Troncy R 2003 Integrating structure and semantics into audio-visual documents *2nd International Semantic Web Conference*, pp. 566–581.

Troncy R 2008 Bringing the IPTC News Architecture into the Semantic Web *In Proc. of the 7th International Conference on The Semantic Web*, pp. 483–498.

Troncy R, Bailer W, Hausenblas M, Hofmair P and Schlatte R 2006 Enabling multimedia metadata interoperability by defining formal semantics of MPEG-7 profiles *1st International Conference on Semantics And digital Media Technology (SAMT'06)*, pp. 41–55, Athens, Greece.

Troncy R, Celma O, Little S, García R and Tsinaraki C 2007a MPEG-7 based multimedia ontologies: Interoperability support or interoperability issue ? *1st International Workshop on Multimedia Annotation and Retrieval enabled by Shared Ontologies*.

Troncy R, Hardman L, van Ossenbruggen J and Hausenblas M 2007b Identifying spatial and temporal media fragments on the Web *W3C Video on the Web Workshop*. http://www.w3.org/2007/08/video/positions/ Troncy.pdf.

Troncy R, Mannens E, Pfeiffer S and (eds.) DVD 2010 Media Fragments URI 1.0 W3C Working Draft. http://www.w3.org/TR/media-frags.

Tsarkov D and Horrocks I 2006 FaCT++ description logic reasoner: System description *Proc. of the International Joint Conference on Automated Reasoning*, pp. 292–297.

Tsinaraki C, Polydoros P, Moumoutzis N and Christodoulakis S 2004 Integration of OWL ontologies in MPEG-7 and TV-Anytime compliant semantic indexing *16th International Conference on Advanced Information Systemes Engineering*.

Tummarello G, Morbidoni C and Nucci M 2006 Enabling Semantic Web communities with DBin: an overview *5th International Semantic Web Conference (ISWC)*, Athens, GA, USA.

TVA 2005a Broadcast and online services: Search, select, and rightful use of content on personal storage systems ('TV Anytime'); part 2: System description ETSI TS 102 822-2V1.3.1.

TVA 2005b Broadcast and online services: Search, select, and rightful use of content on personal storage systems ('TV Anytime'); part 3: Metadata; subpart 1: Phase 1 metadata schemas ETSI TS 102 822-3-1V1.3.1.

Tzanetakis G and Cook P 2002 Musical genre classification of audio signals. *IEEE Transactions on Speech and Audio Processing* **10**(5), 293.

Urruty T 2008 KpyrRec: a recursive multidimensional indexing structure. *International Journal of Parallel, Emergent and Distributed Systems* **23**, 235–245.

Urruty T, Djeraba C and Simovici DA 2007 Clustering by random projections *ICDM 2007, Leipzig, Germany*, pp. 107–119.

Uschold M 2003 Where are the semantics in the Semantic Web?. *AI Magazine* **24**(3), 25–36.

Utsumi O, Miura K, Ide I, Sakai S and Tanaka H 2002 An object detection method for describing soccer games from video *IEEE ICME 2002*.

van Assem M, Malaisé V, Miles A and Schreiber G 2005 A method to convert thesauri to SKOS *3rd European Semantic Web Conference*, Budva, Montenegro.

Van Gool LJ, Breitenstein MD, Gammeter S, Grabner H and Quack T 2009 Mining from large image sets *Proceedings of the ACM International Conference on Image and Video Retrieval (CIVR)*, Xi'an, China.

van Harmelen F 2006 Semantic Web research anno 2006: main streams, popular fallacies, current status and future challenges, pp. 1–7.

van Ossenbruggen J, Nack F and Hardman L 2004 That Obscure Object of Desire: Multimedia Metadata on the Web (Part I). *IEEE Multimedia*.

Vapnik V 1998 *Statistical Learning Theory*. New York: John Wiley & Sons, Inc.

Vassiliou A 2006 Film content analysis: a text-based approach *PhD dissertation, Dept. of Computing, University of Surrey*.

Vert J 2008 The optimal assignment kernel is not positive definite. Technical Report HAL-00218278, Mines ParisTech.

Völkel M 2005 Writing the Semantic Web with Java. Technical report, DERI Galway.

VRA 2007 VRA Core http://www.vraweb.org/projects/vracore4.

Wan V and Renals S 2005 Speaker verification using sequence discriminant support vector machines. *Speech and Audio Processing, IEEE Transactions on* **13**(2), 203–210.

Wang Y, Makedon F, Ford J, Shen L and Goldin D 2004 Generating fuzzy semantic metadata describing spatial relations from images using the R-histogram. *Digital Libraries, Proc. of ACM/IEEE Conf. on* pp. 202–211.

Weber R, Schek HJ and Blott S 1998 A quantitative analysis and performance study for similarity-search methods in high-dimensional spaces *24th VLDB*, pp. 194–205. Morgan Kaufmann, New York.

Wielemaker J, Hildebrand M, van Ossenbruggen J and Schreiber G 2008 Thesaurus-based search in large heterogeneous collections *7th International Semantic Web Conference (ISWC)*, Karlsruhe, Germany.

Wielemaker J, Schreiber G and Wielinga B 2003 Prolog-based infrastructure for RDF: performance and scalability, pp. 644–658.

Wilkinson K, Sayers C, Kuno HA and Reynolds D 2003 Efficient RDF storage and retrieval in Jena2.

Wold E, Blum T, Keislar D and Wheaten J 1996 Content based classification, search, and retrieval of audio. *IEEE Multimedia* 3(3), 27–36.

Wright R and Williams A 2001 Archive preservation and exploitation requirements. Technical Report D2, PRESTO – Preservation Technologies for European Broadcast Archives.

Wu P, Manjunanth B, Newsam S and Shin H 1999 A texture descriptor for image retrieval and browsing *Content-Based Access of Image and Video Libraries, (CBAIVL '99) Proceedings. IEEE Workshop on*, pp. 3–7.

Xiao X, Chng E and Li H 2007 Temporal structure normalization of speech feature for robust speech recognition. *IEEE Signal Processing Letters* 14(7), 500–504.

Xie L, Xu P, Chang SF, Divakaran A and Sun H 2004 Structure analysis of soccer video with domain knowledge and hidden Markov models. *Pattern Recognition Letters* 25(7), 767–775.

Xu D and Chang SF 2008 Video event recognition using kernel methods with multilevel temporal alignment. *IEEE Transactions on Pattern Analysis and Machine Intelligence* 30(11), 1985–1997.

Xu F and Zhang YJ 2006 Evaluation and comparison of texture descriptors proposed in MPEG-7. *Journal of Visual Communication and Image Representation* 17(4), 701–716.

Xu H, Yu D, Xu D and Zhang A 2008 SS-ClusterTree: a subspace clustering based indexing algorithm over high-dimensional image features *CIVR '08: Proceedings of the 2008 international conference on Content-based image and video retrieval*, pp. 95–104. ACM, New York.

Yee K, Swearingen K, Li K and Hearst M 2003 Faceted metadata for image search and browsing *International SIGCHI conference on Human factors in computing systems (CHI)*, pp. 401–408, Ft. Lauderdale, Florida, USA.

Yegnanarayana B, d'Alessandro C and Darsinos V 1998 An iterative algorithm for decomposition of speech signals into periodic and aperiodic components. *IEEE Transactions on ASSP* 6(1), 1–12.

Yeung M and Yeo BL 1996 Time constrained clustering for segmentation of video into story units *Proceedings of International Conference on Pattern Recognition*.

Yeung M and Yeo BL 1997 Video visualisation for compact presentation and fast browsing of pictorial content *IEEE Transactions on Circuits and Systems for Video Technology*, pp. 771–785.

You W, Lee KW, Kim JG, Kim J and Kwon OS 2001 Content-based video retrieval by indexing object's motion trajectory *International Conference on Consumer Electronics*, pp. 352–353.

Yu C, Ooi BC, Tan KL and Jagadish HV 2001 Indexing the distance: an efficient method to knn processing *In Proc. of the 27th International Conference on Very Large Databases (VLDB)*, pp. 421–430, Italy.

Yuan J, Li J and Zhang B 2007 Exploiting spatial context constraints for automatic image region annotation *Proc. of ACM Multimedia 2007*, pp. 595–604. ACM, New York.

Zaffalon M and Hutter M 2002 Robust feature selection by mutual information distributions *18th Conference on Uncertainty in Artificial Intelligence*.

Zhai C and Lafferty J 2006 A risk minimization framework for information retrieval. *Information Processing & Management* 42(1), 31–55.

Zhang D and Lu G 2003 Evaluation of MPEG-7 shape descriptors against other shape descriptors. *Multimedia Systems* 9(1), 15–30.

Zhang GP 2000 Neural networks for classification: A survey. *IEEE Transactions on Systems, Man, and Cybernetics* 30(4), 451–462.

Zhang R, Ooi BC and Tan KL 2004 Making the pyramid technique robust to query types and workloads *20th IEEE ICDE*, pp. 313–324, Boston.

Zhou S and Chellappa R 2006 From sample similarity to ensemble similarity: Probabilistic distance measures in reproducing kernel hilbert space. *IEEE Transactions on Pattern Analysis and Machine Intelligence* 28(6), 917–929.

Author Index

Multimedia Semantics: Metadata, Analysis and Interaction, First Edition.
Edited by Raphaël Troncy, Benoit Huet and Simon Schenk.
© 2011 John Wiley & Sons, Ltd. Published 2011 by John Wiley & Sons, Ltd.

Subject Index

ABox, 106, 185, 203
   syntax, 190
alignment kernel, 72
analytic documentation, 16
archive, 14
artificial neural networks, 73
ASK query, 113
attributed relation graph, 165
audio, 10

basic graph pattern, 115
Bayes classifier, 66
Bhattacharryya, 62
blank node, 84
broadcast archive, 16

canonical processes, 21
CCA, 57
CeWe Color Photo Book, 27
classifier, 164
ClioPatria
   faceted browsing, 274
   keyword search, 270
CoIA, 57
color, 164
COMM, 145
complementary resources
   primary, 201
   secondary, 201
concept, 108
conjunctive queries, 106
CONSTRUCT query, 113
context, 164, 173, 177

controlled vocabulary, 17, 100
cross-modal fusion, 54
cultural heritage, 18

dataset, 113
description logic rules, 107
description logics, 102, 184, 185
descriptive metadata, 17
digital photo services, 8
discovery, 12
discriminant power, 62
divergence, 62
DTW, 72
Dublin Core, 133

early fusion, 53
eigenvalue decomposition (EVD), 56
entailment
   OWL, 106
   RDF, 89
   RDFS, 91
evolutionary algorithms, 74
EXIF, 7

f-$\mathcal{SHIN}$, 184
   concepts, 184
   reasoning services, 191
   tableaux, 185
feature fusion, 54
feature selection, 61
feature term space, 249
filter, 61
filter expression, 114

Multimedia Semantics: Metadata, Analysis and Interaction, First Edition.
Edited by Raphaël Troncy, Benoit Huet and Simon Schenk.
© 2011 John Wiley & Sons, Ltd. Published 2011 by John Wiley & Sons, Ltd.